7

FOURTH EDITION

THE MICROBREWERS' HANDBOOK

Ted Bruning

First published in Great Britain 2007 by Navigator Guides
Reprinted 2008 (twice), 2009
Second edition published by Paragraph Publishing 2009
Third edition published by Paragraph Publishing 2011

This fourth edition published in 2013 by
Paragraph Publishing,
St Faiths House,
Mountergate,
Norwich,
NR1 1PY

www.paragraphpublishing.com

ISBN 978-0-9562681-9-8

A catalogue record for this book is available on requst from the British Library.

Acknowledgements
The publishers would like to thank all those microbreweries who provided information for their case studies and for usage of relevant photographs, and to Ted Bruning for all his hard work and constant enthusiasm for the project. The publishers would also like to thank Chalk Hill Brewery, Norwich for the cover image.

Publisher: Damian Riley-Smith
Design: Jenny Sims
Cover Photo: Paul Beevis
Publishing Consultant: Rupert Wheeler
Printed in Great Britain by Polestar Wheatons Ltd.

Contents

Foreword

By Keith Bott, Chairman of the Society of Independent Brewers (SIBA) and Managing Director of Titanic Brewery

The Brewing Industry in the UK has changed dramatically during the nearly 30 years that I have had the pleasure of being part of it. My brewing career started when I left school, during the summer holidays after I had completed my "A" levels, and whilst waiting to take up a post as a civil servant, I saw an advert in the local paper for a job as a "brewer". With no qualifications other than a little bit of home brewing and a lot of product analysis I applied, and much to my mothers consternation, got the job. Working for a small company who had decided to open a microbrewery gave me my first access to the friendliest industry in the world. Tasked with learning how to brew I just phoned up and asked all the other breweries in the area if they would teach me. All welcomed me with open arms accepting my free labour in return for a whistle stop guide to commercial brewing and one of those I visited was the newly formed Titanic Brewery. When my original brewery decided to call it a day after only four brews, I called Titanic and became their assistant brewer. Without doubt it was the camaraderie of brewing that led to me to have the opportunity of taking over the brewery 3 years later. Maybe if this guide had been available then my original employer would have stuck at it and life may have taken a different course.

Now the UK brewing industry has over a 1,000 breweries across the country all producing fantastic local beer for local people using, where possible, local ingredients. The beer drinker has fallen back in love with knowing where their beer has come from and delights in a connection to what they perceive as "their" brewery.

If you are reading this book it probably means you are considering joining the ranks of those who have chosen to set up a brewery, whilst incredibly welcoming, most already in the industry will, almost without exception, advise you against doing so. I would also caution against rushing headlong into it. The last 30 years is littered with those who have not carefully planned their business. You will not just need to be a great brewer but will need to understand business, sales, purchasing, retailing, design, engineering, HR, health and safety, floor scrubbing and drain unblocking. The list of what you are expected to know seems endless but if you have buckets of entrepreneurial spirit and a love of learning something new every day you will succeed and love every minute of it. If you think it will be easy, make you filthy rich or that you know everything there is to know about it I would heartily propose googling "other idea."

This guide will undoubtedly provide you with a fabulous reference document and sets out clearly how to go about taking your acorns and moving towards treedom, but never be afraid to ask. Advice is freely available from many different sources, and as chairman of SIBA and with 20 years as a member of Council it would not be surprising if I recommended joining. This connects you to a network of people who have all been in the same position as yourself. It also gives you access to technical, legal, training, commercial and political assistance. That said don't ignore other sources of advice, from the local Chamber of Commerce to the local branch of CAMRA, all will assist to build a picture of where your business will fit in.

Finally when you have grafted your way towards your dream, never forget why you are doing it. It is the pleasure of tasting your own produce, or even better seeing your customers nods of approval as they taste your wares that makes everything seem worthwhile.

Happy brewing

Yours forever thirsty

Keith Bott

Titanic Brewery
www.titanicbrewery.co.uk

Author's Preface

If you've picked up this book, chances are you're contemplating a dramatic change of life. You may be a licensee wanting to add a small brewery to your pub; you may already be working as a brewer and looking to strike out on your own; you may be a home brewing enthusiast hoping to turn your passion into a living; you may be none of the above. You may be a "civilian" fed up with the rat-race, perhaps with equity in your house or a redundancy cheque in your pocket that you're looking to convert into a whole new life. If so, microbrewing might be for you.

Every time I ask an established microbrewer for a word of advice for the aspirant, they have the same one: "Don't!" Maybe they're remembering all the struggles and frustrations of making a success of their chosen line of business. Maybe they don't want the competition. But I don't know one of them who isn't happy with his – or, in a few cases, her – lot, or who would trade it for a steady office job. That goes for even the most inveterate moaners among them – and believe me, there are a few!

But before you read any further, ask yourself this question: have you got what it takes? Have you got the grit it takes to jump the bureaucratic hurdles? Have you got the painstaking, near-obsessive character it takes to brew a cracking pint day in, day out? Have you got the ready wit, winning charm, and bloody-minded ruthlessness it takes to get out there and sell? Have you got the imagination, intuition and insight it takes to spot and seize new opportunities? If so, you're halfway there. Add a single-minded attention to administrative detail and faultless form-filling, an accountant's grasp of sums, and a bottomless well of energy, and you might yet make a microbrewer.

Most of these personal qualities, though, are the basic requirements for any sort of business. In microbrewing, there are compensations. One is that everyone will love you because everyone loves beer – and even those who unaccountably don't will be captivated by your story and, at heart, on your side. Another is that you'll be operating in an almost uniquely successful sector of manufacturing. Don't believe me? Just go out shopping and look for widely available brands of cosmetics, confectionery, meat and fish products, biscuits, condiments, shoes, pharmaceuticals or power tools from small independent makers. You may find a few, but mostly in small independent retailers. Microbrewers, though, have torn themselves a niche in the national retail markets, both on-trade and take-home, that few other small manufacturing industries can rival. Depending on your choice you can operate in the warm bosom of your local community or the cold limelight of the national stage, or both. There may be trouble ahead – okay, there will be trouble ahead – but in choosing microbrewing you are giving yourself a far, far better chance of success than, say, the putative restaurateur.

And the good news from the Society of Independent Brewers, which represents the majority of Britain's smaller producers, is that this seems to be holding good despite the recession. Its Local Brewing industry Report 2010 records despite the recession a 3.75% overall volume growth during 2010, with the average number of pub accounts per member increasing from 79 to 94 and throughput per pub account growing by 9%. Members' bottled sales were also up 16%; and all this in a year when production by the bigger brewers was down by 6.5%. So while it might be overegging the pudding to claim that microbrewing is recession-proof, it certainly seems to be a better bet than, say, electrical retailing. Or construction. Or banking.

This book is more of a road atlas than a street map. It will, I hope, direct you up the motorway of hardship to the edge of Success City while leaving you to find the precise route to Fulfilment Avenue for yourself. More prosaically, it aims to set out the challenges and rewards of a new life in small brewing clearly enough for you to make up your own mind whether it's for you and you're for it. All I hope is that I have supplied you with a good long look before you decide whether to leap.

This book would not have been possible without the help of many people. I particularly thank the brewers who submitted themselves to interrogation to provide the case histories you will find within. Others who have given their time and patience include David Smith, David Porter, Arthur Bryant, Keith Thomas, Brendan Moore, Lucy Hunter, Carola Brown, Chris Garrett, Paul Corbett, Richard Shardlow, Peter Amor and Nina Bates. I have also been helped by the planning and licensing officers of South Cambridgeshire District Council, who not only explained the arcana of their crafts but also offered their best wishes to all would-be microbrewers; their names, alas I did not record. To them all, my thanks. To those I have forgotten, my apologies as well as my thanks.

Ted Bruning 2013

Introduction

How it all Started

In 1965 Peter Maxwell Stuart, the 20th Laird of Traquair, unwittingly started a revolution. He had rediscovered the 18th-century brewhouse that once kept masters and servants alike at his ancestral pile at Innerleithen, Peeblesshire, supplied with ale, and decided to restore it. A period of hard work followed, and before long the beer started flowing again at Traquair House.

Neither Peter Maxwell Stuart, nor anyone else thought at the time that the establishment of the first commercial brewery in Britain in half a century was any more a mere curiosity, a one-off. Then seven years later another former brewery, this time at Selby in North Yorkshire, also recommissioned its vessels. Founded in 1894, Selby had stopped brewing in 1954 to become a licensed trade wholesaler. But, said owner Martin Sykes: "I foresaw a revival of interest in real ale and got in early." In November 1972 the first brew was released; and this time it was no one-off curiosity.

Less than a year later the third of the new wave of "microbreweries" – and the first truly new brewery in over 50 years – was founded by former rocket scientist Paul Leyton at the Miners Arms at Priddy in Somerset. It sounded like a pub but was actually a restaurant, where you could only drink Paul's home-brew with a meal; and all the beer was bottled rather than draught. The year after that a fourth micro – and nobody now remembers who coined the term – was set up at the Masons Arms at South Leigh in Oxfordshire, and the microbrewing revolution was well and truly on.

True, it got off to a faltering start. Nearly all of the country's 55,000-odd pubs were either owned by existing breweries or were loan-tied, having accepted the exclusive supply agreement that came with low-interest brewery loans. Half of the 300 small breweries started in the 1970s and 80s failed, most often for want of free-of-tie pubs to sell their beers to.

Campaign for Real Ale (CAMRA)

But the 1970s was also the founding decade of the Campaign for Real Ale (CAMRA), fomentor of the great consumer revolt against the weak, gassy, keg beers the brewing industry was foisting on the public.

CAMRA and the microbrewers were symbiotic from the start — in fact Martin Sykes, the refounder of Selby Brewery, was a member of CAMRA's first National Executive Committee. CAMRA's Good Beer Guide pointed drinkers to pubs that stocked traditional beer, including real ale from new brewers; and the beer festivals staged by its rapidly expanding web of local branches allowed the public the opportunity to try the new brewers' ales for themselves. So, with growing support from enthusiastic beer-lovers and from the handful of publicans who could and would stock their beers, the luckiest and pluckiest of the pioneers clung on. Slowly their numbers grew, and within a decade local independent brewers had become an established, if often penurious, fixture of the nation's beer and pub scene.

Who were they, though, these pioneers? Not the counter-cultural hippy good-lifers you might suppose. Most of them, in fact, were already brewers.

Throughout the 1950s and 60s big breweries had been buying up smaller ones and closing them down. This concentration in the industry was the delayed effect of the economic conditions of the preceding four decades: war followed by depression followed by war followed by rationing had left scores of small family breweries hopelessly uncompetitive. Their pubs and breweries hadn't had a bean spent on them in generations and were in many cases near-derelict; and the owners were desperate to sell, sometimes for ludicrously small sums. Bass once recouped the entire purchase price of a brewery it had bought simply by auctioning the contents of the new acquisition's wines and spirits warehouse. In other cases, buyers found that their acquisitions hadn't had their pub estates revalued since before the First World War , and covered their outlay just by sending round a team of surveyors.

That is, perhaps, another story. The upshot, for our purposes, was that hundreds of experienced staff – executives as well as brewers – were thrown out of work as old-established breweries were rationalised out of existence. Most of them simply disappeared into the general labour force; but a few decided to carry on making a living at their chosen trade and, following the lead of Traquair, Selby and the rest set up breweries of their own.

The Pioneers

And some of them were pretty senior figures. Bill Urquart founded the Litchborough Brewery in a barn at his Northamptonshire home in 1974 having been laid off by Watney after 40 years in the business. His last assignment had been to run down the old Phipps Brewery in Northampton, which was earmarked for demolition to become the site of Carlsberg's brand-new lager plant. Part of his job was to decide who would be made redundant and who would get a job at the new plant. At 58, he had a feeling that his own name was on the list for the chop, so he seized the opportunity to acquire the skills he would need to set up on his own. At Litchborough he not only brewed a bitter so popular that his own golf club sold it, but he also helped many other new brewers set themselves up.

Another senior brewer who set up for himself but spent as much time helping newcomers as running his own brewery was Peter Austin, founder of Ringwood in Hampshire in 1978. Peter had just retired as head brewer of the Hull Brewery, but his new life took him all around the world, installing small brewing plants as far afield as China.

Other highly experienced pioneers included Simon Whitmore, managing director of Courage's western region who founded Butcombe Brewery in Somerset in 1978 after being made redundant. Peter Mauldon, who in 1982 refounded his family brewery in Sudbury, Suffolk (established by his great-grandfather in 1795, bought and closed by Greene King in 1960), had been a senior brewer at Watney in Mortlake, Surrey. David Bruce, who opened the first of the Firkin chain of brewpubs in 1979, had worked in management and brewing at both Courage and Theakston's. Martin Ayres, Ted Willems, John Bjornsson, Chas Wright, Pat Glenny, James Johnstone, Gerry Watts, Roger Catte, Geoff Mumford, Bruce Wilkinson, Tony Allen, Peter Yates, Peter Amor, Bryan Wilson and John Gilbert were other industry veterans who set up on their own in the 1980s and whose ventures are still going strong today, often under second- or even third-generation ownership.

Despite the newcomers' resumés, the established brewers – both the surviving older-established family firms and the new Big Six national brewers – were at first rather scornful of their potential. It was hard to believe they could produce anything decent on brewing plants thrown together from scraps of old dairy equipment and the like; and the older firms were armour-plated, or so they believed. Their tied estates, and the number of free houses they controlled through low-interest loans, surely meant that the upstarts would never gain more than a toehold in the retail trade and could never become a threat. A semi-cordial relationship based on amused condescension soon developed. Pitfield Brewery used to get its fresh yeast from Charles Wells of Bedford, turning up at the back gate with a galvanised bucket which a worker would obligingly fill with barm scooped straight from the fermenter. No-one saw Pitfield as anything but a harmless eccentric – until 1987, that is, when its Dark Star strong old ale became the first microbrewer's beer to win the Campaign for Real Ale's Champion Beer of Britain competition.

For these chaps could brew, whatever the mainstream thought. Brewing is, after all, not that complicated a process (although brewing consistently good-quality beer is extremely painstaking and demanding). A chef-turned-microbrewer I know tells me it is far easier than cooking in a restaurant. Nor does the equipment itself have to be expensive and gleaming, so long as it's well-maintained and only has the holes in it that you actually want. And anyway, most of the new wave had actually been brewing all their working lives, so to find them consistently turning out beers of seriously commercial quality should have come as no surprise.

Actually, in terms of quality they had some key advantages over their much bigger competitors. With no accountants counting the farthings, they didn't have to skimp on the raw materials they used. With no tied estates to guarantee a base level of sales, they could only survive by brewing consistently better beer. And with no marketing departments telling them what the public supposedly wanted, they were free to brew whatever they fancied. Some rifled the recipe books of the past and reinvented vanished beers such as porters and old ales (like Dark Star); others experimented and came up with entirely new styles. The cumulative effect was to arouse in the beer-drinking public an appetite for novelty and variety (and, paradoxically, so far had mass-produced beer fallen by then that a thirst for the traditional and a thirst for novelty were virtually identical) that older, more staid, breweries couldn't or wouldn't satisfy – or in many cases even acknowledge.

Throughout the 1970s and 80s the trickle of newcomers accelerated thanks to a house-price boom that gave most middle-aged people the equity to invest in a new life of their choice, and an economic recession that left many middle-aged people with no choice but to invest in a new life. Fewer and fewer of the second wave came from within the brewing industry itself; but many of them were experienced businesspeople of considerable acumen, so although the retail trade remained largely closed to them, the rate of business failures slowed even as the rate of start-ups increased. By the end of the 1980s the number of new breweries had reached three figures.

Society of Independent Brewers (SIBA)

But while the older brewers stubbornly defended their position – and the Big Six between them owned well over 30,000 pubs – the moderate success the microbrewers had achieved only made them more frustrated. It was this frustration that transformed them from an assortment of very different people with very different backgrounds and ambitions into a movement. SIBA, the Small Independent Brewers Association (now the Society of Independent Brewers, but maintaining the original acronym), was formed at a meeting at a pub in Wiltshire in 1980 after the Brewers' Society (BS) had refused to admit new members who brewed less than 200 barrels a week. (This was regardless of the fact that perhaps a dozen existing BS members brewed nothing like that amount.) It was run from a ground-floor flat in Highbury by a couple of enthusiasts, Elisabeth Baker and the late Tony Williamson, who as well as scraping a living from running the small brewers' magazine, The Grist, were SIBA's occasionally-paid. And if SIBA was a ramshackle, cash-strapped organisation, it included some surprisingly high-powered people. Dave Roberts of the Pilgrim Brewery of Reigate, for instance, was a former senior civil servant in the Welsh Office who knew intimately how government worked; and so SIBA soon became a potent lobbying force.

Monopolies and Mergers Commission

In 1988, as a result of pressure from SIBA, CAMRA and licensees' associations, and with the help of a handful of supportive MPs, the Office of Fair Trading ordered a Monopolies and Mergers Commission (MMC) enquiry into the supply of beer. The mainstream brewers had been here before and had always escaped with no more than cosmetic changes. But this time their complacency proved ill-founded, for much to their surprise the MMC found there was a "complex monopoly operating against the public interest" and recommended measures to smash it. To the brewers' even greater surprise and indeed fury, the then Trade Secretary Lord Young proposed to implement measures in full. He quickly disappeared in a reshuffle, and although his successor, Nicholas Ridley, watered down the MMC's proposals considerably, the 1990 Beer Orders still had the desired effect.

The Beer Orders

The Beer Orders contained two key clauses. The first was that the Big Six (Bass, Allied, Grand Met, Whitbread, Courage, and Scottish & Newcastle) must either sell or lease free of tie half of their pubs above a 2,000 ceiling; the second was that their tenants should be allowed to stock one guest real ale of their choice. The first measure led, eventually, to the demise of the Big Six – only the smallest of them, Scottish & Newcastle, survives in anything like its 1980s form – and the emergence of giant non-brewing pub companies such as Punch Taverns and Enterprise Inns. The second led to a huge increase in the number of microbreweries, thanks to the opportunities it opened up in the retail trade.

The early 1990s were halcyon days for the micros as they saw their beers go on sale in pubs owned by the Big Six (and served, to add insult to injury, through handpumps and beer engines also owned by the Big Six, much to the Big Six's disgruntlement). One of the country's most successful local brewers, Wickwar of Gloucestershire, was founded in 1990 by Courage tenant Ray Penny expressly to supply himself and fellow tenants with guest ale. It wasn't long before Ray got rid of his Courage tenancy to concentrate on running his brewery, which has since moved to bigger premises and supplies around 350 free houses in the region. He is also developing a small tied estate of his own.

End Product Duty

Small brewers were boosted still further three years later when the archaic system of levying duty on the unfermented wort was scrapped. This system, which dated back to Gladstone's introduction of malt duty in 1880, put the smaller brewers at a serious disadvantage. All brewers – however large – were allowed 6 per cent duty-free wastage allowance, which meant that the more efficient big brewers were in effect selling 4 or 5 per cent of their beer tax-free, whereas smaller brewers were sometimes charged duty on beer they couldn't actually sell. In 1993 this system was replaced by end-point duty calculated according to the beer's percentage alcohol by volume. It meant there was no longer any need for a wastage allowance: brewers were only charged for what they had actually sold, and could even claw back the duty on beer spoiled or returned as unfit for consumption. So end product duty meant a more level playing-field as fas as prices were concerned.

The party didn't last, though. By the late 1990s the Big Six had disappeared, and the guest ale market had disappeared with them. But the microbrewers had made plenty of hay during the brief period of sunshine. Consumers had got used to the idea that Courage pubs didn't only stock Courage beer and Whitbread pubs didn't only stock

Whitbread beer. Choice had been put in front of them, and they liked it. The new non-brewing pub companies saw this and started stocking cherished traditional ales from big-name family brewers such as Adnams, Fuller's and Young's alongside national brands such as Bass and Tetley. Once the idea of choice had caught on it gradually became easier for microbreweries to find outlets, and by the time the Beer Orders were finally scrapped in 2002 as no longer relevant, they had trebled in number.

Progressive Beer Duty

The scrapping of the Beer Orders was mourned by CAMRA but not by anyone else. For as luck would have it, the very same year saw the next great legislative boost for microbrewing in Britain when after years of hard lobbying by SIBA (aided by CAMRA) the chancellor introduced the same system of progressive beer duty that was already common in Europe. Progressive beer duty (PBD) or "sliding scale" allowed a 50 per cent duty rebate to producers of less than 5,000 hectolitres or 3,000 barrels of beer annually (a brewer's barrel is 288 pints or 1.63 hectolitres). Brewers of more than 3,000 but less than 18,000 barrels a year got smaller rebates. It was a step that enraged established brewers; and even those few micros that had grown to near the 18,000-barrel ceiling were less than pleased, describing it as a disincentive to growth. But the proof of the pudding is in the eating: PBD has led to a further doubling in the number of micros in Britain, of which there are now approaching 600.

Direct Delivery Scheme

Following the hat-trick of the 1990 Beer Orders, end product duty, and PBD, SIBA scored again with its Direct Delivery Scheme. The withering away of the guest ale market as the Big Six were one by one snuffed out threatened to close the pub trade to small brewers again thanks to the appearance of non-brewing pub companies or "pubcos", as they are inelegantly abbreviated. Most of them were founded with estates of discards bought wholesale from the Big Six. Often they were set up by ex-Big Six personnel, financed by Big Six loans, and stocked only Big Six beers. But one by one they were bought out by aggressive predators funded by venture capitalist houses, in many cases foreign-owned, and very soon the links between the pubcos and the old national brewers were severed. Next, the Big Six themselves were targeted; as we have seen, all of them except S&N have now gone, bought out by Carlsberg from Denmark, Coors from the USA, and InBev, the Belgian-based Stella Artois giant.

At the same time, many old-established regional breweries were succumbing to a fad for separating their brewing operations from their pub estates. The near-universal business model in the industry had always been "vertical integration" – that is, the ownership of both manufacture and retail. One of the effects of the Beer Orders, though, had been the driving down of wholesale beer prices as the pubcos shopped around for their beer. In the light of this, the City of London decreed that brewing was no longer profitable, and that the regionals should become pure retailers. Those that gave in to the siren song and got shot of their breweries – Usher's, Gibbs Mew, Morland, Greenall Whitley, Morrells, Brakspear, Eldridge Pope, Burtonwood and others – have by and large been taken over, either by other pubcos, or by those rival regionals that still believe in vertical integration. For its opposite, as things turned out, has proved to be total disintegration.

Today's big pubcos source their beers more widely than either the national or regional breweries to whose estates they succeeded ever had. But they have always found it difficult to deal with local microbrewers. They own no drays or depots of their own; their pubs are supplied either directly by the brewers themselves or, more commonly, by third-party wholesalers. Micros have generally been too small to deal direct with such giants nationally – Punch Taverns owns 9,000 pubs, and Enterprise Inns has almost as many; and few of the new wave of brewers could supply even a fraction of such vast estates. And even when microbrewers' beers have been listed by a pubco for regional distribution, it has always been galling to have to deliver to a depot 100 miles away or more to get into a pubco-owned pub just down the road – and even more galling to have to return to the depot for the empties!

SIBA's Direct Delivery Scheme has created software that allows local brewers to deal directly with pubco-owned pubs that want to stock their ales. The scheme may appear cumbrous in that it's still the pubco that actually buys the beer from the brewer and sells it to the pub; but the transaction is entirely electronic, and in practical terms it's the brewer who makes the sale, delivers the beer, and collects the empties. (No small matter when the 72-pint firkins most commonly used cost £80 apiece, and the longer they stay out in trade the more you need to own.)

The Future of Microbrewing

In a nutshell, then – rather a large nutshell, granted, but the story of 40 years can't be told in 40 words – that's how we got where we are. The question is: where do you fit in?

That depends very much on your ambitions and your access to capital. Microbrewers cover such a huge span these days that the very term "microbrewer" is no longer applicable to them all. The biggest of them are very big indeed: Sharp's of Rock in Cornwall brews more real ale than the duchy's established regional, St Austell. Hop Back of Salisbury expanded through two Business Enterprise Schemes and now has a dozen pubs of its own and, in Summer Lightning, a genuinely national brand. St Peter's of Elmham, Suffolk, was founded by multimillionaire John Murphy of Interbrand expressly as an exporter and now sells its beers in their distinctive green flasks all over the world. At the other end of scale there are dozens, if not scores, of publicans who have installed tiny brewplants in their outbuildings just to supply their own pub. And the Jenkins brothers of Bullmastiff Brewery in Cardiff have won award after award but resolutely refuse to expand to meet the huge demand for their beers because they've got things just how they like them.

With such a wide range of sizes and styles of operation, microbrewers have experimented with a number of descriptors that would encapsulate them all. "Craft brewers" was tried and rejected as being too evocative of the kitchen sink. "New brewers" seems out of date when some of them are celebrating their 30th birthdays and have even changed hands as their founders retired. "New wave brewers" is accurate enough, but again, many of them don't feel that new any more. So they have settled on "local brewers". It may not always be accurate when many of them have national and even international distribution, but it feels right. It positions them in the local food movement and creates an identity that separates them from the regionals and nationals.

That's the hope, anyway; it doesn't always work out, but for many local brewers it has come true. It can come true for you, too.

Chapter One
How to Brew

CASE STUDIES

Brewing beer, as I mentioned in the introduction, is not actually all that complicated. Not in theory, at any rate. You simply steep malt in hot water until its starch content turns into sugar and dissolves; you strain off the resulting thin syrup and boil it up with hops; you add yeast; and when it all stops bubbling and foaming, you drink it. (Or, in your case, sell it for someone else to drink.) How hard can it be? But as with any craft, success depends on absolute mastery of materials and processes; and within the simple framework outlined above, the possible variations are almost infinite. So before you can present the public with a product of consistently good quality (and achieving consistency will be one of your key skills), you're going to need a lot of training, both theoretical and practical.

You can learn the brewing skills you will need on courses run by formal training providers such as **Brewlab (www.brewlab.co.uk)**. You can learn the basic engineering skills you will need to keep your plant running (if you don't already possess them) from the fabricator who installs it. But before you take either of those risky and expensive steps, you can teach yourself most of what you will need to know about beer by reading, tasting widely and intelligently, visiting working breweries, and by brewing at home. And you can start the reading part of your home-learning course right here.

In nearly 40 years of pub-going, I have never got over my surprise at how little the public actually knows about its beer. Many regular drinkers don't even know what it's made of; and public ideas about alcoholic strength are hopelessly confused. So let's start with a potted description of what beer is and how it's made.

The Basics of Brewing

Most alcoholic beverages have three components. The first is a liquid medium. The second is the sugar that the yeast will digest, alcohol and carbon dioxide being its waste products. Finally, there are the aromatic or flavour components. If you're making wine, all three come in one handy little package – the grape (or the strawberry or blackberry or whatever soft fruit you're using). All you have to do is crush your fruit, add yeast – or let naturally occurring wild yeasts do the job for you – and let it ferment. Hard fruits such as apples and pears are harder to crush – you have to pulp them in a mill before you can press the juice out of them – but the principle is the same. Everything you need is in the fruit. With beer, things aren't so simple. In fact, cereal grains are so hard to ferment it's a wonder that beer was ever invented. For in their natural state, grains contain almost none of the required components: no liquid, no sugar and not much in the way of flavour. All three have to be painstakingly brought together before you can even think about introducing yeast.

Malt

Malt is usually made from barley, but brewers also use other grains, both malted and unmalted. Wheat you will probably already be familiar with as a brewing grain, but oats, rye and even rice are also added for particular purposes. In essence, malt is cereal grain that has been tricked into germinating by being steeped in warm water to mimic spring weather. During germination, enzymes in the grain start to convert its insoluble starch into sugar, which the plant needs in order to fuel its growth. This sugar, maltose, is soluble, and is what you want as your fermentable material. So, soon after the grain has started to germinate, it's dried in a kiln to halt the process before it goes too far. Different kilning times and temperatures produce pale, amber, chocolate, crystal, roast and other types of malt, each with its own properties. You will almost certainly be using blends of malts, so you will need to be familiar with all their characteristics. You may also want to use brewing sugars in some brews: these have flavour characteristics of their own and are perfectly respectable adjuncts in many brewing traditions.

Liquor

Once the little sprouts have been shaken off, the malt is ground into a coarse flour ("grist") ready to be "mashed" in hot water (always referred to as liquor – water, in a brewery, is for washing and cooling). Mashing both completes the conversion of starch into sugar ("saccharification") and dissolves the sugar to produce a thin malt syrup or wort. The mineral content of the water is critical: gypsum-rich water from the artesian wells beneath Burton-upon-Trent produced the classic pale ales that succeeded the darker stouts and porters brewed with London water (well-water, not Thames water!) in popular favour in the 19th century. British ales require hard water while continental lagers require soft water, for reasons that will be explained later on. Some breweries are lucky enough to have their own wells or boreholes; you will probably have to settle for mains water, but you will want to analyse it and alter its mineral content first.

Hops

Next, the boil; and the addition of hops. Hops contain acids that kill bacteria and protect the beer from infection, and were introduced to British brewing in the 14th or 15th century chiefly as a preservative. But they are vital for flavour, too: their acid may give beer its bitterness, but they also contain oils and resins that produce much of its taste and aroma. Some hops are more acid than others; some more aromatic. And the aroma characteristics vary hugely from strain to strain: Goldings, for example, produce a citric tang, while Bramling Cross are said to give an aroma of blackcurrants. Again, you will need a thorough knowledge of hop varieties if you are to become a master of your craft.

Yeast

A microscopic fungus, yeast digests the maltose in the wort and excretes alcohol and the CO^2 that puts the fizz into the beer. Every brewery has its own strain, and every strain is different both in its handling characteristics and in the flavours it produces. Yeast is tricky stuff to handle and is very prone to infection; at the University of East Anglia in Norwich is the National Yeast Bank, where brewers send samples of their yeast from which to propagate fresh supplies if problems occur at the brewery. Brewers are very proud of their yeast strains, saying they give their beers their house character; so the strain you choose at the outset will define your brews for years to come.

Other Ingredients

More and more these days, you will come across other ingredients, especially fruit. Fruit has been used for centuries by brewers, especially in Belgium, where some beers are fermented by naturally occurring wild yeasts rather than cultured brewing strains. Wild yeasts and the other microfauna and flora that accompany them often produce sour flavours which are not to everyone's taste: adding pulped cherries or blackcurrants to the mash of sour brown ales and lambic beers is a traditional method of tempering their sourness. In the last 20 years or so these Belgian fruit beers have become more widely available in Britain, and many British brewers have been inspired to experiment with fruit as well. Grapefruit beer from the St Peter's Brewery of Suffolk is one unlikely but surprisingly palatable example

Almost as important as the raw materials to the character of the beer are the processes by which it is made. The grist must be mashed at a specific temperature for the starch to be fully converted and the sugars to be fully dissolved. The boil can be long or short – a long boil caramelises some of the sugars in the wort and produces a richer, darker beer – and hops are added in different quantities and at different points in the process. The hopped wort has to be cooled before the yeast is pitched, and there are various methods of fermentation which each produce different results. All of this you must learn thoroughly.

Vessels

The vessels can be critical, too, especially the shape and size of your fermenters and conditioning tanks. We'll come to the number and type of vessels you'll need in a later chapter. But as an example, take fermentation. It's a biochemical process which, like all biochemical processes, generates a certain amount of heat. This can dictate the speed of fermentation and help or hinder the creation of by-products such as phenols and esters, all of which will affect the finished product; and the thermodynamics of different-shaped fermenters will give different results. Modern cylindro-conical lager fermenters produce more heat and work more quickly than old-fashioned horizontal ones, so a lager conditioned in a cylindro-conical will be detectably sweeter and less clean than one aged in a horizontal. There are plenty of other similar examples; so even before you build your brewery, you will have to have made critical decisions about what sort of beers you want to brew. And it's important to make these decisions from a position of knowledge, which you can get right now – assuming, that is, that your local is open. Because the best way to learn about beer is to drink it, but to drink it intelligently. That means sampling as many different examples of as many different styles as you can get hold of. And sampling is not the same as swigging!

Tasting Beer

The first time I was invited to an organised beer-tasting, on a press trip to Bruges nearly 20 years ago, I thought someone was taking the mickey. Arriving at a bar called 't Brugs Beertje, the party was shown into the function room in which was a huge table garnished with an astonishing array of glasses (you shouldn't use the same one too often), and bottles of mineral water and dishes of dry crackers to cleanse our palates between samples. Once seated, we first held up each sample to the light to check its colour, clarity and head. There followed a thorough swirling and nosing à la Jilly Goolden, with tasters detecting aromas of leather, tobacco, new-mown hay et al. A meagre mouthful was then ritualistically swilled; air was sucked; and mouthfeel (ie residual sugars, if any, coating the tongue) and flavours were pronounced upon. Finally, we swallowed (you don't spit out beer as you do wine, since the bitterness receptors are at the very back of the tongue). A long hush ensued, with expressions of intense concentration while the length and bitterness of the finish (or aftertaste) were thoroughly cogitated. Finally, there followed a brief but intense discussion of each sample – was it true to style? Were there any off-flavours? How complex was it? – and each beer was marked.

Well, it seemed daft to me at the time. But after many years of attending such tastings, I have discovered the value of really concentrating on what you're drinking; and at tastings you have to strain your tastebuds in exactly the same way you have to strain your ears to hear a distant melody. Conveying your experiences is difficult, because you have to describe the flavours you detect in terms of others that are commonly recognisable; hence the descent into vocabulary such as "green fruit", "treacle toffee", "marmalade" and other comparables. It all sounds incredibly pretentious, yet the language of tastings is expressly not intended to exclude the uninitiated. In fact the very opposite is true: it's meant to allow a group of individuals (normally brewery staff) to describe their personal experiences of a range of sensations in a format that can be shared. But the difficulty of translating tastes and smells into words even affects skilled and experienced brewers. You might expect, perhaps, that professionals who live and breathe diacetyls, esters, phenols and sulphites would describe their beers in those terms. They do, but only to a limited extent. The flavour wheel used by brewers all around the world since the 1970s to delineate the aromas and tastes of their products includes those terms, but also uses expressions such as liquorice, chocolate, butterscotch, toffee, pineapple, catty, papery, leathery, grassy – even Horlicks!

It struck me quite early on in my beer-tasting career that there's a metaphysical dimension to it, especially to the attempt to explain your findings. You're trying to do two things, both of which are impossible: to objectify the subjective, and to describe one set of sensory experiences in terms of another. You might as well try to paint Beethoven's Fifth, or describe red. Nevertheless, you have to do it. You have to explore all the different possibilities of beer with your intellect as much as with your tastebuds, and for two good reasons. First, having chosen what you want to brew you have to be able to recreate your choice exactly, every time you brew it. And second, you need to upgrade from mere organolept – that is, a dilettante who just likes beer – into a ruthlessly efficient sensory analyst whose finely honed papillae (Latin for "tastebuds") can instantly and accurately detect and identify imperfections and variations from brew to brew.

From a practical point of view, though, wide-ranging tastings of a good selection of beers aren't that easy to arrange. To be meaningful, tastings need to be comparative: you need to be able taste and decide between a range of at least half-a-dozen bitters or best bitters or old ales or stouts or whatever. Very few pubs stock more than one example of any one beer style other than British-brewed lager, of which they might have two or three. Even the handful of real ale Meccas that carry 10 or 12 cask beers at a time will try to stock as wide a spread as possible. The same is true, if to a lesser extent, of supermarkets. They may carry a bewildering variety of pilsner-style beers, from cheap "slab-packs" of canned French or Belgian lager brewed mainly from corn syrup to top-notch Czech beers such as Budweiser Budvar, but of bottled ales they will have only one or two examples of each style. But by shopping around you can assemble wide enough selections of styles to hold meaningful tastings of your own, possibly inviting friends round to make an event of it. And don't exclude women! It's long been known in the industry, and recently confirmed by medical research, that by and large they have more accurate sense-memories than men. So a serious beer-tasting is not merely an excuse for a night in with the lads.

You should also join the **Campaign for Real Ale (CAMRA) www.camra.org.uk**. Each CAMRA branch is supposed to field an official tasting panel to help select entries for the *Good Beer Guide* and to judge the branch's annual beer competition. Members of these panels receive free expert training, which will be invaluable to you later on. But for some reason, branches are often short of volunteers to serve on their tasting panels, so if you put yourself forward you will undoubtedly be a very welcome member of the team.

Further Reading

Membership of CAMRA will also give you access to lots of useful reading matter. Its monthly newspaper, *What's Brewing*, is sent free to all members and carries interesting features on other breweries as well as beer-related news from around the country. The organisation is also a book publisher and seller. There is a host of good and informative books about beer styles and traditions by authoritative writers such as Roger Protz and Michael Jackson on the market; you may not find them in bookshops, but CAMRA stocks most of them for sale by mail order at a discount to members. There's also a commercial bookseller called Beer Inn-Print **www.beerinnprint.co.uk** which, despite its title, has a good selection of out-of-print titles.

You should also visit as many breweries as you can. Ten or fifteen years ago it used to be almost impossible to tour a working brewery unless you were in the licensed trade. Visitors were generally considered a nuisance who only got in the way, and most old-established breweries were such a jumble of pipework, narrow spaces, uneven floors, and steps and stairs, mostly awash with water, that tour parties were a health and safety nightmare. Today, it's all different: most of the older regional breweries run conducted group tours, and some even have visitors' centres with shops, bars and museums or, at least, displays of antique brewing equipment and brewery ephemera. This change of attitude has been driven mainly by marketing considerations. Smaller breweries with little to spend on advertising have come to see the benefit of drawing the public in. As a result, most brewery visitor centres are aimed more at the tourist than the would-be brewer; even so, they will acquaint you with the sights, sounds and, indeed, the smells that will surround you every day once you have opened a brewery of your own.

For more informal tours, often conducted by the brewer himself, which gives you a chance to ask questions and get full and informed answers, your CAMRA membership will come in handy yet again. All local branches organise regular trips to breweries large and small in their areas and sometimes further afield. You may even get the chance to do a bit of real hands-on brewing: most branches have a special beer made by a local brewer for their annual beer festivals, and sometimes branch members are allowed to do the actual mashing. This could be you! And if you get friendly enough with a local brewer, he may even allow you to help out occasionally (don't expect to be paid, though!).

Brewing at Home

Would-be brewers are luckier in one enormous respect than people considering new careers in other industries. For brewing is one of the few trades, along with baking and one or two others, in which you can coach yourself almost up to professional level in your own home, without giving up your job, and without investing too much of your savings.

Many of today's local brewers started off as home-brewing enthusiasts who finally tipped over and turned their hobby into their living. But even if you're not a frequent and regular home brewer yet, you should become one: it's the single most important step in your preparations for your new career. If you are already a home brewer, you'll understand the advantages: for instance, that it's possible to start slowly, with a kit that requires little more than the addition of hot water to make a passable beer. From the all-in-one kit you can graduate step-by-step through the various phases of difficulty until you're quite at home with all-malt, whole-hop brews. Home-brewing kit can also be upgraded piece by piece from a bucket in the airing cupboard to a miniature tower brewery complete with mash tun, copper, hop back, wort cooler, fermenter and conditioning tank, all of gleaming steel and copper, perfect in every detail. One is always reading of enthusiasts whose garden sheds or garages are home to lovingly constructed brewing plants that need only an excise licence to start operating commercially – indeed, there are one or two cases where this has actually happened.

Brewing at home will familiarise you with brewing materials and techniques, with faults and their cures, and with different beer styles. It will also give you the chance to experiment with recipes; and I have heard of local brewers who perfected their entire ranges during their amateur days and were able to turn professional with enviable smoothness. But there's more to it than that. Home brewing will help you in practical terms, yes, but also in psychological terms. Have you really got what it takes to be a brewer? Are you a true enthusiast, or merely a dabbler? If you find your interest in brewing as a hobby palls after a few months, you can be sure it's not going to be a wise move to try and make a living at it. By the same token, if you find the bug is really biting, you can be more certain than many others in your position that the step you're about to take is the right one.

Since Boots stopped stocking home-brewing and wine-making equipment and materials some years ago, they are not as easy to come by. However, the Craft Brewing Association (CBA), founded in 1995, can fill the gap. Its website, **www.craftbrewing.org.uk**, lists local home-brew supplies shops and, for those who don't have one in their area, mail-order suppliers too. It's a mine of information in other regards, as well: anyone with even a passing interest in beer could browse its technical pages for hours. In addition to that, the CBA offers training for beer tasters, runs competitions, helps with recipe designs, has beer style guidelines of its own, and has a quarterly newsletter, *Brewer's Contact*. It's divided into local groups covering most of the country; if there isn't one near you, use the link on the site to the **National Association of Wine and Beer Makers (www.nawb.org.uk/)**, which is more wine-oriented but still lists many home brewers among its membership. You can now buy home brewing kits online at sites such as **www.brewingathome.co.uk**.

Formal Training

So you've drunk intelligently; you've read up on the subject; you've toured breweries; you've brewed your own beer. Now you're ready for your formal training. But where are you going to get it?

Fundamentals of Minibrewing

The obvious answer is that you'll get all the information you need from the person or company who (assuming you're not going to do it all yourself) designs and installs your brewery. And in fact the man recognised as one of the country's leading microbrewery designers started out 17 years ago as a trainer. David Smith of York had spent over 20 years brewing at the legendary Sam Smith's in Tadcaster before branching out as a consultant on his own in 1989, just as the microbrewing boom was about to go into overdrive. Teaching people how to brew was his next venture; but it soon became clear that obtaining small-scale brewing equipment was one of the new entrant's biggest problems, so David branched out into sourcing and fitting the kit as well as imparting the knowledge needed to operate it. Since then he has designed and installed nearly 100 small breweries, but he has never lost sight of the fact that the equipment isn't much use without the know-how. So, twice a year, he teaches his four-day fully residential Fundamentals of Minibrewing course, using the conference suite at York Brewery as his classroom and York Brewery itself for the practical side of the course. See details at **www.brewingservices.co.uk**.

Fundamentals of Minibrewing, which costs £695 plus VAT, is firmly rooted in the practical experience that David and his guest lecturers from other areas of the brewing industry – including maltsters and hop merchants – have gained in their working lives. The purpose of the course is to instill the basics of good brewing practice and a greater insight into the art and science of brewing.

It combines lectures and workshops, run by a number of specialists from the industry, with visits to a maltings, other microbreweries and retail outlets. For those who may not have any practical experience there is the opportunity to help with a complete brew at York Brewery before the course gets under way.

Heriot Watt University

At the other end of the scale is the Bachelor of Sciences course Brewing and Distilling at Heriot Watt University **www.hw.ac.uk** in Edinburgh, one of the world's leading brewing schools since 1903. The BSc course has an annual intake of 10–14 undergraduates who in their four years (this is a Scottish university, not an English one) at Heriot Watt will learn not only how to make beer but how to turn it into whisky as well. It is a complete course training graduates for employment in both the craft and larger brewing companies.

Heriot Watt is also world-renowned for its Master's degree in brewing and distilling – so world-renowned, in fact, that usually less than a quarter of the annual on campus intake of 35-45 are British. If you fancy the letters MSc after your name, you need a first degree in chemistry, chemical engineering, biochemistry, microbiology or something similar, and if you are applying from outside Scotland a substantial sum for the fees. Again, it is a complete course training graduates for employment in both the craft and larger brewing companies.

Heriot Watt's Postgraduate Diploma distance learning programme, though, is a different matter altogether. Consisting of 8 courses at £1000 each, it's tailored for busy working people and you can take up to four years to complete it. One can choose only the courses that seem most useful to you, getting a separate certificate for each course you complete, or a Postgraduate Certificate on completion of 4 courses. On completion of the PG Diploma it is possible to continue to complete the MSc. Seven courses are taught entirely electronically via an online portal and email and telephone contact is encouraged with course directors. The cohort of approximately 100 attracts brewers from all over the world and all students attend a 2 week practical workshop to gain hands-on experience in brewing, distilling, malting and microbiology during their studies. For more details, go to **www.postgraduate.hw.ac.uk/prog/119**

Brewlab

Brewlab (**www.brewlab.co.uk**) is a centre for brewing studies and analytical, microbiological and consultancy services, based within Sunderland City boundaryb and teaches courses such as Start up Brewing and British Brewing Technology, and also supplies specialist training in taste evaluation, microbiology and small scale bottling services. Courses range from one day introductory workshops in bottling, tasting and practical brewing to three week and three month

CASE STUDY

Home Brewer Turned Professional– Hoggley's, Kislingbury, Northants

It was a burning desire to find out what people really thought of his beer that started Roy Crutchley on the path to full-scale commercial brewing. And the path was, quite literally, his garden path.

Youth worker Roy had been brewing his own since his student days in the 1970s, but only using kits and malt extract. Then he got hold of one of the many CAMRA-published books by home-brew guru Graham Wheeler, which explained the mysteries of full-mash brewing – and from then on he dreamed only of turning pro.

Partner Julie Hogg gave him another powerful push in the right direction when she evicted him from the kitchen of the couple's cottage in the Northamptonshire village of Kislingbury. For full-mash brewing is a very different proposition from boiling up a dollop of extract in the old Burco.

"Mashes will occasionally boil over however careful you are, and wort is the stickiest substance known to man," says Roy. "The whole process was incredibly messy, and eventually Julie chucked me out of the kitchen.

"Fortunately there's a patch of ground at the bottom of our garden just big enough for a shed – so I built one."

It wasn't quite that simple, though: as Kislingbury is a conservation area, and the shed Roy had in mind had to be 10-foot tall to accommodate a three-stage gravity-fed or "tower" brewery, he needed full planning permission. And he was sure he wasn't going to get it – until, that is, he got a visit from a sympathetic environmental health officer.

"His dad had been a full-mash brewer, and he seemed to be the only council officer with a grasp of what I was proposing," says Roy. "I can't say for sure it was down to him, but after his visit the whole thing seemed to go a lot more smoothly."

The equipment itself was home-assembled and had a capacity of 100 litres, and Roy put it together with a lot of help from the Craft Brewers' Association.

"The advice of the CBA is essential for anyone like me who is evolving rather than jumping in at the deep end," he says. "They were incredibly helpful."

Roy always planned to go commercial, as much to massage his ego as anything else. "I'd brewed for family and friends but you can never be sure how good you are when you give people free beer, because they're always going to say nice things about it," he says. "What I wanted was to see a pumpclip with my name on it on a beer handle in a pub, with people I didn't know buying a pint of my beer – and then coming back for another."

There were other issues as well – Roy often felt frustrated that as a youth worker he couldn't see the results of his efforts until years later; and anyway, he felt he had reached a career plateau. But chiefly, he says: "Running your own brewery is every middle-aged man's dream."

Actually turning pro, Roy and Julie found (the brewery is called Hoggley's, although Crutchley & Hogg might have had a certain traditional ring to it), wasn't that difficult: change of use permission went through on the nod, and Customs couldn't have been more helpful. Roy did forget at first that he needed to talk to the water company as well, but that was soon put right.

Anyway, in March 2003 the first pint of Hoggley's Northamptonshire Bitter went on sale at the Alexandra Arms in Kettering, and Roy's dream of seeing people actually buying his beer came true. A year later, it was formally confirmed that people he didn't know liked his beer when Hoggley's Mill Lane Mild was named joint Beer of the Festival award winner at Milton Keynes.

All this was achieved without Roy giving up his day job or moving out of the 10-foot tower brewery at the bottom of his garden. But in 2006 he was made redundant, and the money went into equipping a new eight-barrel brewery on an industrial estate in nearby Litchborough. And he's a happy man.

"If any home brewer is thinking of going all the way, I would say: DO IT!" is his advice.

www.hoggleys.co.uk

programmes in brewing theory and practice. It also runs a range of 'after sales services' to assist brewers with quality control requirements, tax and excise compliance, yeast storage and propagation. Due diligence packages can be tailored to individual brewery needs, helping to maintain quality and compliance with the industry codes of practice.

Brewlab was founded at the City of London Polytechnic in 1986 to provide laboratory analysis for brewers large and small, the training side of the operation was only developed later to enable staff in the microbrewery sector to understand and interpret analysis results.

In 1992 Brewlab moved to the University of Sunderland to take advantage of it's extensive biotechnology laboratory facilities.

In 2008 Brewlab moved out of the University into temporary brewing and educational facilities, and then in 2011 Brewlab moved into a modern purpose built headquarters incorporating facilities for teaching, analysis, brewing and consultancy services.

Its laboratories provide up to date facilities for practical work alongside dedicated equipment for detailed studies. There is a fully operational 100 litre (22 gallon) pilot training plant and small 25 – 40 litre kits for recipe development and training. Small scale bottling facilities have been recommissioned allowing practical experience and the opportunity for students to brew beers for sale in local outlets. Students also have access to other laboratory and research facilities.

Brewlab also operates a fully commercial three barrel showpiece brewery which can provide an industrial setting for student placements and allow them to experience full scale production. There is also strong links to local microbreweries allowing for the opportunity for student placements in a variety of plant types, sizes and settings.

Managing the teaching programme and laboratory activities are Dr Keith Thomas, Arthur Bryant, Mike Hitchen and Alison Hedley, with many years industry experience. In addition to the scheduled Brewlab courses, they occasionally lecture on local university courses and practical sessions such as microbiology, biotechnology and food science. Dr Thomas's specialist research areas are yeast physiology, beer analysis and bioremedioration. He is often a judge at major beer festivals and competitions both within the UK and internationally and a regular contributor to the SIBA journal. All of the above team are Members of The Institute of Brewing and Distilling (IBD).

Additional staff provide specialist support for both teaching and analytical services, while students also benefit from the expertise of visiting lecturers, who work in the brewing industry or allied trades.

The Practical Brewing and British Brewing Technology courses are accredited by the Open College Network (OCN). This follows official assessment of Brewlab's course structures and teaching quality. Brewlab students may also study for external qualifications through the Institute of Brewing and Distilling notably the General Certificate and Diploma exams. Students can also focus on developing practical skills using the varied brewing capabilities available on site.

Brewlab Training Courses

- Brewing Skills Development for production personnel
- Start Up Brewing
- Certificate in Practical Brewing
- Diploma in British Brewing Technology
- Microbiology Workshop
- Advanced Craft Brewing Workshop

For further information contact **info@brewlab.co.uk** or go to **www.brewlab.co.uk**

The course options include one day workshops for craft brewers who want to develop their skills and knowledge, three or four day start-ups, residential three week and three month courses for those who wish to make a career in brewing, either to manage their own brewery, to seek employment in the industry or to further their career prospects.

For many of the courses, groups are limited to 12 students to enable staff to spend more time with individual students. Although some knowledge of maths, chemistry and biology would be beneficial on the more detailed courses, Brewlab provides full student support and additional tuition.

There is a range of professional and development courses for those who already work in the brewing industry wanting to improve their technical and brewing skills.

Students successfully completing course assessments and examinations, will receive a Certificate in Practical Brewing or a Diploma in British Brewing Technology accredited by the Open College Network equivalent to NVQ level 3.

For the following courses, all dates and fees are published on the Brewlab website.

Brewing Skills Development for Production Personnel

Brewlab has created this course following requests from clients for training to enhance the knowledge of existing staff. The course is aimed specifically at staff involved in the brewing and production processes and those wanting to improve their understanding of:

- The key factors on which quality ale production depends
- The brewing process – from raw materials to final product
- Production problems, enabling them to be addressed on site, leading to more consistent quality products

The course starts with targeted sessions on the four main raw materials – liquor, malt/adjuncts, hops and yeast. Each session discusses the raw material, its specifications and commonly encountered problems with its usage. The fundamental brewing processes are covered in brewer's terms, along with current industry best practice.

The processes covered include malting, mashing, sparging, boiling, cooling, fermentation, maturation, finings, racking, casking, cleaning, hygiene, quality control and tasting.

Course Programme

- Day 1 - Recipe formulation, brewing liquor, malt, mashing and sparging
- Day 2 - Hops, boiling, cooling, yeast, fermentation, maturation, cask racking and fining
- Day 3 - Quality Control, food safety and beer flavour profiling
- Day 4 - Food safety and practical brewing

It is not necessary to attend the full four days and participants can choose which days they wish to study.

Start Up Brewing

Start Up Brewing is a comprehensive three day course with an optional fourth day hands on brewing in a working micro brewery. The course provides a realistic introduction to the brewing process and the requirements of commercial brewing. It is suitable for potential or recently engaged brewers and those looking at brewing as a possible career change or business opportunity.

No background knowledge of brewing is required for the course and the aim is to provide you with a sound overview of the industry and an understanding of the requirements of operating a commercial brewing venture. Opportunities are offered to view the stages of a commercial brew, to study the basic steps of brewing and to discuss business start up issues with working brewers and specialists.

Certificate in Practical Brewing

A detailed three week examination course in the theory and practice of brewing. The Certificate in Practical Brewing has an emphasis on traditional procedures and provides up to date knowledge, with a strong practical element through laboratory work and placement brewing opportunities. Course work is extended with a series of site visits and external placements.

Additional subsidiary components cover:

- Basic microbiology
- Taste training , cellar craft
- Food safety
- Marketing, sales and business start up

Students completing the course will be awarded a Certificate of achievement comparable to a NVQ level three. Accommodation is provided at the University of Sunderland halls of residence as part of the course fees.

Diploma in British Brewing Technology

The three month examination course offers extensive practical experience with regard to recipe formulation and practical brewing, allowing in depth utilisation of theory gained on the course. External placements are timetabled throughout the course and additional time is dedicated to self study and research for assessments where relevant. Students are also extended the opportunity to study marketing, business planning and sales techniques and experience group presentations.

Assessments include Brew to specification technique, Quality Assurance plan construction and brewing log records.

The module supporting the course, Brewing Technology, has no formal prerequisites for study, but students are advised to anticipate some basic biology, chemistry and maths, also some experience of brewing on a home brew or commercial plant would be beneficial.

Students completing the course will be awarded a Diploma of achievement comparable to a NVQ level three.

The majority of students attending this course are successful in either starting up their own business or gaining employment in the industry.

Accommodation is provided at the University of Sunderland halls of residence as part of the course fees.

One day Microbiology Workshop

Brewlab's dedicated Microbiology Workshop offers a comprehensive introduction to the use of microscopy and microbiology techniques. The course can be used to develop practical skills in laboratory analysis as well as to instigate a quality control programme for due diligence purposes. This is a one day workshop where students can learn to use a light microscope to carry out basic microbiological techniques. Some understanding of the brewing process, along with basic maths and chemistry would be beneficial but not essential.

By the end of the course attendees will have gained knowledge on how to set up and implement your own microbrewery hygiene and quality control procedures.

Advanced Craft Brewing Workshop

Advanced Brewing Workshops provide opportunities for brewers with amateur or small scale commercial micro-breweries to investigate their brewing operations in technical detail. The sessions run over an intensive weekend based at Brewlab's headquarters.

Full laboratory support is provided including technical staff, brewing facilities and a dedicated analytical laboratory. Participants are provided with opportunities to develop analytical skills including microscopy, basic chemical procedures and flavour evaluation. Some experimental work is conducted by Brewlab on relevant brewing issues. The sessions are intensive but rewarding and excellent value in helping to improve your products or to provide quality assurance.

Analytical Services

Brewlab provides extensive analytical and investigative services to the brewing industry, ranging from routine microbiological and chemical testing to detailed fault finding and product evaluation.

CASE STUDY

Brewlab Graduate – Potbelly, Kettering, Northants

For Northamptonshire businessmen Toni Hooper and Glenn Morris, starting thier own brewery was a sternly practical exercise.

Toni and Glenn run a factory in Kettering where they employ 40 people making leather fashion accessories. But increasingly cut-throat competition from cheap imports prompted them to consider diversifying, and setting up brewpubs seemed a likely venture. A one-day seminar on the pub trade soon changed their minds, though. "We went along to have our choice confirmed or to be put off and when we came away we'd decided we didn't want to run pubs after all," says Toni.

They hadn't given up on the idea of brewing, though, and Toni booked them in for Brewlab's three-day Start-Up Brewing course in early 2004.

"We went on the course for the same reason we attended the seminar – to confirm our choice or to put us off," says Toni. "It confirmed our choice."

Following the course the two men set about installing a 10-barrel plant at the back of their factory, and Toni immediately booked himself in for the two-week Practical Brewing course that July.

"It covered much the same ground as the Start-Up course, but in much greater detail and with hands-on training that included a placement at York Brewery," says Toni. "I was never a home brewer and we decided to go into brewing for purely commercial reasons. Our intention was always to go to market with products of consistent high quality, and Brewlab helped us achieve that."

The proof of the pudding is in the sales ledger – and in the trophy cabinet. Within a year of going into full production, Toni and Glenn's Potbelly Brewery had won the Society of Independent Brewers' 2006 National Beer Awards with Beijing Black. Potbelly is now brewing twice a week and Toni has had to co-opt one of the factory staff, Ben Bulcock, to do the brewing while he hits the road to drum up more sales. Ben, too, was trained on Brewlab's Practical Brewing course.

If Toni has one criticism about the training he's received, it's that it doesn't tackle the nuts-and-bolts engineering side of the business.

www.potbelly-brewery.co.uk

Microbiological Services

- Quality control and fault finding
- Analysis services

Chemical and Physical Analysis

- Due Diligence analysis packages for legislative compliance

Yeast

- Yeast supply
- Yeast safe deposit service
- Yeast analysis and characterisation

Product Development

Brewlab can assist with recipe formulation, water analysis and product specific liquor treatment advice, as well as providing access to pilot brew plant of 25 litres through 500 litres. Our analytical/microbiological laboratories have access to a variety of instruments to analyse any new products:

- Gas and HPLC chromatography
- Mass spectrometry
- X ray analysis and electron microscopy

Sensory analysis services include a panel of fully trained and experienced beer tasters. Where appropriate advice on how to conduct taste analysis on your site and are pleased to work with breweries to develop quality assurance programmes involving both analysis and preparation.

Brewing Consultancy

If you require advice on simple recipe or raw material uses, through to a full turnkey project, Brewlab's technical staff can assist you through the process. Our brewers have extensive consultancy experience, contact **mike@brewlab.co.uk**.

Metal-bashing

Learning to make beer is not the end of the skills you'll need as a full-time brewer. Big breweries have their own engineering services departments; you'll have to be your own engineer, because brewing on a small scale is a very hands-on business. Assuming that you can't afford just to hire contractors and sit back and watch them work, you'll

CASE STUDY

Blue Monkey Brewery

The Blue Monkey brewery was established in 2008 by former IT worker John Hickling. The initial intention was to keep things small. John was finding working in IT for a bank very unrewarding, and wanted to do something creative, hands on and give him some independence. He spent a good year or so coming up with ideas for businesses, and of all the ideas he came up with, setting up a brewery was the one that stuck. He had always been a fan of good beer, so it was very appealing.

The first few months had their highs and lows – John was very fortunate in that his beers and branding were an immediate hit and selling them wasn't too difficult, but what he had not prepared for was the challenge of spending much of his working day alone in a cold and draughty brewery. This all changed though once he had managed to build the business up to a big enough size to justify hiring his first employee. For him, that was when he started really enjoying running a brewery, and now there are five employees working and there's some real energy in the place, which he thrives on.

After quickly winning a clutch of CAMRA and SIBAS awards, Blue Monkey's growth rate has been phenomenal, with sales going from 25 to 250 firkins a week in just over two years. Sometimes success can cause its own problems though. The one major regret that John had was that the first brewery he built was just too small. The brewery was creaking at the seams less than a year after he had started. He then went through the expensive and disruptive process of building a new and much bigger brewery.

Above all, John believes that what really counts in the market place is the quality of the beer you brew. Producing great beer is absolutely critical and should be at the heart of everything you do. If you get the beer right then everything else will fall into place.

The final piece of advice that John has is that brewing is a business, not a hobby. Brewing beer is hard, physical work. Selling beer, day in day out, is exhausting. Doing your accounts and administration is a daily grind. Drinking beer is the fun bit, and he would strongly recommend that you save that for your well-earned evenings off!

www.bluemonkeybrewery.co.uk

need both building and project management skills to prepare your premises even before the brewery equipment arrives.

Knowing one end of an oxyacetylene torch from another, unfortunately, doesn't form part of any known brewing course; so some experience of practical fabrication will be a must when the plant is being installed and assembled. Your consultant, if you use one, will oversee the project; but you'll need to play an active part because once the contractors have left it will be up to you to run and service the plant. And as there's a lot of pipework in a brewery, there's no better way of gaining the intimate knowledge of your plant that you'll need to keep it running smoothly than actually rolling up your sleeves and helping install it. And once it's installed, you'll need the skills and experience to make alterations and to carry out running repairs. Some people delight in metal-bashing and take to it naturally; others don't. Which category you fall into is something you'll only discover when a crucial valve springs a leak in the middle of the night...

Chapter Two
Where to Brew

CASE STUDIES

One of the first, most important, and in some ways most difficult tasks you are going to face is finding a home for your brewery. In some cases it won't be such a problem because the premises will have come first. You might be a publican installing a brewery in an outbuilding, for instance, or you might even have been inspired to start brewing commercially because the ideal site has come up. But whichever direction you decide on, finding a home is the first major practical step.

And once you've found the premises, there's a long road ahead before the first brew is loaded on to the dray. Not only does the building have to be converted and equipped, but there are more regulations to satisfy and procedures to go through than you can shake a stick at. More of these later, though; for now, let's stick with finding the right place to make your brewing dream a reality.

The first new brewery in post-war Britain, as we have seen, was founded in a Scottish baronial pile, Traquair House, when the laird restored the 18th-century brewhouse that had originally kept the castle's many inhabitants and visitors supplied with ale. By a strange coincidence, the first new brewery in Northern Ireland was also founded in the outbuildings of a grand country house. Hilden Brewery was set up by Anne and Seamus Scullion in 1981 in the stables at Hilden House, which was built in 1837 by the owners of Barbour Threads in nearby Lisburn, one of the biggest thread mills in the world. The immediate advantage was the instant availability of a suitable building; but Hilden House also had plenty of room for a bar and restaurant and for an annual beer and music festival, which becomes more popular every year.

Few would-be local brewers have a Traquair or a Hilden House to base themselves in, and the vast majority find homes on industrial estates. But there's an important question to answer here, because your first brewery may have to last you for years. Do you see brewing as purely an industrial activity, something you just want to get on with as efficiently as possible? Or is there more to it than that? Is the brewery itself part of the retail strategy with tours, a shop, a bar, maybe a restaurant?

Both strategies have their pros and cons. Making the brewing process as efficient as possible, without the distraction of tour parties and retailing, will hold your costs down and give you more time to spend out on the road selling your beer. On the other hand, welcoming people into the brewery generates cash and can also be a useful marketing device. But whichever model you choose, you will be bound by it for some time. If the former, you will opt for a unit on an industrial estate which no tourist will ever see. If the latter, you need to find a site that is attractive in itself thanks to its location, or to its architectural and historical value, or even both.

Historic Properties

The local brewing revolutions in mainland Britain and Northern Ireland started in grand country houses, as we have seen, and historic properties have proved popular sites for small breweries ever since. But by historic properties, I don't only mean grand country houses like Traquair and Hilden, but also units in regenerated industrial areas – 18th-century canal wharves, 19th-century cotton mills – and one-off conversion opportunities such as watermills, railway sheds, smithies, bakeries and so on. What these have in common is that they make the brewery itself a focus of interest, with the potential to generate considerable income from shoppers and tourists. This category could also be extended, therefore, to cover breweries in caravan parks, garden centres and other locations that attract paying customers.

Country Houses

Let's deal with country houses first. Ironically, perhaps the most ideally sited country house brewery remains resolutely off the tourist trail. Stanway House, near Winchcombe in the Cotswolds, is a fabulous confection of late Elizabethan and Jacobean stone with a baroque water garden that boasts the world's highest gravity-fed fountain and, for good measure, a 14th-century tithe barn. But the house is no museum, and the family that still lives in it after nearly 500 years only opens it to the public on Tuesday and Thursday afternoons in June, July and August. The brewery founded by Alex Pennycook in 1993 in Stanway's original brewhouse doesn't even unbend that far: if you want to see it and its two ancient wood-fired coppers, you have to content yourself with a virtual tour on Alex's website.

Another country house brewery; another wood-fired boiler. With the help of Keith Bott of Titanic Brewery of Burslem, Staffordshire, the National Trust has restored the old brewhouse at Shugborough Hall near Cannock. Keith oversees the brewing there, and says he has a lot of fun with that boiler. For getting the temperatures right can be a nightmare: on calm days it's a struggle to get it to boil at all, he says, while on windy ones it's a struggle to get it to stop!

Then there's North Yorkshire Brewery which, like Hilden, makes good use of its very stately home. Pinchinthorpe Hall near

Guisborough is also a bistro, restaurant and hotel. Mention might also be made here of St Peter's Brewery, tucked away deep in the Suffolk countryside. It's not quite a stately home, but a Tudor moated manorhouse that was built using medieval gothic stone doorcases and huge ecclesiastical windows salvaged from the nearest dissolved monastery. The brewery is housed in former farm buildings in the grounds; St Peter's Hall itself, decorated with the owner's collection of French medieval ecclesiastical artwork and other antiquities, is surely the most unusual and beautiful restaurant and functions room in the country.

Another country-house-based brewery is Elvedon, set up in the old stableyard at Lord Iveagh's stately home near Thetford as part of a larger development of visitor attractions. And although these are exceptional cases it might still be worth your while knocking on the doors of country houses in your district and asking if all their old stables, coach-houses, kennels and so forth are in use. After all, the worst they can say is "'Fraid so, old boy".

Restored Industrial Areas

A better bet, if you have decided on a visitor-based "heritage" dimension to your business, is a unit in a restored industrial quarter. Rents in these kind of developments are likely to be high, but they attract shoppers and tourists, they ought to be well suited to the purpose, and planning permission should be fairly straightforward – either they're designated as light industrial land will already have all the utilities and access you need, or they're unsuitable, in which case you'll be told at the outset.

The list of local breweries that have found themselves a home in some gem of industrial archaeology is a long one. Grainstore is in a Victorian railway granary in Oakham. Stonehenge is in the turbine-house of a hydroelectric power station that originally supplied the Royal Flying Corps aerodrome in Netheravon, Wiltshire. Big Lamp occupies a former steam-powered pumping station in Newburn, Tyneside. Archers, Atlas, Beckstones, Empire, Howard Town, Millstone, Old Mill, Saltaire, Shardlow, Slaughterhouse, Stirling, Strathaven – the list goes on.

Surprisingly often, the old industrial buildings in question were once breweries, maltings or hop kilns themselves. Exmoor of Wiveliscombe, Somerset, Wickwar Brewery of Wickwar, Gloucestershire, Wychwood of Witney, Oxfordshire, Tower Brewery of Burton on Trent, Uley Brewery of Uley, Gloucestershire, and Shaw's of Dukinfield, Greater

Manchester, all occupy former breweries. Frog Island of Northampton and Black Sheep of Masham, North Yorkshire, are both housed in former maltings, and Marches and Mayfields of Herefordshire are in old hop kilns. This kind of location adds general interest to any brewery lucky enough to find one, and sometimes makes a quirky and attractive shop, brewery tap and/or visitors' centre too.

As well as stately homes and historic industrial buildings, local breweries are to be found in a wide range of often bizarre locations. Plassey and Haywood Bad Ram are on caravan parks whose bars sell their beers. Bell's is in a garden centre. Nelson is in the historic dockyards at Chatham, Kent. Country Life is in the Big Sheep visitor attraction near Bideford, Devon. Brew Wharf is attached to the Vinopolis permanent wine exhibition in Southwark. Barearts is part of an art gallery in Todmorden, West Yorkshire, which is devoted to the human nude. Locations like this are one-offs, and finding one depends on your local knowledge. If your business model involves retailing from the brewery, though, finding a site that already has a substantial footfall of people who come prepared to part with cash could be well worth the effort.

Farm Buildings

Farm buildings have always been popular locations for new breweries, and in fact more than 100 of today's 600-odd local breweries are sited on farms. (Another half-dozen or so are on vineyards, which is pretty much the same thing.)

Usually, these are straightforward commercial arrangements where new brewers looking for sites have found farmers trying to get a rent out of older buildings that don't suit modern agricultural needs. In many cases, farmers with surplus buildings and homeless would-be brewers have been brought together by good fortune, sound local knowledge and a wide network of contacts. In others, old farm or estate buildings have been converted into straightforward industrial units for rent just like any other. This accounts for the location of Oldershaw's on the Harrowby Hall estate near Grantham, and Wentworth Brewery in a former generating room on the Wentworth estate near Rotherham.

Often there's a conservation aspect to these arrangements. Old brick or stone dairies, cowsheds, hay barns, granaries and the like may be picturesque, and may even be listed; but if they're to be preserved they need a commercial raison d'être, and brewing is in many cases the ideal use. Two such breweries are sited in redundant farm

CASE STUDY

Brewery in a Former Agricultural Building – Peak Ales, Pilsley, Derbyshire.

Rob and Debra Evans were teachers living in Worcester and working in the Black Country. But Rob knew what he really wanted to be – a brewer. And he knew where he wanted to brew – the Peak District National Park, near his native Chesterfield. So Rob and Debra sold the family home and moved with children Lucy and Matthew to a rented house in Bakewell so they could pursue the dream.

It took a good six months to find the right place to base their brewery – and much, much longer than that before Peak Ales beers began to flow. At first, with York-based brewing consultant David Smith as their guide, the Evanses trawled the usual round of industrial estates. One promising unit fell through, and David urged them to reject a second because if they'd ever wanted to move out they would have had to surrender it as they'd found it. That would have meant removing such necessary works as sloping floors, so moving out would have been as expensive as moving in.

Rob had already decided that the project needed a more characterful setting than an industrial estate, anyway. If the brewery was going to establish a connection with the Peaks, he felt it needed a more traditional building. He had decided to start asking farmers about surplus buildings when the landlord of the Devonshire Arms at Beeley, where Rob and David were having a meeting, suggested they approach the Chatsworth House estate office. Rob wrote a letter on the spot and dropped it round to the office, where he happened to meet the estate's deputy manager. And yes, the estate had the perfect building, quite nearby, the right size, a good sound roof; it just needed a bit of restoration.

Well, that roof (recently retiled) and four stone walls were just about all the barn had. It didn't have luxuries like floors or water or electricity.

It also had a yard that was several feet deep in the litter of generations of cows. But the estate had already decided to restore it and had in fact tried to get permission to turn it into housing or a shop. No way, said the Peak Park Authority. It could only be agricultural or light industrial. So Rob came along at just the right moment.

The restoration was a partnership. Rob applied and – after much form-filling – got a Rural Enterprise Grant from DEFRA (which is what the Ministry of Agriculture is called these days). The Chatsworth estate matched the grant and recommended a specialist firm of builders to do the work. It also appointed a project manager.

"It was very much a two-way street," says Rob. "I got my business in a beautiful location; the estate got a good building with offices and storage and a tenant paying a commercial rent."

Despite the estate's co-operative attitude, though, it wasn't all plain sailing. The work included building internal walls and floors, digging drains, laying cable and laying on a water supply – and it wasn't a help when a brood of baby barn owls was discovered in the building. Barn owls are protected, and work had to stop for three months while they grew up. And all that was before the brand-new Ten-barrel brewery from German manufacturer Moeschle was installed.

Finally, though, in February 2005 Peak Ales was open for business.

"It was a bit tricky and there were plenty of headaches," admits Rob. "It was two years behind schedule and all our contingency funds had disappeared, and I'm not sure I'd recommend it to anyone else.

"But in the longer term the Chatsworth connection will be invaluable. It's an idyllic setting and one that epitomises everything craft brewing is supposed to be about."

There's a short-term advantage to the Chatsworth connection, too. Both the pubs owned by the estate – the Devonshire Arms at Beeley, where it all began, and the Devonshire Arms at Pilsley – stock draught Peak Ales beers, as does the estate workers' social club. And both the Chatsworth gift shops stock the bottled versions. In crudely commercial terms, identification with Britain's premier stately home has instantly given Peak Ales sales volume and brand recognition that goes right off the scale.

The brewery now employs seven people, has a very good working relationship with the Chatsworth estate and is looking to export bottled beers. Its main business is still centered around supplying the free trade in and around the Peak District.

www.peakales.co.uk

buildings owned by the National Trust. Branscombe Vale was founded in 1992 in two former cowsheds on the quaintly named Great Seaside Farm at Branscombe, Devon, while Westerham was founded in barns at Grange Farm, Great Cockham, Kent, in 2004. An increasing number of new brewers are finding derelict farm buildings to renovate, and in some cases the farmers themselves are converting their disused buildings into breweries on their own account. For many, though, it's more than just a sideline aimed at making some additional profit out of disused buildings; it's actually a way of giving some life and purpose to an economic activity that has become almost marginal.

Farm-based Breweries

Making the leap from growing the ingredients to brewing the beer is nothing new. Paine's of St Neots and Ridley's of Chelmsford, two old-established regional brewers that have disappeared within the last two decades, started as farmers; a branch of the Paine family is still in the flour-milling business. Following their lead, Larkins Brewery was set up by the Dockerty family on their hop farm at Edenbridge in Kent in 1986; while Rother Valley Brewery was set up on a Sussex hop farm in 1993.

More recently, though, this tendency has gained new impetus. Many growers are not only fed up at the rock-bottom prices paid for top-quality produce – premium malting barley, for instance, has sold at little more than feed prices in recent years – they are also keen to join the environmental movement. So we have farm-based breweries such as Atlantic in Cornwall, High House in Tyne & Wear, Wagtail in Norfolk, and Wold Top in East Yorkshire growing organic barley and having it specially malted for them to reduce food miles; disposing of waste water in natural reedbeds; generating the electricity to power their breweries with wind turbines; feeding livestock with their spent grains; and using spent hops as mulch. If you see yourself and your brewery as part of this movement, you could do a lot worse than seek out a like-minded farmer to team up with. In purely commercial terms, environmental awareness is no longer a mere fad but an important and growing aspect of the market; and of course there is also great satisfaction to be derived from the knowledge that you are doing your bit.

The attractions of life on the farm are many and compelling. But, realistically, for most would-be local brewers there are two alternatives: an industrial estate or a pub.

Pub Breweries

Once upon a time almost all pubs brewed their own ale, and in many parts of the country pub-breweries were still relatively common until the late 19th century. By the early 1970s, though, there were just four left: the Three Tuns at Bishop's Castle and the All Nations at Madeley, both in Shropshire; the Old Swan at Netherton, West Midlands, universally known as Ma Pardoe's; and the Blue Anchor at Helston, Cornwall. (All four are still brewing, although Ma Pardoe's did actually close for quite a long while and has recently started brewing again.)

Reviving the tradition of pub brewing has been one of the driving forces of the real ale revolution, and not for romantic reasons alone. With most of the pub trade firmly closed to the new brewers, ownership of the means of retail has seemed a sine qua non to many, and buying a brewery tap has been an important goal. There have even been some cases where the brewer has bought a pub expressly as a base for the brewery: Wensleydale Brewery in North Yorkshire, for example, won awards for its Rowley Mild when it was Lidstone's, based in Cambridgeshire. The owner then decided to move north, and spent a long time looking for a suitable pub to base production in before settling on the Foresters Arms in the Yorkshire Dales National Park. It was a bold move that paid off, and within a year the brewery was forced to relocate for a second time to a farm near Bellerby.

Buying a pub expressly as a home for a brewery is probably a bit extreme for most start-ups, though, especially if you have no experience in the licensed trade; but before moving on, mention should be made of the most extreme variant of this type of venture: the boutique brewery.

Boutique Breweries

Boutique breweries are much more common in America and on the Continent than in Britain, but there are examples. Essentially, the boutique brewery is a bar or beer café where the brewing equipment itself occupies centre stage, normally behind a huge sheet of glass. Big vessels and pipework of gleaming brass make a spectacular backdrop for an evening's drinking; not surprisingly, the few boutique breweries in Britain are all urban and often very chic. In London there's the Porterhouse and the Bünker Bar in Covent Garden, Mash in Great Portland Street, and Zerodegrees in Blackheath (branches also in Bristol and Reading); in Glasgow there's Clockwork Beer Co; and in Peterborough there's Oakham Ales' Brewery Tap in a converted 1930s dole office opposite the railway station. They were all founded by people with seriously deep pockets, which is perhaps why there are so few.

More commonly, breweries have been installed in cellars or barns by publicans who feel that the wholesale price of beer is simply too high, and to whom brewing on their own account looks like cutting out the middleman and pocketing the extra profit. It also adds a unique selling point to the pub and attracts real ale fans from all over the country, and there's more money to be made by selling ale to beer festivals and other pubs. The success of the best-known figure in the history of microbrewing in Britain, David Bruce, who chose pubs as the sites for his Firkin breweries, merely proves the point. He put his first brewery into the cellars of the Duke of York in Southwark in 1979. Renamed the Goose & Firkin, it was the first in a chain of brewpubs that eventually (and under different ownership) covered the country.

Actually, it's something of an illusion that you need to own a pub or pubs in order to make a go of a microbrewery. Many of the best-known survivors from the earliest days – Chiltern, Larkins, Cotleigh, Exmoor and others – have done perfectly well without trying their hand at retail. Conversely, many of the casualties – Trough, Hull and very nearly Banks & Taylor, which went into receivership but was bought by a white knight and is thankfully still with us – only got into difficulties through ownership of pubs. For the truth is that running pubs demands a completely different set of skills from running a brewery; and possessing the latter doesn't necessarily mean you possess the former.

Brewing on the Premises

Be that as it may, surplus space at a pub is an obvious location for a brewery; and more than 130 micros, or about a quarter of the total, give their address as a pub (or in one case a workingmen's club, and in a couple more a restaurant). Many more of today's crop of local brewers started out in pubs but have since had to move to larger premises. There are dozens of highly successful publicans turned brewers, some of whom have gone on to sell their pubs to concentrate on wholesale brewing. Many small pub-breweries, on the other hand, brew pretty much exclusively for their own use: Front Street Brewery at the Chequers at Binham, Norfolk, is one example; Leith Hills Brewery at the Plough at Coldharbour, Surrey, is another. Often their breweries are squeezed into the tiniest of spaces – at Front Street and at Gwynant Brewery at the Tynllidiart Arms in Capel Bangor, Ceredigion, the brewing equipment is housed in the old outdoor loos.

They might let the odd firkin out for beer festivals, but given the price of casks and the difficulty of getting their empties back, they have decided to supply their own pubs but not the wider free trade. Some of these pub breweries have achieved cult status, perhaps thanks to the very fact that their beers are so hard to find. The Railway Tavern at Brightlingsea, Essex, is a fine example: its Crab & Winkle Mild regularly wins awards at beer festivals, and although one or two favoured stockists might have it, to be absolutely sure of getting any you have to go to Brightlingsea. As indeed many beer-lovers do.

Partnerships

Less common are partnerships between publicans with spare space to let and would-be brewers looking for homes. This is a very good option for the brewery, since the host pub invariably stocks some or all of its products. Perhaps the best-known example is Hesket Newmarket in the Cumbrian village of the same name, where the brewery and its host pub, the Old Crown, are owned by two entirely separate co-operatives of villagers (many of whom, though, are members of both). The miniature tower brewery at the famous Three Tuns at Bishops Castle, Shropshire, is also now under separate ownership from the pub across the yard, and there would be a national outcry if the pub didn't stock the brewery's beers. Other similar examples include the Alexandra Arms in Rugby, whose plant is shared between two local CAMRA members going under the name Atomic Brewery and the landlord, who brews his own ales under the pub's name. If you're looking for a home for your brewery, sounding out local free trade licensees (although not the tenants of tied houses, of course!) is a very good first step.

Industrial Estates

If your hopes of brewing in the stables of a stately home, or in a converted woollen mill, or a former dairy or cowshed, or in the yard of a friendly pub come to nothing, there is always the industrial estate; and the fact that maybe half of all new brewers operate from units on industrial estates speaks volumes.

The advantages are many and obvious. They're often fairly cheap to rent; there's little conversion work to do; and they're comparatively easy to find. Considerations such as adequate access for delivery vehicles, industrial-grade flooring, three-phase electricity and proper drainage can pretty much be taken for granted (in theory, at least), while all the necessary planning permissions ought to come with the

CASE STUDY

Renting Buildings From a Pub – Farmer's Ales, Maldon, Essex

Brewing may not be rocket science – but it's pretty close to it when you're talking about Farmer's Ales of Maldon, Essex. For founder Nigel Farmer used to be an engineer in the nuclear power industry, and worked at Bradwell power station until the opportunity of redundancy came up.

Nigel had been home brewing since his teenage years and had progressed as far as installing a full-mash brewery in his shed. There, his friends told him, he turned out beers that were better than some of East Anglia's leading brands. So with the prospect of spending more time with his mash tun looming, he decided to look for somewhere to set up a more serious operation.

Keen to find a home with a bit of character to give his beers a unique selling point, his first choice was an old shop at the dusty end of Maldon's high street. But a local Businesslink adviser soon put him right.

"He was asking questions I simply hadn't entertained, questions about footfall, about parking, about access, about services, and by doing that helped me avoid a costly disaster," says Nigel. "With his help I put together a checklist of criteria, and set about looking for somewhere that satisfied each one.

My advice to anyone would be not to jump at the first place you see. I nearly made that mistake and I'm lucky I didn't."

The disused outbuildings of a pub always seemed like a good prospect to Nigel, not least because the host pub would be bound to stock his beers. Several landlords were interested at first, but problems always cropped up that stopped the project going through. Then a friend suggested the Blue Boar, an imposing 18th-century coaching inn at the top of Maldon's high street, with a run-down stable-block that – with love and attention would be ideal. Nigel approached the owners, Old English Inns, which to his delight the company was very interested indeed.

"I think they saw it as adding interest to the hotel, and the deal was that I would restore and repair the stables to an agreed standard in return for a lower rent," says Nigel. They shook on it, and Nigel applied for planning permission to restore the Grade II listed building – the hotel's glory hole, he calls it, full of broken furniture and ancient VAT records – and change its use.

Getting all the necessary consents wasn't at all straightforward – although along the way Nigel found he was due a 40% English Heritage grant towards restoring the exterior, provided he used authentic materials such as cast-iron guttering – and it took nine months for the council to give him the go-ahead. A few days later, Old English Inns sold out, and the Blue Boar's new owner, Greene King, wasn't interested in Nigel's proposal.

"I went from elation to dejection in the space of a week," says Nigel. "Greene King just didn't want to know. I wrote, I rang, I got shunted from department to department – it was all very dispiriting."

A few months later, though, the Blue Boar appeared on the list of former Old English Inns that Greene King didn't want to keep, and was sold to local landowner John Wilsdon. Nigel approached him with the idea; he liked it; the planning permission was still in force; and in 2002 the first beers flowed out of the hotel's old stables.

That heart-stopping moment when Greene King took the hotel over illustrates the downside of basing your brewery in someone else's back yard: you're not really in control. But for Nigel, the advantages still outweigh the drawbacks.

"It made moving in easier because all the mains services were already in place, and I actually buy my electricity and water off the hotel which means I don't pay any standing charges," he says.

"The biggest advantage to me, I suppose, is that I have a brewery tap, and the biggest advantage to the hotel is that having a beer of its own has hugely increased its bar trade. There used to be four casks on stillage behind the bar – now there are 12.

"Another advantage to the hotel is that I manage its cellar, and when I first started trying to sell beer to other pubs I had someone to talk to who knew the licensed trade and could tell me where I was going wrong."

"The hotel manager sometimes calls it 'his' brewery, but I don't mind," says Nigel. "I'm not sure a working relationship like this would suit everybody – but it's certainly worked for me."

www.maldonbrewing.co.uk

lease. Often, too, a base on an industrial estate offers the option of quick and easy expansion into neighbouring units – Hop Back Brewery of Salisbury, for instance, now occupies almost the entire industrial estate onto which it moved 25-odd years ago. (It was actually founded in a pub.)

But although they are usually both well equipped and cost-effective, industrial estates do have their disadvantages. They are never going to figure on the local tourist trail, for instance; and an integral brewery shop or tap is pretty much out of the question too. So choosing a home on an industrial estate is not just a question of cost and convenience but has longer-term implications for how you plan to run your business. Unless you decide to relocate, you will inevitably remain first and foremost a manufacturer.

Size and Services

The choice of premises will not only be one of the first decisions you make, but also the hardest to go back on; and in large part the style and scope of your future operations will be determined by what you find now. So having made the choice between an industrial estate and something more characterful and picturesque, start your hunt for premises with a clear idea of where you'd like to be in five or ten years' time.

Size, especially, is going to matter. You can brew in a broom cupboard if you have to, but it's difficult and uncomfortable and leaves no room for expansion when your beers become a cult hit, as they undoubtedly will. Many's the brewery that has chugged along quite happily in a cramped and crowded basement until a major award came along, and with it a deluge of orders. Three such breweries, Mordue, Kelham Island and Coniston, found that winning the Champion Beer of Britain competition – held every August at the Great British Beer Festival – brought as many headaches as it did joys. They all had to farm out part of their production, and two of them then moved into bigger premises.

And although the brewing itself can be made ergonomically efficient almost to the point of absurdity – Rob Jones of Sussex's Dark Star Brewery, for instance, sells a prefabricated plant dubbed the Russian Doll, whose vessels fit inside one another – other activities, notably cask storage and washing, can't. So although you may be impressed, on your brewery visits, by miracles of compact engineering such as Front Street or Green Jack in Lowestoft, do get as much space as your budget allows. Even if the building itself is small, a good-sized yard can always be partly covered over to shelter a cask-washer – and additional fermenters, when you need them.

If you're looking at older premises for conversion, be sure to keep a project manager's eye out for practical pitfalls. Brewing in a largely residential area is possible; but some people object to the smell, and getting your plans past them can be difficult. While the fabric of the building itself may be sound or, at least, not unfeasibly unsound, you also have to think about things like flooring, drainage and water treatment (essential – you can't just pour waste water down the drain), and ventilation (equally essential: you're going to be filling the place up with CO^2 which, believe me, you don't want to breathe in). Can your building be adapted, within your budget, to meet all these needs?

Planning Permission

Having found the perfect building – picturesque, perhaps; on a well-trodden tourist trail; in reasonable condition; suitable for adaptation; with space for future expansion, yet still miraculously within your budget – you have to discover whether you're actually going to be allowed to brew there.

Getting planning permission can be drawn-out, time-consuming, complicated and expensive. But it's not half as drawn-out, time-consuming, complicated and expensive as not getting it. Councils vary in their attitude to enforcement, some being absolutely rigorous and others more flexible, but the last thing you want is to have to undo work you've already done. And if you don't have planning permission, or if you do have it but then breach the terms and conditions set by the council, enforcement is a very real danger.

Class B2 Planning

If you've gone for the industrial estate option, getting permission to brew in your unit should be relatively straightforward. The site will already have permission for general industrial use – or use class B2, in planning jargon. However, there will still be some issues to settle with the local council – city, borough or district – before you start grinding your grist. If B2 use exists – and your solicitor should check this with the vendor or landlord of the site before you exchange – then the exact details of what you propose to do will be classed as "reserved matters" (another useful piece of planning jargon!), which you will need to agree with the council's Development Control department.

Other necessary compliances involving building regulations,

health and safety matters, and waste water disposal will be dealt with in Chapter Four, Installation. However, you may find that the council won't give you the final go-ahead unless these compliances, especially health and safety and waste water disposal, are fully dealt with in your application.

Change of Use Permission

If you propose to brew in a historic property of some sort – or a farm or pub, that isn't already classed as B2, you will need change of use permission. You may very well need another, more difficult, permission, too: older buildings are frequently protected by statutory listing and need listed building consent before any alterations, however minor, can be made. Strictly speaking, this is not part of your change of use application as it's covered by different laws; but in practice the two applications can be submitted at the same time. They remain separate applications, though, and the outcome of one doesn't affect the outcome of the other. You can be granted listed building consent on the grounds that your proposals don't affect the building's character, but still be refused change of use permission on the grounds that, say, vehicle access is inadequate. Refusal of either application can be challenged; so if you're knocked back on the one but get through on the other, the game's not necessarily up.

Planning Applications

Getting your planning application under way will involve you in two processes: evaluating your building itself, and evaluating its surroundings. Obviously, it's wise to carry out this exercise as thoroughly as possible before signing a lease or making a purchase – you don't want to be stuck with premises you can't brew in, or face a long hard battle to get your application through before you can start trading.

As far as the building is concerned, there's quite a long list of questions to answer:

- If it's an older property, is it listed?
  (The local council will give you this information.)
- Can it jump all the relevant health and safety hoops?
- How good is the vehicle access?
- Is there room on the site for all employees, visitors and deliveries to park off-street and for vehicles to manoeuvre, i.e. for articulated lorries to reverse?

As for its surroundings, is it in an area where general industrial use is permitted by the Local Development Framework (LDF)? And while you're at the town or city hall or on the council's website checking the LDF ask yourself the following questions:

- Will you in future be likely to require a shop (use class A1) or a bar (use class A3)?
- Are permissions for these uses likely to be granted, or does the LDF frown on the volume of traffic a shop and bar are likely to generate?
- Is the premises in a conservation area, an Area of Outstanding Natural Beauty or a National Park?
- How many traffic movements on and off site do you envisage?
- Can you control noise and emissions to the satisfaction of local residents, if there are any?

Getting planning permission is – supposedly, at least – somewhat more straightforward these days than it used to be; and thanks to the web, it's also much easier to get the information you need than it was. At **www.planningportal.gov.uk** you'll find a pretty exhaustive guide to planning matters; it even enables you to apply online and includes a fee calculator to help you work out roughly how much the council will charge you. It also has links to other sites detailing, for instance, national policy on converting surplus agricultural buildings to other employment-related uses (Public Planning Guidance note or PPG 7). Since 2002 there has been a presumption that councils should approve "well-conceived farm diversification proposals, particularly involving the re-use of existing buildings for business purposes". Another useful link is to the Royal Town Planning Institute's **www.rtpi.org.uk**, which includes a list of planning and development consultants. Hopefully you won't need one, but if your application is any less than perfectly straightforward, you very well might!

In law, councils have to determine your application within eight weeks of receiving it; but they have the power to extend the permitted period under certain circumstances. If they do, you can appeal to the Secretary of State, but it's rather self-defeating as the appeal will almost certainly take longer than the council would. So if you want your application to be determined as quickly as possible – and if you want to be sure of getting the right result – then everything hinges on thorough preparation and good communication. What this means, in effect, is that the whole planning process will take longer – a lot longer – than eight weeks – eight months, more like! – but the vast bulk of the work will be done before your application actually goes in.

Preparing the Planning Application

The important thing is not to be afraid. Some people are: they think the council is there to frustrate them, or tie them up in red tape, and the only sure way to get an application through is to sneak it under the town hall's radar somehow. And while it's true that councils can be awkward if they want to, and that there are more formalities to go through than at the State Opening of Parliament, in essence the council is there to help you and even has an obligation to make suitable provision for industrial developments such as yours. The grounds on which applications can be refused or substantially varied are finite; and if you discover in advance what they are, you shouldn't find it too hard to put together an application that is more or less bound to succeed.

First, check the relevant sections of the LDF thoroughly. This is available from the city, borough or district council, or on its website. Unfortunately it's not a single document, but a collection of them: however, the site should be fairly easy to navigate. In the case of my local district council, South Cambridgeshire, the home page has a link to "planning", which has a further link to "district planning policy". Finally, there's a link to "Local Plan 2004" (as this book goes to press, old-style local plans are being replaced by new-fangled LDFs, but in most cases the old plan is still in force). The Local Plan is divided into two sections: general policy across the whole district, and more specific policies location by location. Armed with this information, you should at least be able to tell whether the building you have your eye on is likely to be acceptable for B2 use. Other council websites I have checked are less clear than South Cambridgeshire's, but still far from impenetrable; and by and large I have found that, once all the documents are located, LDFs are fairly straightforward and easy to understand.

However, the Planning Portal site and LDFs are very general documents and paint rather a rosy picture of the planning process. Things can and do go wrong. It took Oakham Ales two frustrating and interminable years to get permission to move from its original home in Rutland and convert the former DSS office in Peterborough it inhabited until only recently. It later transpired that discussions aimed at razing the whole area to extend the neighbouring shopping mall had been opened, but the project hadn't made it as far as the Local Plan when Oakham put in its application. Two years later the Brewery Tap had a compulsory purchase order slapped on it, and has lived under the shadow of demolition ever since. (At the time of writing it's still trading, and indeed still brewing, although the company has found a new and much bigger production base on an industrial estate.)

Other breweries could doubtless tell planning horror-stories just as hair-raising, but in the main, where applicants have run into really serious planning delays the premises have been in urban centres where there is strong competition for development sites. (And often, councils can't reveal details of negotiations such as those in Peterborough because of commercial confidentiality.) On other occasions, applicants have tried to get their putative breweries approved as B1, which covers light industrial use in residential areas and imposes stricter conditions than B2 on noise, fumes and traffic movements. Then again, attempts to install breweries in quirky or unusual premises have often run into problems with planning officers who argue that the use is unsuited to the building. And another big stumbling block that has faced pub-breweries in the past has been the thorny question of mixed used. A lot of local planners have simply been unable to get their heads round the concept that a single business premises might want A4 use for part of the site and B1 or B2 for another. This was part of the problem that caused so much delay for Oakham, but with the huge increase in the number of pub-breweries round the country, local planners seem to be getting less resistant to the idea than previously. If mixed use does seem to be presenting a problem, you can usually overcome it by arming yourself with a good number of precedents, preferably fairly local.

The Planning Officer

There is no way of guaranteeing that these kind of problems won't arise; but one way of reducing the risk is to open a good channel of communications with the planning officer working on your proposal. You will have to pay an administration fee to cover the costs of your application anyway, but some councils will charge extra for an initial consultation with your planning officer.

However, as it's this very officer who will recommend acceptance or rejection of your plan to the council's Development Control Committee, the extra charge is probably money well spent. Involving this officer in the evolution of your brewery makes it almost as much his or her baby as yours; in fact one planning officer I spoke to while researching this chapter told me: "We always like getting a brewery to work on – it's a lot more interesting than most of the cases we get!" So by the time the application is ready to be formally considered, the two of you should have made it almost watertight. Of course, this process may itself have its ups and downs; but remember, if your planning officer flatly vetoes an idea of yours, it's not out of bloody-mindedness (probably not, anyway) but out of a deep knowledge of local policies and a very shrewd understanding of what will or won't get through.

The planning officer's main concern, after the suitability of the building itself, will be the impact of your business on its neighbours, especially if there are any residents in the immediate area. This concern is partly met by the LDF's approved uses for the locality. But the officer will also be interested in noise, smells and other emissions, waste discharge and any changes you propose to the appearance of the building, particularly if you will need equipment such as extractors, flues and tanks.

Consultation

A large part of the process involves consultation, and not just with the neighbours. Of course, you will have to advertise your application both in the local press (another expense!) and on the application premises, to give them the chance to object if they fear your plans will threaten the "amenity" of the area. But the grounds on which they can object are limited by statute, and the planning authority is specifically instructed (see the Planning Portals website again) that the mere number of objections is irrelevant.

But the main consultation will be with other statutory bodies:

- The local council's own environmental health department
- The county council as highways authority
- The parish council
- The Environment Agency
- The Health and Safety Executive

This is the part of the process that really takes the time, and it's these agencies that are likeliest to have objections. If your application conforms to the LDF and is made out thoroughly and in proper form, you should hopefully clear this last hurdle and start brewing. But what if you fall at this stage?

Appeals

The obvious thing is to appeal, and you have six months from the rejection of your application to do so. Once again, the Planning Portal contains all the details you need – you can even appeal online!

Appeals are handled by the Planning Inspectorate, which appoints an independent inspector to whom you can state your case and comment on the planning authority's statement of case. The Inspectorate also appoints a case officer who will help by telling you what documentation to present and when, although when dealing with unfamiliar, complex and jargon-ridden rules and regulations you may well feel you need the (expensive) advice of an experienced surveyor or planning consultant.

An appeal is the only way to overcome refusal of listed building consent; but if you are refused change of use permission, a quicker and cheaper option is to find out what the council didn't like about your application, amend it accordingly, and resubmit. If you do this within 12 months there will be no fee and as objections to the original application are considered to have been dealt with already they can't be submitted again. This is the way large corporations wear down local objectors – they simply present slightly altered applications again and again until their opponents either run out of new objections or are simply too worn down to continue. That's perhaps not quite what you have in mind but if you have been refused on minor or technical grounds this is the simplest solution.

Once you have been granted planning permission, you have three years to act on it before it expires.

Chapter Three
Planning Your Finances

So you've done your training and you've found your premises. Your next step is to move in and install your brewery. But not just yet. First, you've got to make sure you have enough money not just to set yourself up but also to keep yourself afloat until the business starts paying its way.

This is not the place to tell you how or where to raise the money you'll need. You've doubtless got some capital already: the proceeds of the sale of your house, perhaps, or a redundancy cheque, or your life savings, or a bequest. You'll probably need more, though, and the obvious place to go is the bank. But before it parts with any money, the bank is going to ask some searching questions and will expect to see a detailed business plan.

The Business Plan

A business plan, as you undoubtedly already know, is far more than just a rosily optimistic cash-flow projection cobbled together to lull the bank into underwriting a daydream. If carried out thoroughly, it will confront you with the mathematics of your project coldly and in stark detail. And if it doesn't make you run away screaming, it will help ensure that your newborn business will be a sturdy infant capable of making it through to adulthood.

Your business plan will force you to answer hard questions about practical issues such as your premises, your products, your personnel, your market, and indeed your very purpose in starting up your brewery. But the final and most important section will be a cash-flow projection that should tell you, just as it will the potential lender, how long it's going to take to get into the black, and whether your brewery really has a financial future. Now, a business plan is a very specific document – a highly detailed and indeed compendious questionnaire that leaves nothing out. But to put a proper business plan together you don't have to reinvent the wheel and set yourself, as it were, your own exam. Rather, use a specimen business plan supplied either by your bank or by a Local Enterprise Agency (LEA); visit the National Federation of Enterprise Agencies (NFEA) at **www.nationalenterprisenetwork.org**. to find your nearest.

LEAs are just one source of invaluable start-up help and advice available to all businesses. In fact there's a staggering amount of information on the web from a surprisingly wide variety of agencies. Plunder them all.

Online Advice

An especially useful link on the NFEA site is to the Small Business Advice Service (SBAS) **www.sbas.org.uk**, which has more than 200 trained and accredited business advisers all of whom can be consulted free online on topics ranging from starting a business while on benefits to international copyright protection. The SBAS site also has a start-up guide you will, I guarantee, find indispensable.

Visit **www.gov.uk** for a guide to everything from starting up your business to selling it on, and also for a directory of grants and other forms of assistance. There's a surprising amount of grant schemes, soft loans, training programmes and other practical and financial help out there if you know where to look; the site includes an online directory, and your Local Enterprise Agency will give you practical assistance in the tedious and time-consuming business of applying. And don't imagine that because you're not restoring a historic building, or creating jobs in a deprived area, or planning to export, you don't qualify for help. The chances are that you do, and while your bank may be able to help you source grants (and there are commercial agencies that will do so for a fee), with your Local Enterprise Agency you have a LEAs are not only useful as sources of business advice and factual information. They also hold regular seminars and other events covering all sorts of topics from environmental considerations to employment law that are not only useful in themselves but are also opportunities to meet other local businesspeople. It's called networking, and it's not to be sneered at!

But every type of business has peculiarities that will affect its financial prospects, and this chapter aims to identify some of the peculiarities of the brewing business.

Capital Expenditure (Capex)

Your financial planning should fall under two headings: capital expenditure and running costs. Let's start with the capital you'll need.

By the time you get round to writing your business plan, you will know the exact purchase price (or premium, if leased) of your premises. But making your new home ready to receive your brewing equipment will, as we saw in the last chapter, entail costs that other businesses might not have to bear. Breweries use water, lots of it, not just to brew with but to heat up and cool down the various vats and vessels and, of course, to keep everything squeaky clean. In terms of capital, using so much water means you must have adequate drainage including sloping floors (which you may have to lay at your own cost even in a brand-new unit on an industrial estate) and washable surfaces. It may also mean you will have to treat effluent before disposal, another capital item to consider. The cost of non-slip flooring, essential on safety grounds, should also be built into your capital estimates; and, again on safety grounds, so should adequate ventilation. Carbon dioxide, one of the products of fermentation, is deadly not only to the microbes that would otherwise colonise your beer, but also to you and your staff. Your council's environmental health department will insist on the prevention of concentrations of CO^2; so – and especially if you plan to use traditional open fermenters – you will have to factor in the cost of whatever ventilation and/or extraction measures (as well as monitoring equipment) it stipulates.

Casks

After your premises, your equipment will be your largest item of capital expenditure, and it can be anything from £7,500 (plus installation costs) to – well, the sky's the limit. One item of capex that is frequently and sometimes disastrously underestimated in brewery start-ups, though, is the stock (or population, to lapse into jargon) of casks that you will need. The most common size in use today is the nine-gallon firkin, and orthodoxy has it that a free trade brewer whose beers are distributed widely and whose empties trickle home with agonising slowness needs around 30 firkins for every bulk barrel, or 36 gallons, of production capacity. A five-barrel brewery should therefore own a stock of at least 150 firkins, which cost around £80 each new; so put down at least £12,000 in your capital estimates for casks. The temptation to skimp on casks is almost irresistible, especially if you're on a tight budget and other costs are running above your estimates. But you really will need that many: it's easy to be caught short by even a modest surge in demand, and disappointed customers tend not to come back. And do buy new: there are cheap second-hand casks around, but ask yourself why they're being sold. Unless they are the stock of a brewery that has closed, the answer, very probably, is that they're "leakers", and therefore worth only their scrap value. No publican or wholesaler likes to have to deal with a leaker, nor has any affection for the brewery that sends them one.

A vital note regarding casks is that they're extremely vulnerable to theft. After use they tend to be stacked at the pub's back door until the owner comes to collect them; so if you're talking about an irregular account that only orders every month or two, your expensive firkin could be sitting out in the open for weeks. The temptation to swipe it soon becomes irresistible to the more light-fingered of the local inhabitants (casks do have a scrap value), or alternatively it might be "uplifted" by another brewer collecting his own empties. For this reason the British Beer & Pub Association (**www.beerandpub.com**) allocates a unique colour band to any brewer, whether a BBPA or SIBA member or not the joining kind, who cares to apply for one. The banding consists of three hoops chosen from 10 colours, normally painted round the middle of the cask between the rolling rings. For more details of the scheme visit **www.beerandpub.com** and follow links to its colour band register. The BBPA register is free to join and is an entirely voluntary scheme, so brewers are at liberty to pick their own colours if they prefer. However, if you're unregistered and your banding duplicates that of a registered brewer, you may find that helpful people keep sending your casks back to the wrong place. So you can either register, or choose a colour scheme of your own that's so unusual it's bound to be unique. And your unique identifying mark doesn't have to be in the form of bands, either. Some brewers I know paint both ends of their casks; one paints the entire vessel pale pink! It's also a good idea to stamp your name indelibly on the cask end.

The Delivery Van

Another vital piece of equipment that's easy to get wrong is the delivery van. A full firkin of beer doesn't take up much space but weighs about 50kg, so unless you don't mind being done for overloading your van from time to time, you need to choose with an eye to axle-weight rather than capacity. You also need to think about what sort of licence and tax-disc you will need. An ordinary Transit-style van should carry 20–30 full firkins without straying into the "Larger Goods Vehicle" vehicle excise licence category (TC55), and you will be able to drive it on your ordinary driving licence. This should be perfectly adequate for local deliveries; if you're planning anything more ambitious you may need a vehicle whose total laden weight or maximum authorised mass exceeds 3.5 tonnes. Visit **www.gov.uk** and follow the links to "motoring" for more details.

Finally, managing the start-up will not be cost-free. For instance, there are the administration fees referred to in the previous chapter, payable to the local council merely for the privilege of applying for planning permission. Your consultant will also want paying: factor his fees into your financial plan. Find out in advance when these payments, as well as those due to builders and fabricators, will fall due. Knowing when you need to have funds available gives you two benefits:

you won't be caught short when tradesmen demand payment; and you can organise one of those loans which you only draw on (and start paying interest on) when necessary.

Running Costs

Your brewery will have to bear many of the same general running costs as other businesses: utilities; insurance; business rate; rent or mortgage repayments; repayments on other borrowings; wages (if any) and associated costs; marketing and advertising; and deliveries. The information you will need to compute all these should be available from your bank or accountant, from one of the agencies listed above, or even from the websites of the relevant government departments.

However, as with any business sector, brewing has its peculiarities. One, in particular, might just help you with your business rate.

Your business rate, which is the property tax levied by your local council, is perhaps the easiest of your regular outgoings to calculate. It is based on a valuation carried out in 2003 (and due to be updated soon) aimed at establishing the rent your premises would command on the open market – exactly the same system, in fact, that used to be used to calculate domestic rates. This notional rent is known as the property's rateable value or **www.gov.uk/calculate-your-business-rates**, and what you actually pay is calculated as a percentage of it. The difference between modern business rates and old-fashioned domestic rates is that the council doesn't decide what percentage of your RV it may levy. Instead, the government sets it each year: for 2008–9 the standard percentage, or "multiplier", was 46.2p in the pound, dropping to 41.4p for 2010–11. There's a small concession, though, for businesses with an RV of £15,000 or less (£21,500 or less in Greater London): 45.8p for 2008–9, and 40.07p for 2010-11. So make it a priority to find out the RV of your premises before you move in – if you find a site with a lower RV, it could save you quite a bit. Another relief that may help you with your business rates is that any part of your premises that is empty and unused can be rated separately, with a 50 per cent reduction. For up-to-date information visit **www.gov.uk/calculate-your-business-rates**.

Finally there's a third, rather complicated, relief aimed at protecting threatened village amenities, and which you might just be able to benefit from. The only pub in a village of 3,000 inhabitants or less is entitled to a business rate rebate of at least 50 per cent (or up to 100 per cent at the council's discretion) provided its RV is less than £14,000. If you base your brewery at a qualifying pub, and it doesn't push the RV over the £14,000 ceiling, you'll still get the full rebate. Even if it does, you might be able to have the brewery and the pub rated separately; in which case the pub will still qualify for its 50 per cent relief, while the brewery might qualify for the small business multiplier.

Excise Duty

When you pay your excise duty, and how much you'll pay, are explained in detail in the appendix. To give you some idea, though, duty is calculated as hl% per percentage point of alcohol by volume per hundred litres or hectolitre. (A 36-gallon barrel is 163hl). The current (2013) rate is £19.12 per hectolitre per degree alcohol; so the full rate of duty on a barrel of a 3.8 per cent ABV session bitter would be £118.43. However, if you're a small brewer producing up to 3,000 barrels a year you're entitled to 50 per cent relief and would only pay £59.21. There is also a sliding scale of relief for brewers producing more than 3,000 barrels a year; the complicated equations by which this is calculated are set out in the appendix, see page 194. Duty must be paid by direct debit on the 25th of the following month – again, details are in the appendix on page 178.

That leaves your ingredients – malt and hops – which are not going to be the biggest items on your list of outgoings. In fact, and to many people's amazement, they could well amount to less than £11 per firkin of standard-strength bitter.

Hops

The amount of hops you will need depends on your recipes, but a very rough guide for budgeting purposes is a maximum of 500g of aroma and a slightly lesser quantity of bittering varieties per barrel of a pretty hoppy beer. A five-barrel brewlength would therefore require about 2–2.5kg of each or 4–5kg all told. The price, of course, varies from harvest to harvest and according to the type of hops you use but is usually in the range of £5–10 per kilo: at the time of writing, the Worcestershire hop merchant Charles Faram was charging £9.65 per kilo for Goldings; £8.25 for Fuggles. Your hop bill is therefore likely to be in the region of £20–50 per five-barrel brew; £5–10 per barrel; or £1.25–2.50 per nine-gallon firkin.

You will need to be more precise than this when making the cash-flow projections for your business plan, but Faram's produces a brochure for start-up brewers that will give fuller details. Faram's will also accept payment over the phone by credit, which – assuming you pay your credit card bill in full every month – is the equivalent of 30 days' free credit.

Malt

Calculating the amount of malt you will use is more complicated, but a rough rule of thumb (again, very rough) is that a tonne will make 40 barrels of 4 per cent ABV bitter. A tonne of pale ale malt, at time of writing, is upwards of £600 – again, depending on what type of malt you are using. So that's about 125kg costing £75.00 per five-barrel brew; 25kg costing £15.00 per barrel; or 6.25kg costing £3.75 per firkin. So your ingredients, excepting the liquor, will have cost only £4.77– 5.62 for 72 pints of best bitter. The duty, with the 50 per cent small brewery relief, will be £13.37 for a 4 per cent ABV beer – a total, then, of about £25 a firkin. Throw in the cost of your utilities, and you should be able to produce a five-barrel brewlength for about £550; which is £110 a barrel or £27.50 a firkin.

You may have to pay a delivery charge, but even that is only about £45 for a one-tonne pallet – enough for eight five-barrel brews – carried by Palletline, Palletways, or a similar group haulier. (Most merchants will make up a mixed pallet for you, so you don't have to buy the crystal and other specialist malts you will only use in small quantities in one-tonne lots: they come in 25kg bags. Many malt suppliers will also deliver packs of hops on the same pallet.)

These are, as I have said, very rough costings intended only to give you an overall picture of what you are likely to spend. Your brewing consultant will go into more detail with you, but before you reach that stage I strongly recommend you to visit **www.murphyandson.co.uk**. Murphy & Son is a very old-established firm of brewers' suppliers based in Nottingham, which stocks all the consumables and process additives that you will ever need, from liquor treatment to finings. More than that, though, its site is an absolute treasure-trove of highly detailed technical and other information including extraction tables for all varieties of malt. In fact Chris Garrett of Warminster Maltings in Wiltshire recommends the Murphy's site not only to start-up brewers planning their first year's outgoings, but also to more established brewers working out hop and malt grists for new recipes.

Pricing

So much for your outgoings. Now we come to the more pleasant side of the equation – namely, your likely income.

The most important thing you'll want to know is the price you can expect to charge per firkin – the nine-gallon barrel in which most cask beer is delivered these days. This should be determined based on your costs of production plus the profit margin you need to stay in business. There is an old industry adage which states "any fool can give beer away, a wise brewer sells beer at a profit"

Many Brewers swap beer and to enable brewers to swap different-strength beers with the minimum of fuss: instead of invoicing and paying each other in full every time, the brewer of the weaker of the beers being swapped simply paid the other brewer the strength-difference. It is important to remember that if you work on reduced margins swapping beer you need to make that margin when you sell the other brewers beers. There are many different formulas in use but most multiply the ABV by a set factor (no lower than 25) and then add a set amount per barrel (often £100) thus a barrel (36 gallons) of a 4% ABV beer would cost £200.00 (4% * 25 + £100) or £50 per firkin.

Note, though brewers who use a formula almost always add a margin of their own when selling the beer into trade – although wholesale prices have actually been declining recently thanks to the number of new breweries opening, and there are brewers who will indeed sell at the formula price or even lower. A great deal depends on how efficient your brewery is, and how much of your beer you retail yourself. If you produced 2,999 barrels a year at an average of 4 per cent ABV and got the Bristol formula for all of it, your annual turnover would be £570,000. Your ingredients and duty would account for £300,000 or thereabouts; only you know, having totted up your rent, rates, repayments, utilities, wages, marketing costs and so on whether there would be enough change out of the remaining £270,000 for you to live on. Probably not; so you'll doubtless be looking to charge your own margin on top of the formula price.

SIBA's Direct Delivery Scheme (DDS)

A good idea of how big a margin you can reasonably expect is provided by the prices agreed with pubcos and supermarkets by SIBA's Direct Delivery Scheme (DDS). Like the Bristol formula, these prices are based on alcoholic strength rather than the cost of ingredients, which you might perhaps see as an incentive to pad out your malt with as much rice and corn syrup as you can get away with. But don't give in to temptation. As a craft brewer, the quality of your beer is your greatest strength, and if it's no good no one will buy a second bottle or pint. Anyway, the DDS-agreed prices aren't always that bad: with one company SIBA has agreed a price of £48.53 per firkin at 4 per cent ABV. That's £24 a barrel extra margin over the Bristol formula price, or an additional £72,000 a year.

The problem is, though, that thanks to the recent proliferation of brewers and the introduction of new tiers of middleman in the form of wholesalers and non-brewing pubcos, the brewer's slice of the cake is actually dwindling. A pub in south-eastern England will charge, at time of writing, around £3.10 including VAT per pint of session-strength bitter – call it £2.58 net. Publicans typically add 100 per cent to the wholesale price to cover their own not inconsiderable overheads, which suggests that they're paying at least £75 a firkin for a beer of less than 4 per cent ABV. That price comes, typically, from a pubco or a wholesaler; you can bet the brewer isn't getting much more than 60 per cent of it. Unless, that is, the brewer in question happens to be one of the shrinking band of established regionals with tied estates whose tenants have to pay whatever they're asked.

In fact tied tenants are famously the only people in the pub trade who pay list-price for their beer: anyone with any bargaining power at all gets a discount. Often, small brewers deliberately ramp up their price-lists so they can appear to be generous with discounts; I have often heard of free trade publicans who negotiated discounted prices from microbrewers that were actually higher than the pre-discount prices offered by competing regionals. What's important in all your price negotiations, though, is that you should keep a firm idea in your head of what your lowest economical prices are, and should never give in to the temptation to take less for the sake of volume.

The thought of all that extra margin has lured many small brewers down the path of pub ownership. Whether you should think of owning a pub or pubs yourself will be explored in Chapter Eight, see page 158.

Cashflow

No matter whether you have agreed comparatively low prices for bulk orders, or are charging a premium for single-firkin orders of ales that have just won major awards (always a good opportunity for a temporary price uplift), actually getting paid can be a nightmare. I have even heard stories of small, and generally new, breweries that went under because their credit with their suppliers had run out, their overdraft facilities had been exhausted, and they were missing loan repayments – and all the while, they were owed enough to cover the problem and more but simply couldn't get their money.

Whether these stories are true or not I don't know, but poor cash-flow is a perennial problem for local brewers whether it's large customers whose accounts departments aren't as sharp as they might be – later on in the book you will meet one brewer whose regular pubco cheques kept going to a competitor with a vaguely similar name – or small one-off accounts who had ordered a couple of firkins as a guest ale and never got round to paying for it.

The smaller debts used to be the bigger headache – you could be sending out invoices and reminders for months and then get an unsigned or undated cheque and have to start the whole procedure again. These kind of accounts need not be a problem any more, though: if it's a small enough account you can demand payment in advance by credit card either via your website, by email, or simply over the phone. If customers proves awkward, remind them that provided they clear the card in full every month, this amounts to 30 days' interest-free credit. Your bank will advise you on "customer not present" card payments, but there are other providers who you may find are cheaper.

Paypal (**www.paypal.com**), for instance, charges no start-up or one-off fees and takes a 20p fee plus 3.4 per cent commission on all transactions up to £1,500 per month. You can take payment on your website or, if you don't have one, by email or even by phone. The commission goes down in stages to 1.4 per cent, so you might even arrange for your bigger customers to pay you in this way. If they pay the whole month's bill in one go at the very end of the month, they can actually get two months' free credit while you pay just a single transaction fee. And if the month's bill is more than £1,500, the commission you pay goes down as well. Play the system – charge really trusted regular accounts every two months or even quarterly, and as long as your cash-flow will stand the wait, everybody gains.

There are, of course, other providers such as WorldPay, and you can spend many a happy hour trawling the net to find the provider that suits you best. None of them, though, can help you with big customers – supermarkets and pubcos – whose alleged credit periods of 60 or 90 days are honoured more in the breach than the observance. And once a glitch enters the system, it can take months to iron out. One way round this is to join SIBA's Direct Delivery Scheme which ensures, among other things, regular and reliable payment; however, there is a joining fee and some brewers aren't too happy with the prices they get through DDS.

Exit Strategy

At some point in the future, the likelihood is that you'll want to sell up. You might fall ill, you might fall out with your partners (a very common cause of the closure of small breweries), or you might just want to retire. It may, when you're planning to get started, seem a bad omen to be planning your demise as well. But remember, this is not only your living but your pension. The question you'll want to ask, then, is: how much is my brewery worth? And the simple answer is that it's worth what somebody will pay for it.

The actual value, however, is not the same thing as reaching a valuation, which depends very much on what you're selling. If you boil the brewery down to its constituent parts, there's the site, the equipment; and the brands and goodwill. The valuation of the site depends on whether it's freehold or leasehold and is something that only you can determine. A freehold you can have valued by any commercial property broker. A lease may or may not be "heritable" – that is, it may revert to the landlord, or you may be able to sell on its unexpired term. If the latter, the same commercial property broker will determine it for you. Your equipment undoubtedly has a second-hand price, which will depend on its condition and fluctuate according to market conditions; any brewing consultant can assess it for you. The valuation of the brands and goodwill will depend on how well you're trading, but a figure of three times net profit is often mentioned in valuing businesses without assets.

The same could be said of the value of your business as a going concern. It's such a rare occurrence in the small brewing world that there are no reliable guidelines: whether you can sell it as a business at all depends very much on whether there's someone willing to buy, and also on the nature of the sales. One brewery that I know sold recently for 2½ times its gross profit of £60,000. It had a turnover of £200,000 but was in leased premises, and sold relatively easily because it had top-class, fairly new equipment and a good steady trade with plenty of repeat custom. It may not seem a huge sum, but it was still considerably better than the break-up value. However, the deal depended very much on the fact that owner was introduced to a waiting buyer by a brewing consultant; and given that mature small breweries sell relatively rarely, perhaps the best way to get your business valued and sold is through one of the established consultants.

Chapter Four
Brewing Equipment

CASE STUDIES

There's many a good tune, my father used to remark, played on an old fiddle. And although he didn't actually have brewing equipment in mind, he might as well have. For a brewery doesn't have to be brand spanking new, all gleaming steel and copper and costing a small fortune to produce perfectly good beer. It can equally well be second-hand or even – if you really know what you're doing – cobbled up from scrap.

And indeed in the early days of small brewing, it had to be. There were, of course, long-established firms of brewers' engineers, mostly based in Burton-on-Trent, which could design and install a world-class brewery for you at the drop of a blueprint – provided, that is, that you wanted to brew a hundred barrels at a time. But if you only wanted to brew five, you had to forage for your vessels. In fact many small brewing pioneers – especially Peter Austin, founder of Hampshire's Ringwood Brewery – were just as well known for helping other start-ups as they were for their own. And I well remember a visit as late as 1993 to Central Bottling International (CBI) of Doncaster, which originally specialised in selling reconditioned second-hand bottling lines but found a new lease of life once the real ale revolution started. Microbrewers used to swarm all over its huge yard looking for usable vessels and pipework, and my visit coincided with Scottish & Newcastle's decision to take the old-fashioned cellar tanks out of its big managed houses and workingmen's clubs and replace them with conventional keg dispense. There they all stood in rows, fully insulated freestanding, five-barrel tanks with pipe-stubs at their tops and bottoms and even little viewing panels so you could see what was going on inside – 1,800 of them. They made absolutely perfect fermenters and/or conditioning tanks, and CBI sold the lot in a few weeks of getting them in.

Constructing your own Equipment

You could, if you had to, follow the example of those pioneers and cobble up your brewery from scrap, and not just out of parsimony, either. Tim Dunford of the Green Jack Brewery of Lowestoft did just that as recently as 2003, but only because he had to. Green Jack occupies the back part of a house that was, rather weirdly, left standing when the front was demolished for road-widening, and it's such an odd shape that he couldn't get what he needed off the peg and had to fabricate his own. His tanks were all originally used to make onion flavouring for the crisps and snacks industry, and when he first got

CASE STUDY

Monty's Brewery, Montgomery, Powys

Pam Honeyman was looking for career change having previously been a teacher and then working for a local charity. She had always been interested in real ale. With her husband Russ they run a successful beer wholesale business in Montgomery. They noticed a gap in the market in that Montgomeryshire did not have a microbrewery of its own.

She signed up for the two week practical brewing course at Brewlab in June 2008 to see if she enjoyed the experience and to find out if she could work as a brewer. After successfully completing the course they then set out to find equipment and somebody to put it together for them. From local brewers they sourced two second hand FV's and after spending a day with Dave Facer at Facer's Brewery in Flint, he recommended contacting Johnson Brewing and Engineering.

The order was placed for a 12 Barrel LPG fired Brewhouse built from new at the start of November 2008 and the first brew was put through the new equipment on 28th January 2009. Vincent from Johnson Brewing & Engineering tells of how most of their business is based on word of mouth and clients recommendations. 'The best advert for us, is our equipment being successfully used to make top quality beer in an efficient manner, we set our breweries up to be easy to use and easy to maintain. We are often surprised in that clients expect pretty low levels of efficiencies in that they have been told 80% extracts are the norm. They are pleasantly surprised that a minimum of 90% is what we expect from the equipment. This adds up to lot of extra beer in the long term.'

Pam is looking to add more beers to her range as she gains experience of the market. From the initial decision to set up to the completion of a working brewery took a little over seven months, which as Pam says felt like an age, but is a lot quicker than most from conception to start up.

www.montysbrewery.co.uk
www.johnsonbrewing.co.uk

them they stank of it. They all had to be altered in various ways, but Tim sought the aid of a local fabricator who instructed him in the finer arts of the angle-grinder and the oxyacetylene torch, and he did a brilliant job – as the rows of award certificates decorating one wall of his pub, the Triangle, demonstrate.

A word of caution here, though. The pioneers of the 1970s were able to make their Heath Robinson breweries work because, almost to a man, they were themselves refugees from the mainstream brewing industry. They'd been working with mash tuns and hop backs and paraflows since their teens and they knew everything there was to know about brewing equipment. Similarly Tim Dunford, although he has never worked for a mainstream brewery, had run another micro for a long time before founding Green Jack. He too knew from workaday experience what vessels he needed, what was going to be expected of them, and how to connect them all up. You, even if you're a veteran home brewer, are probably not quite so intimately familiar with the hardware you're going to need; and you're also not going to be quite as expert in keeping the whole rig running once it's actually set up.

A wise maxim here, then, is that you should spend as much as you can on your equipment, even if it means skimping elsewhere. Once you're up and running you're going to be more than grateful for well-designed and well-built plant that works smoothly and efficiently, using up a minimum of energy and a minimum of labour and never breaking down. Something you can take for granted, really. And that, for most mortals, is going to mean a purpose-built brewery from one of the main consultants – people like ABUK, Mossbrew, David Porter, David Smith and Brewing Design Services.

Brewing Consultants

If your budget isn't big, you may at this point be getting a bit nervous about how much you're likely to have to spend on your all-singing, all-dancing, cutting-edge computer-controlled brewery. Well, it depends on who you talk to. Some consultants can do the job for less than £20,000. Then there are others who would advise you to spend not less than £100,000. That doesn't mean that the former are cheapskates and the latter are con artists, though. All it means is that there's a broad spectrum of people coming into the industry, all with different philosophies, ambitions and levels of resource, and the various consultants specialise in catering for different segments of the market. They will all, by and large, do you proud; just make sure, though, that you talk to three or more of them before deciding which you want to hire. Whichever of the consultants you opt to put your business with, however, you will soon discover one unexpected fact: at bottom, the equipment you will need is surprisingly simple, and the actual pieces of kit are surprisingly few.

Opting for purpose-built brewing equipment rather than improvised kit made from vats with former lives in other industries (and one brewery has vessels originally used to mix lipstick!) doesn't necessarily mean buying new, though. There's a lively trade in second-hand kit too. Something like 20 breweries close every year, some of them tiny, some of them (like Gale's and Ridley's, which both closed in 2006–7) pretty big. Much of the equipment that comes on the market as a result is fairly well worn, but then much of it is brand new.

Well-established family brewers aren't above picking clean the corpses of their fallen rivals; if you're lucky enough to be shown round by the brewer himself when you visit an old family brewery, he will inevitably and often rather gloatingly point out plant and vessels rescued from recently deceased competitors. And if it's good enough for them, it's good enough for you!

But most of the second-hand kit you or your consultant are likely to pick up won't come from closed breweries at all, but from successful breweries that are expanding and replacing outgrown gear.

Let us now pretend that we're walking through a classical tower brewery – that is, one where the malt and water are hoisted or pumped up to the top, and all the subsequent processes are on successively lower floors to allow gravity to do the work. We are in the company of Richard Shardlow of Brewing Design Services of Tring, to whom I am indebted for this virtual tour. As we go, you can tick off what you're going to need piece by piece

Fulton Boiler Works

Fulton have been associated with supplying steam boiler plant for the brewing industry for more than forty years. As the larger breweries have closed and merged and the move to smaller micro breweries has developed the Fulton range of small vertical steam boilers has proved to be ideal for the application. Compact with a small footprint, the fuel fired and electric boilers are suitable for even the smallest brewery. Reliability,efficiency and longevity are essential considerations for the brewery owner when selecting steam boiler plant and Fulton

Brewing Consultants

Most start-up brewers these days will use the services of the brewing consultancy sector, which has evolved as the small brewing industry has grown over the years. In fact, given today's regulatory complexities it's a brave would-be brewer who doesn't exploit the vast expertise that is available from the various consultants, most of whom have huge experience and access to a network of other support services that you will find indispensable. However, they all have different specialities and ranges of services, so shop around!

Below are listed details of the best known of them.

AB UK: The full name of AB UK, based in Misterton, Nottinghamshire, is Advanced Bottling, and it supplies and installs bottling equipment as well as breweries. Specialises in reconditioned equipment.
Tel: 01427 890099; **Email:** info@abuk.co.uk; **Web:** www.abuk.co.uk

Brewery Services: Brewery services offers work place based training and general consultancy for the microbrewing industry.Chris has particular experience with small scale bottling machines and with water management as well as some years spent writing and teaching brewing courses when not making beer at a commercial brewery.
Contact: Chris Holliland; **Tel:** 01777 703074;
Email: info@breweryservices.co.uk; **Web:** www.breweryservices.co.uk

Brewing Design Services: Founded as Inn Brewing by former Whitbread senior brewer David Shardlow in 1982, the company changed its name 10 years later to broaden its appeal beyond the brewpub and is now run by David's son Richard, who also runs Tring Brewery. BDS specialises in larger projects such as Butcombe's brand-new 60-barrel brewery in Somerset.
Tel: 01442 890721; **Email:** info@tringbrewery.com

Iceni Brewery: Iceni is owned and run by Brendan Moore, founder of the East Anglian Brewers co-operative. Brendan also installs both brewing and bottling equipment and advises on marketing techniques.
Tel: 01842 878922; **Web:** www.icenibrewery.co.uk.

Johnson Brewing & Engineering Ltd: Vincent Johnson set up Johnson Brewing & Engineering in 2003. Time served as a boilermaker working on

M.O.D contracts, the end of the cold war meant redundancy and a career change into brewing. Successful completion of a BSc (Hons) in Brewing & Distilling at Heriot Watt in 1997 followed a year working for S&N as a Project Engineer. Then followed five years working for brewing maverick Brendan Dobbin, installing brewplant all over Europe and the UK. Since 2003 he has concentrated mainly on the UK market and has supplied & installed new many breweries from 4 barrel to 50 Barrel in size. Cask washing and keg washing equipment is also a speciality. With his previous experience in Europe they have help set up lager brewing plant and filtration equipment in Sweden (Jamtlands Bryggerri) and the UK (Taddington Brewery).
Tel: 01204 887754; **Web:** www.johnsonbrewing.co.uk

Mossbrew: As well as running Mossbrew, Graham Moss owns the Ministry of Ale brewpub in Burnley. A biochemist by training, he also holds master's degrees in Malting & Brewing Technology and Business Adminstration. He founded Mossbrew in 1994 and supplies everything from half-barrel pilot breweries upwards, with full installation, training, and consultancy services thrown in.
Tel: 0114 2492309; **Web:** www.mossbrew.co.uk

David Porter: Founder of Porter Brewing based at the Griffin Inn at Haslingden, Lancashire, David once owned a chain of five pubs but slimmed down this side of the business to concentrate on his installation and consultancy business. David has installed over 250 breweries in Britain and many overseas, and runs a full service including training and troubleshooting.
Tel: 07976 845705; **Web:** www.pbcbreweryinstallations.com

David Smith: York-based David Smith had brewed at Samuel Smith's in Tadcaster for over 20 years before establishing DH Smith Brewing Services in 1988, primarily to supply microbiological testing and other technical services. In 1989 he branched out into training, and then, as a natural next step, into sourcing and installing breweries as well. He still runs his twice-yearly Fundamentals of Minibrewing course at York Brewery. David has designed an installed over 100 breweries, including Le Brewery at la Ferté-Macé in Normandy.
Tel: 01904 706778; **Web:** www.brewingservices.co.uk.

provide equipment with a national after sales service support to meet these expectations.

The Fulton reputation has been built on providing a "total service" for our customers. Whenever possible, we like to be involved at the initial planning stage with the customer or design team when our expertise in steam boiler application is available at no cost. As a British manufacturer with a factory in Bristol, we can then provide the boiler and ancillary plant as individual components or as an off site pre-fabricated package as a turnkey CE certified assembly which minimizes the site work.

We welcome enquiries to either the company website or to the Bristol office.

www.fulton.com

0117 972 3322

Utilities and Services

Water

Beer is mostly water, and you're going to use an awful lot of it. What with cooling (and automatic temperature control at every stage of the process means a more consistent product and a less exhausting life) and cleaning, in fact, you're going to use about five times as much water as you will actually brew with. So if you produce 3,000 barrels of beer at 36 gallons per barrel, you're actually going to get through more than 400,000 gallons of water a year. This will put you on your water company's industrial tariff, details of which you will need anyway for your business plan. If you are lucky enough to have your own well or borehole, however, your water will be free. The Environment Agency allows the abstraction of up to 20m^3 or about 4,400 gallons of groundwater per day from wells and boreholes without a licence. That's more than 1.6 million gallons a year, or nearly four times what a 3,000-barrel brewer is likely to need. You can even have your own well sunk: numerous commercial borehole diggers can be found by surfing the web.

(The water company doesn't just supply water; it also takes it away. However, to take advantage of this service you will need trade effluent consent, of which more below.)

Electricity

As well as water you will also need energy to move it around, heat it up and cool it down. You can have a gas-fired copper, if you like; but

CASE STUDY

Expert Help with a Difficult Installation – Le Brewery, La Ferté Macé, Normandy

On the D908 between La Ferté Macé and Joue du Bois is an ancient cider farm that is home to a former teacher from Peterborough, Steve Skews, and his wife Jane. It has also, since 2001, housed Le Brewery, a project the Skewses cooked up with Jane's sister Sue Green and her husband, another Steve, after the winter storms of 1999 destroyed two-thirds of the orchards. It was a blow that the Skewses knew would take years to overcome.

Back in England, the Greens had seen not one but two microbreweries sprout in their home village of Wickwar, near Bristol. And although there were no breweries in Normandy, says Steve Green, there was no shortage of interest in beer. So installing a micro looked like a short-cut to getting the Skewses back on their feet.

"Normandy had a large British population both of residents and tourists," he says. "Local people were also known to have an interest in a good half-litre, and there were bars around the region that sold premium Belgian beers and would surely be interested in a good local product. We discussed the idea over a long evening and decided to investigate the feasibility of setting up a brewery at the farm."

After visiting the nearest brewery – 200km away – the partners decided their idea was worth pursuing and sought the professional services of York-based consultant David Smith, who visited the farm in late 2000.

"His report listed a number of hurdles but showed that a brewery could be installed and operated in some of the buildings," says Steve. "Considerable work would be needed to make them suitable, but it was not impossible."

While the Skewses set about tackling French officialdom – no easy task, since the authorities themselves seemed to have little idea of what was necessary – Steve Green and David Smith started the hunt for suitable second-hand kit. After visiting a number of breweries with plant for sale, they finally settled on the Worldham Brewery, located in an old oasthouse on a farm near Alton, Hampshire, and recently closed by owner Hugo Sharpe, who was calling it a day after 10 years .

"It was of the ideal brewlength and in a good location for transport to France," says Steve. "It was a 10-barrel plant with extras that allowed its use as a five or possibly 15-barrel plant – a flexible future-proof solution for a new company setting up in a new market in a new country."

Getting the brewery to Normandy wasn't so simple, though. In fact, getting it out of its oasthouse wasn't so simple.

"A team of engineers, electricians, and other helpers assembled one cold weekend in mid-January to dismantle the brewery," says Steve Green. "I could come up with many quotes in connection with moving the plant, but they'd all be unprintable! "The brewery had evolved over many years and it was hard to believe the building wasn't built around the pipework rather than the pipes fitted into the building."

The main difficulty was shifting the heaviest items such as the copper over floorboards of insufficient loadbearing capacity.

"We had to lift the vessels down using a site trolley and various pieces of wood, all of which broke during the process," says Steve. "No one was seriously hurt, but we came close. It was also icy during most of the work, which made things more difficult still."

Although the stables at the farm had been in a state of dilapidation, one feature proved ideal for a brewery: they're in a row. This made it possible to house each stage of the process in a separate temperature-controlled compartment, in a series, with the malt store and mill in the old haylofts above.

Test brews were soon completed, and the first batches of Norman Gold, a blonde beer of 4.9 per cent ABV, and the hoppier, darker Conquerant at 5.5 per cent were soon ready for drinking. They caught on, too – not only are they stocked in local supermarkets, but Le Brewery now owns two English-style pubs.

"I don't think we could have done it without David," says Jane. "He found us the ideal brewery and then came over here to inspect the site and work out how it could all be set up in the buildings we had. Then he planned the installation and supervised it, and he wasn't above rolling his sleeves up and mucking in.

"He's still a good friend, and he comes over to see how we're getting on with the brewery every now and then. He's even brought his family over, too."

www.le-brewery.com

even if you do, you are still going to use electricity on a literally industrial scale. You are therefore recommended to install a three-phase supply, even in a five-barrel plant, both as an economy measure (a single-phase supply, will draw 120–130 amps to generate 36kw of heat, whereas a three-phase supply will draw only 63 amps) and so that you don't keep blowing fuses.

Laboratory Tests

A key heading under services is the laboratory. The number and sophistication of tests and checks required out these days places a fully functional lab beyond the resources or expertise of the smaller brewery. But there are plenty of contractors who will undertake them for you; your consultant will provide some of these services and will be pleased to put you in touch with a specialist such as Brewlab who can do the rest. Before you start brewing, you'll need to have your liquors analysed, and it's a good idea to carry out regular checks in case of changes to the water supply. Your consultant can arrange these. The regulations concerning sampling for alcoholic strength are set out in the appendix; again, a specialist such as Brewlab can carry these out, but nearer to home there are many commercial providers of laboratory services. If your consultant hasn't recommended one, check in Yellow Pages under "chemists – analytical and research" or "laboratory facilities". Local laboratories can also carry out the regular microbiology tests recommended in the Society of Independent Brewers' code of conduct. If you sell beer to large retailers such as supermarkets, you will also be required to check regularly for heavy metals such as lead and arsenic, aflatoxins produced by fungi, and toxins produced by nitrites in the water supply.

Pipework

Finally, not exactly a utility or service, but an essential part of the brewery that won't fit under the heading of any particular vessel, is the pipework. This is something you will become intimately familiar with very quickly, and Richard's advice is not to cut corners by choosing flexible hoses over solid pipes. There are two reasons for this: hoses won't last and will have to be replaced; and there will be an irresistible temptation to tuck them away where you won't see them. Actually, you're better off with highly conspicuous pipes in places where they can easily be inspected, cleaned and, if necessary, repaired or replaced.

Malt Mill and Hot Liquor Tank

The brewing process starts when you fill your mash tun, so up in the very attic of our tower we find our malt mill and hopper and our hot liquor tank. Only Richard's advice is: don't bother with the malt mill. Years ago, maltsters expected their customers to grind their malt to their own specification and were rarely called on to do it themselves; and when they did, the grists they came up with weren't necessarily exactly what the brewers wanted. Nowadays, though, with so many micros not possessing mills of their own, grinding is all part of the service; so why spend good money on kit you don't need? (You'll still need the hopper, though.)

On the right, an example of a mash tun

The other constituent of the mash is, of course, hot liquor. A fairly plain insulated stainless steel tank with everyday immersion heater elements will do the trick, says Richard. It needs to heat the liquor to 72°C but can be allowed to do so overnight; an energy-saving thermostat and a timer switch (so you don't have to slip back to the brewery last thing at night to turn it on) might be useful extras.

Also in the attic we're likely to find the cold water tank. This is going to supply the four-fifths of your water requirement that you don't actually brew with, and it's got to be both refrigerated and well insulated. This is because the hopped wort hot from the kettle is going to have to be cooled rapidly to 18–19°C before it can be pitched with yeast; given that the cold water tank is right up there in the attic where the sun can beat down on it uninterrupted, it will need refrigerating down to about 12°C before it can do the job.

Mash Tun

The mash tun is one of the most important vessels in the brewery, for it's where the malt is made to yield up its fermentable sugars to create the sweet wort that will, eventually, be beer. But it's also one of the simplest – basically, it's just an insulated stainless-steel cylinder with a tight-fitting lid on top and a sieve in its bottom.

To get the best extraction, your mash tun will need to be well enough insulated to hold the mash at 72°C for an hour or an hour and a half. It also needs a sparging arm built into the lid: this is rather like the spray arm in your dishwasher at home, and once the wort has been run off it sprays the grains with hot water to get the last little bits of sugar out of them. The grains lie either on a stainless steel plate drilled with just the right number of right-sized holes to let the wort run through once mashing has been satisfactorily completed, or on a wedgewire mesh that does the same job.

Emptying the mash tun is quite a job: a five-barrel brew of session bitter requires something like 120kg of malt, which will by this stage be sodden and therefore even heavier. It all has to be dug out by hand, which is made more difficult by the fact that a five-barrel mash tun isn't actually all that big, so it's very awkward to get at its contents. There is a tilting mash tun on the market that makes the job easier, although digging out the mash tun is the brewing industry's time-honoured way of keeping apprentices and juniors fit, and it's certainly a good work-out.

Spent Grain

What do you do with all that spent grain? If you're brewing every day it will soon mount up, so the best thing to do is to sell it as animal feed. Since July 2005, though, thanks to swine vesicular disease and foot and mouth, British brewers have had to conform to European feed hygiene legislation (although you can bet that French brewers don't!) as they affect "co-products". In practice this means registering with the Brewing, Food & Beverage Industry Suppliers Association (BFBi) Feed Assurance Scheme, going on a one-day training course and paying a one-off fee of up to £350 depending on how big a brewery you are. It may seem just another bit of bureaucratic nonsense, but given the devastation caused by foot and mouth it is perhaps a forgiveable precaution. Anyway, the sale of co-products is so important for your cash-flow (the first payment you ever receive will almost certainly be not for the sale of beer, but for the sale of spent grains), and any other means of disposing of spent grains would be so difficult, that registration is indispensable. Visit **www.bfbi.org.uk** for more details.

The Copper and Hop Back

The next stop for your wort on its journey towards becoming beer is the copper, where it will be boiled up with hops. The wort will already be pretty hot when it is run into the copper, so a couple of thermostatically controlled immersion heater elements should suffice to bring it nicely to the boil and keep it there for as long as need be. But although this is the cheapest and easiest way to do the job, it's by no means the only way. A handful of breweries, notably Caledonian, have coppers fired directly by gas-jets, which is picturesque and produces caramel that gives the beer a highly distinctive taste, but there are health and safety issues. Another more expensive method is a gas immersion coil, which is highly efficient. Then there's steam either in a coil or an external jacket, which is very expensive to install and highly dangerous if not assiduously maintained, but is both cheap to run and very flexible, since the same steam can also be used for cleaning. The copper needs to be vented externally to get rid of the steam it produces, and it can have a sprayball built into its lid to allow thorough cleaning with detergent and rinsing with boiling water after use (although you can just hose it out with boiling water, since it should already be sterile).

Perhaps the hop back should come under the same heading as the copper, although strictly speaking it's the next stage in the process. Essentially it's just a huge sieve, like the steel plate or wedgewire mesh in the mash tun, only in this case it's the "trub" or used hops that it's there to collect so the hopped wort can run clean into the fermenters. Richard, who runs Tring Brewery as well as Brewing Design Services, actually uses the hop back for the last addition of hops. Instead of putting the late hops into the copper itself, he puts them in the hop back, dampens them a little, and then lets the wort sit on them for a while before it drains through them. Adding the late hops directly to the boil, he says, evaporates many of the aromatic resins. Doing it his way gives him more aroma from fewer hops. Be that as it may, the used hop litter or "trub" has to be got rid of. Pelletised hops, says Richard, tend to disintegrate in the boil to the extent where (with the water company's express approval) they can simply be hosed down the drain. Used whole hops have to be dug out by hand (not a very onerous task, since they're not used in any great bulk) and make excellent mulch.

Fermentation

This is where your plentiful supply of refrigerated water comes in, since the hopped wort leaves the copper having just boiled and has to be rapidly cooled to 18–19°C ("collection temperature") before it can be pitched with yeast. The wort therefore needs to be run through a heat exchanger, of which there are a number of types, into the fermenting vessel, whose temperature also needs to be controlled (preferably automatically, so you don't have to alter it manually every two hours day and night) by means of a cooling jacket. Fermentation is an exothermic reaction – that is, it generates its own heat – and the temperature of the fermenting wort will gradually creep up as the process goes on. Too hot a reaction will generate off-flavours, so the fermentation temperature needs to be kept down to about 21°C. Different shapes of fermentation vessel produce different temperatures, too; consult your installer as to which shape – cylindrical, cylindro-conical or square – will suit you best. The cooling jacket also needs to be efficient enough to bring the temperature of the wort down sharply to 8–12°C within 24 hours of fermentation ending.

Open Fermenters

There is some argument as to whether fermenters need lids or not. In some parts of the country, especially in the north of England, open square fermenters made of slate are traditional; and being able to peer into the fermenting vessel as the yeast begins to grow its thick, rocky head is perhaps the most aesthetically pleasing part of the whole operation – visitors love it. But there are drawbacks, says Richard. Open fermenters demand excellent ventilation and extraction equipment, since the CO^2 given off during fermentation is lethal even in fairly low doses. CO^2 in high concentrations burns the nose and eyes; but it will cause unconsciousness at much lower levels than that, and will then asphyxiate you. Your environmental health officer will probably stipulate what monitoring and ventilation equipment you will need at the planning stage; some EHOs have been known to veto open fermenters altogether.

Closed Fermenters

Closed fermenters are also more hygienic, says Richard – which is vitally important, as this is the most delicate stage in the proceedings. Historically, it's infection when fermenting that is the brewer's biggest danger. Improved hygiene and handling mean that spoilt brews due to yeast infection aren't as common these days as they were;

Pitching yeast into a fermenter

nevertheless, you need to be ruthlessly disciplined in your hygiene procedures at this stage. Not only do closed fermenters mean there is no risk of birds, rodents or brewery visitors making unwelcome deposits in them, but they are less vulnerable to passing bacteria and spores and can have "cips" or cleaning in place systems built in. And opting for closed fermenters means that your entire brewery is virtually a closed system from mash tun on. Almost the only time it needs to be opened during the process is for the addition of hops – which are, of course, added to the wort while it is boiling and therefore safe from infection.

Once fermentation is complete – and it can last for anything from a day to a week – there are really only two processes left for most microbrewers: collecting the yeast, and racking the beer into casks. Your yeast will grow by five or six times during fermentation; you can skim excess barm off the top of the wort with a sterile bucket, and there's a trap known as a thimble that fits into the base of the vessel to collect the clumped or "flocculated" yeast cells that sink down to the bottom. Once you've harvested enough yeast for your next brew, the rest will go as highly nutritious pig-feed – provided, of course, that you have signed up to the BFBi's Feed Assurance Scheme.

CASE STUDY

A Brewery Put Together from Scrap – Green Jack Brewery, The Triangle, Lowestoft, Suffolk

East Anglian brewers scored a resounding double when Ripper from Lowestoft's Green Jack Brewery won the CAMRA's Winter Beer of Britain award in January 2007. For the overall Champion Beer of Britain title was already held, for the second year running, by Crouch Vale of South Woodham Ferrers, Essex, with Brewers Gold. And as it happened, CAMRA's reigning National Pub of the Year, the Swan at Little Totham, was also in Essex.

Not only was Ripper's victory a historic hat-trick for the region, but Green Jack itself must be one of the smallest breweries ever to win a national title. Its 10-barrel plant isn't unusually small in terms of capacity; but it's squeezed into the tiniest space imaginable, and with only two fermenters it can only turn out three brews a fortnight. "We're part-time brewers looking to go full-time," says boss Tim Dunford, who is permanently on the lookout for bigger premises.

What Tim has got is an odd-shaped outbuilding across a tiny yard at the back of the Triangle, a pub in a Victorian residential quarter on the edge of Lowestoft town centre. But the pub known today as the Triangle actually used to be called the Oddfellows Arms; and the outbuilding that houses the brewery is all that remains of the original Triangle. The back-to-back pubs coexisted perfectly happily for many years until the Triangle was partly demolished – to widen the street it fronted. Somehow its name migrated to the Oddfellows; and it was universally known as the Triangle when Tim embarked on his drinking career – slightly but not dramatically underaged – just over a quarter of a century ago.

The story of Green Jack starts in 1988 when one Charlie Forbes started brewing at Oulton Broad just outside Lowestoft. Tim, a local lad who was first a signmaker and later an antique dealer, was an enthusiastic customer and a pal of Derek Longman's, who actually ran the brewery while Charlie lived in London.

"I thought to myself, if Derek can do this I could probably do it better!" says Tim. "So I got hold of Charlie and started hassling him, and eventually he sold me the brewing equipment and the lease on the premises."

That was in 1993, and – Charlie having taken his surname with him – Green Jack was born, with Tim brewing in partnership with his sister. That arrangement lasted until five years ago, when the partnership broke up and Tim took over the Triangle. The constraints of the new site made buying a standard brewing plant impossible. Instead Tim and local fabricator Richard Payne – a childhood friend – purpose-built the kit themselves.

From a local scrap-yard they bought four small steel tanks that had once held onion flavouring for crisp manufacturers. Once the lingering odours had been scrubbed away, one of them became the mash tun, while Tim and Richard welded a metre and a half of extra capacity to the tops of the others to become the copper and two fermenters. The whole process is digitally temperature-controlled using equipment which had to be sourced from a German firm because the brewery doesn't have three-phase electricity, which Tim says was required for all the domestically-produced equipment he looked at. It was all rigged up by a local refrigeration engineer who is also a friend of Tim's. And cleaning in place was provided by the installation of simple pumps and sprayheads – they cost about £30 each, says Tim – in all the relevant vessels.

The extempore nature of the kit hasn't affected the quality and consistency of the beers it produces, either. Almost as soon as it flowed from the fermenter, Ripper was winning awards: it was champion barley wine at Peterborough Beer Festival in 1996, '97, '98 and '99. The summer seasonal beer, Summer Dream, has won a few trophies too, and Orange Wheat – "a bastardisation of Hoegaarden", Tim admits – was champion specialist beer at Peterborough and Norwich in 2006 and overall winner at both Norwich and Leicester in 2005. But it was Ripper, a dangerously drinkable, hoppy, pale ale of 8.5 per cent alcohol, that notched up Green Jack's first national title, much to Tim's delight.

A word of caution, though. You might opt for a brand-new brewhouse, or prefer to put one together yourself from second-hand vessels of respectable provenance. Tim was able to create an award-winning brewery from used vessels employing local tradesmen with no prior experience of brewery equipment because he already knew his brewing inside out. Do you?

www.green-jack.com

Casks

As we discovered in the previous chapter, most brewers reckon they need a ratio of seven casks for every one out in trade. The commonest casks these days are nine-barrel firkins at around £80 each, so they're not cheap, and you might also want a handful of 4½-gallon "pins" for very strong beers or 18-gallon "kilderkins" for fast-selling beers.

Cask racking at St Peter's Brewery

(The full 36-gallon barrel is almost obsolete as an actual cask and is really only used as a unit of account these days. That's not because few pubs or clubs can sell beer quickly enough to warrant such a large container: there are probably thousands whose throughput of lager would justify the use of full-sized barrels. But they are nightmares to handle, weighing 200 kilos or more when full. I do believe there are still a handful of 54-gallon hogsheads in circulation in the north-west, but I have never seen one. And you may also come across metric barrels of 11 and 22 gallons, which were once quite commonly used for lager but are now pretty much obsolete. If you are offered any, check them scrupulously: they are not only second-hand but probably very old and likely to leak.)

Alternative Barrels and Casks

The price of new barrels may tempt you to investigate second-hand ones; but do be careful, especially if there are any visible mends, for a "leaker" will cost you dear. Not only does it have to be replaced, but if it fails all over a publican's cellar, it may well cost you his business. Similarly, plastic casks have attracted considerable criticism. On paper, they're the answer to the small brewer's prayer – lightweight, rigid, durable and cheap. But some say they're prone to sudden and dramatic failure, that they don't stand up to the stresses of frequent rough handling, and that the plastic is degraded by washing at high temperatures. And whether the criticisms are justified or not, there are many licensees (including beer festival organisers) who simply won't accept them, not because they've had bad experiences themselves, but because they've heard of others who have. Still, plastic casks have their uses. If you're a brewpub or have a retail outlet on site, you can use plastic casks and reduce the risk of failure by handling them carefully. And if one does fail, the only person you've antagonised is yourself. You can also keep a stock in reserve for emergencies and for special events such as brewery open days that might make too much demand on your population of steel casks.

Whatever casks you use, cleaning them to sterility is absolutely essential. Rinse them out thoroughly with a power-washer; use a sprayball with detergent; and then steam them. (In fact ease of cleaning is one reason why the industry has resolutely refused to countenance cubic containers. Barrels are traditionally barrel-shaped so they can be rolled round the brewery yard and up and down the ramps in pub cellars quickly and easily. Cubic containers would be more efficient, stacking into a smaller space, and could easily be moved around on tail-lifts and similar devices. But the corners would be havens for bacteria, which single disadvantage outweighs a multitude of advantages.) Let discipline in your cask-washing procedure slip even once and you risk sending out spoilt or infected beer, for which the retailer on the receiving end may never forgive you. Let it slip more than once and you risk being labelled as a dirty brewery, a reputation that is hard to shake off and that could kill your business (and I do know of one brewery near my home where this actually happened; other brewers will doubtless have other examples).

Bottling and Kegging

Not all your beer, of course, will go into casks. Depending on your business model, you might well sell as much in bottle as on draught. Bottled sales will almost certainly be very important to you, and the ramifications of bottling your beer will be dealt with in a later chapter. But there's a small but growing band of brewers interested mainly in foreign beer styles whose draught volume is not cask-conditioned but keg.

One of the biggest changes in the microbrewing sector since the first edition of this book was published has been a growing affinity for keg. For many years – decades, in fact – microbrewers by and large stood shoulder to shoulder with the Campaign for Real Ale in loathing the stuff, remembering unfondly both the ghastly almost non-alcoholic national brands that prompted the consumer backlash of the 1970s (Watney's Starlight at 2% ABV!) and the second generation "smooth" keg brands of the '90s (whatever happened to Boddington's?).

But as long as 20 years ago there were those who broke ranks. Inspired more by the cafe culture of the Continent than by the British pub tradition, Freedom and Porterhouse breweries set up more or less all-keg boutique bars in the West End of London in the mid-1990s, and the Porterhouse in Covent Garden is still with us. Others followed, notably West Brewing in Glasgow and Camden Town Brewery in north London.

The earliest micros to adopt keg did so in order to recreate the best beers from Europe and North America, where serving cold, filtered, and sometimes pasteurised beer under CO^2 is too natural to be worthy of comment. Indeed the French don't even have a word for cask-conditioned beer: the equivalent of "draught" is "sous pression"; its opposite, "sans pression", would be meaningless. And legions of British travellers to Belgium – even those who fancy themselves real ale diehards – have discovered to their delight that the method of dispense isn't really all that important: it's the quality of the liquid that counts.

Now, however, there's a new impetus that is driving more and more British microbrewers to look seriously at kegging their ales: mathematics.

Five or six years ago there were 500 independent brewers and 60,000 pubs, or 120 potential accounts per brewer. Today there are only 48,000 pubs and more than 800 independent breweries, or 60 potential accounts per brewer; and the ratio is getting worse every day. Unanswerable maths dictates that the old-school pub trade can no longer sustain the number of brewers clamouring to supply it. And the bottled trade, monopolised as it is by the big supermarket chains, is for most independent brewers a marginal alternative at best. Few microbrewers make their living selling bottled beers to specialist off-licences and at farmers' markets!

But there is one huge opportunity for keg sales which the microbrewing sector hasn't had to examine before but is now beginning to investigate. All over the country, thousands of licensed bars in village halls, leisure centres, sports clubs, theatres, arts centres and community centres can't stock cask beer but are intensely local in their outlook. Often owned by parish or district councils, they would willingly, were it possible, stock a beer from a brewery that employed local people and paid their business rates to the local town hall. In terms of overall annual barrelage many of these accounts might appear too small to bother with. But many of them, even the most modest village halls, operate as functions rooms that can shift impressive volumes per session – and are in fact just the kind of venues for which keg was developed back in the 1930s.

There are other keg opportunities to exploit, too. Such workingmen's, comrades, and social clubs as survive are in many cases keg-only but – if the price was right – their committees would stock locally-produced keg beers if they could. Breweries with intensely rural trading areas, especially in Scotland, are looking to kegging to give their beers the shelf-life that small isolated pubs with big seasonal fluctuations in throughput require. Keg can also be more convenient than cask for outside bars – no settling time, so no need to move on to the site 48 hours in advance of the event itself. And even dyed-in-the-wool real ale pubs will often stock a craft-brewed keg lager, stout, or Belgian-style blonde ale to underline and enhance their support for microbreweries

But isn't kegging hugely expensive? Well, not necessarily. An expert tinkerer who already filters and carbonates beers for bottling can adapt the bottling line to keg short runs for individual accounts. But it's unbelievably time-consuming, and any microbrewer who is serious about the opportunity would probably want to install a keg washer and filler such as the trolley-mounted 4-nozzle DME from ABUK of Misterton, Nottinghamshire, which would come to about £7,000 new.

If keg sales take off, there are two alternative routes to expansion. One is to take the plunge and invest in in-line carbonation (much more reliable than carbonating in tank, according to ABUK), sterile filtration, and a larger manual filler starting at around £20,000 the lot (depending on the strength of the pound: much of the equipment is imported from Germany or the Czech Republic). You pretty much have to filter, since the method of dispense rouses the beer; but there's no need to pasteurise: sterile-filtered and carbonated beers even of moderate ABV can have a six-month shelf-life without being boiled!

If, however, £20,000 is rather a large step you might always consider contract-kegging with a regional brewer as an intermediate phase. Everard's of Leicester, for example, has a minimum kegging run of 60 barrels (or about 200 11-gallon kegs), which is probably fairly typical. Expect to pay £3-4 per keg for filling, or up to £800 for a minimum run, to which you would have to add the cost of tanking your beer to the brewery. This could be in the region of £500 for a 180-barrel (600-keg) tanker, depending on the price of diesel this week; the cost is likely to be the same even if the tanker isn't full. If you don't have your own kegs and pallets you'll have to hire at around £6 per keg. The total cost, therefore, of contract-filling 200 kegs will be around £2,300 or £11.50 a keg, so it might be just a stopgap while you consider your longer-term options.

There are other costs, too: temperature-controlled conditioning tanks, for instance, especially if you want to brew an authentic lager. Large ones cost a lot, but the German fabricator Speidel makes freestanding 240-litre chiller tanks at around £800 each (its UK agent is Vigo Ltd of Devon). And then there are the fonts themselves, and here the sky is the limit. But you don't have to include flashing lights and a brushed titanium finish: a modest font will cost no more than £70 and if even that seems a lot, there is a hidden advantage. Once landlords have allowed them to be installed on their bars, it seems, they regard them as permanent fixtures (provided your beer sells, of course), which gets you off the guest ale rollercoaster.

To offset all these costs is the 50p-a-pint premium that keg lagers, stouts, Belgian-style blonde beers and even British-style golden ales can command at the bar, which should translate into a premium of £20 a keg for the brewer. There's also the fact that new 11-gallon kegs from a company like Kammac of Burton-on-Trent cost £6 less than 9-gallon firkins, which all adds up in the long run. And there could be a sales advantage, too: real ale publicans who won't extend their range on grounds of throughput and ullage might very well consider adding a long-lived, waste-free, keg beer to their offering.

To many microbrewers, of course, the very idea of keg is anathema. But these are hard times and getting harder, and there can be few microbrewers nowadays determined enough to dismiss a business opportunity on grounds of ideology. And as for quality, anybody who has enjoyed a superb draught de Koninck in Antwerp or Gouden Carolus in Mechelen will surely agree that a squirt of CO^2 won't harm a good beer – even as being pulled through a handpump won't improve a bad one.

Water Disposal

The last step in the brewing process is getting rid of the four-fifths of water you used but didn't brew with safely and hygienically. This water is going to be full of all sorts of pollutants, some of them organic – bits of hop-litter, protein, suspended solids – and some of them chemical, in particular, industrial-strength detergents. It may also have a low ph value.

You can't simply flush it down the drain – not, at any rate, without trade effluent consent from your water and sewerage provider. This comes with a detailed questionnaire about what sort of and how much effluent you propose to discharge; the advice is to be scrupulously honest, since if you breach the terms of your consent the sewerage provider not only can but probably will revoke it altogether (it can fine you, too). And since sewerage providers continually monitor discharges, you won't get away unnoticed with craftily tipping it away, either. Especially as it's going to be in the region of 450,000 gallons a year!

If you don't have mains drainage, you might consider planting a reed bed, for which you will need discharge consent from the Environment Agency. It is becoming increasingly popular to dispose of pollutants by creating a reed bed certain reeds can actually digest most if not all of the nasties a brewery generates, and the site itself will be a haven for wildlife and is therefore environmentally friendly. But there are drawbacks: you may not have enough space, and establishing a reed bed is a slow and expensive process. Most brewers who have done so seem satisfied with the result, though, and I have heard of only one instance where the reed bed's performance has been ruled inadequate by the Environment Agency. Two sites where you'll find more information about reed beds are **www.oceans-esu.com** and **www.sheepdrove.com**, and for more details on effluent discharge regulations visit the ever-helpful **www.businesslink.gov.uk** and follow the "health, safety, premises" link to "managing health, safety, environment". The Environment Agency also has a helpline: **Tel:** 08708 806506.

Chapter Five
What to Brew

CASE STUDIES

Wine snobs like to be dismissive of beer. Wine, they say, has infinite variety: red, white, rosé, still, sparkling, dry, sweet, New World, Old World, grown on chalk, grown on flint, grown in a hot climate, grown in a cool climate, grown on a south-facing slope, grown on a west-facing slope – and as for grape varieties, why, the permutations are endless! But as for beer… well, it's just beer, isn't it?

And if the big multinational brewers had their way, the wine snobs' scorn would be justified; 90 per cent of the world's beer is a pale (both in colour and in character) imitation of Czech or German pilsner, brewed (as one disgruntled ale brewer of my acquaintance once put it) from broken biscuits and old bus tickets, and served cold to mask the nasty tinny flavour that comes from too short a lagering period (which is the only flavour most of it has). It's made as cheaply as possible, using the lowest-grade ingredients available, taking every short cut in the brewing chemist's book, and intended to be the same wherever you go, whatever the label. However, some 90 per cent of the world's wine is pretty ropy, too, if you think about it, but as a craft brewer you stand shoulder to shoulder with the makers of the best 10 per cent. That's the bracket you want to be in; it's only a pity that the best of beers can't command the same price premium as the best of wines!

Don't, whatever you do, let the winemakers command the high ground, or get you down, or dent your self-esteem. Beer has just as many variations as wine – more, in fact – and you are setting out to master all of them. And in so doing, you have picked a harder row to hoe than any winemaker, for most of their work is done for them. All their ingredients come in one juicy little package, the grape, the sweetest fruit of them all, with liquor, fermentable material and aromatics all contained in one skin. All they have to do is grow it – and in many cases they needn't even choose which variety to grow, so protected are they by regulation – press it and ferment it. If things go wrong, they can always blame the weather. Even the choices they do have to make – how to prune, whether and how long to macerate, what temperature to ferment at – are largely dictated by tradition and regulation. They don't have to make anything like the number of choices you will – how to treat your liquor; what grains to blend in your grist, and in what proportions; what hops to use for bitterness and aroma, and in what quantities, and at what stage of the boil to add them; what strain of yeast to propagate; even what styles of beer to brew. For you, the choices are infinite, and you are going to have to make them yourself.

Don't be daunted, though. For as a brewer you are free to make whatever you want.

- Pale
- Cloud
- Lemony
- Belgian-style wheat beer
- Baltic-style porter
- Bitter
- Mild
- IPA
- Brown ale
- Stout
- Barley wine
- Pilsner
- Bock
- Kölsch
- Alt
- Bière de garde

And as for ingredients! There's pale malt, brown malt, amber malt, black malt, crystal malt, chocolate malt, unmalted roasted barley, malted wheat, unmalted wheat, torrefied wheat, oats, rye, rice (if you're American), sorghum (if you're African), buckwheat (if you're gluten intolerant), invert sugar, candy sugar, honey... and we haven't even started on the hops yet!

As a home brewer, you may have played around with all of the above, and you might also have experimented with flavourings such as ginger and other spices, various fruits, exotic sugars and even spirits. Some of your experiments may have been, frankly, mistakes; others may have been triumphant successes. But as someone whose livelihood now depends not so much on what you can brew as on what you can sell, you're going to have to think hard about the range of beers you can realistically put out there in the market place.

Being thoughtful and realistic, though, needn't mean being too cautious. Yes, it's a foolhardy brewery that doesn't have a session bitter, a best bitter and perhaps a special bitter in its range. But in order to compete effectively – with the established regionals in the early days and more recently with each other – small brewers have always had to offer something different. And part of the glory of being a micro is that you can create something completely off the wall that will break even at only two or three barrels a week, whereas for someone like Greene King or Marston's it would have to sell two or three hundred barrels a week to be worthwhile.

The ability to brew interesting beers in small quantities has always been one of the small brewing sector's key advantages over the older and larger brewers. By the time the small brewing revolution came along, regional and national breweries had been rationalising their ranges for years. It was the downside of economy of scale: any beer style they couldn't brew in economic quantities was doomed, even though there might still be some demand for it. The tied house system and the closed mentality that it engendered meant there was no chance of two or three brewers combining to continue to offer, say, a single old ale between them: if they couldn't sustain a brand of their own, they wouldn't buy in from another brewer. The exception to this rule was Guinness, which as it was Irish and had no tied estate wasn't regarded as dangerous. So old-established breweries that hadn't had a dry stout of their own for generations would happily stock Guinness; and Guinness got a very healthy trade out of it, commanding five per cent of the entire British beer market. Why the brewers couldn't extend this logic to other beer styles remains a mystery; only recently have some family brewers that no longer produce a dark mild allowed those of their tenants who still have a demand for it to stock Bateman's DM.

Even in the early 1970s, though, most regionals produced quite a wide selection of draught beer. Mostly there was a light mild, a dark mild, a session bitter, a best bitter, a special bitter and sometimes a strong Christmas special. On the bottled side, most brewers, even quite small ones, still had a light or pale and a brown ale of their own (often just the bitter and the mild pasteurised and bottled), a sweet stout and a barley wine. But these were dying out, and regional brewers were increasingly concentrating on their core ranges. It was years before they noticed that the micros were succeeding by experimenting and diversifying; ever since then they too have been experimenting and diversifying, but you can safely say, in terms of beer styles at least, that where the micros have led the regionals have tended to follow.

For having the motive, means and opportunity to experiment has always suited the temperament of microbrewers in general. Few of them turned to brewing for the money; and the love that drives most of them has always been evident not only in the quality of what they brew, but also in their enthusiasm for different styles and genres. There's also, sometimes, a strong sense of fun in evidence. Yes, these people are trying to make a living. But for many of them the conventional ways of making a living have been crushing disappointments, and they're out to enjoy their working hours as much as their leisure time. This is a wholly laudable urge (which you must share, or you wouldn't have picked this book up), and it has produced some wholly laudable, if somewhat eccentric, beers.

For instance, there was the Kitchen Brewery of West Yorkshire, now defunct, whose beers were all brewed with various vegetables. Then there was a hotly contested but strictly unofficial competition in the early 1990s between Parish of Leicestershire and Sheffield's Frog & Parrot brewpub to see who could brew the stronger version of their respective barley wines, Baz's Bonce Blower and Roger & Out. I can't remember who managed to keep their yeast alive and fermenting for longest, but the eventual victor was, as I recall, over 20 per cent alcohol. I never got to try either contender. I did, however, try the ginger wheat beer produced by Salopian, always an experimental brewery. It was a delight, and I wish it were still being produced.

I could mention Hanby's Cherry Bomb; I could mention Nethergate Umbel Ale. I could go on and on, indeed; but in a way it would be misleading. For where the microbrewing industry as a whole has really led the way has not been in the production of eccentric one-offs, fascinating though they are, but in the setting of trends that the rest of the industry has followed.

Themed Ranges and Unusual Ingredients

The creative drive that has allowed smaller brewers to get away with experimenting is based on the inversion of the law of economy of scale described above: micros can make an acceptable margin on volumes that would be loss-making even for the smallest of regionals. This has proved very useful in preserving beer styles, especially the darker styles such as mild, barley wine and old ale that would otherwise have vanished completely or, indeed, that already had. The first micro-brewed beer to win CAMRA's Champion Beer of Britain Competition back in 1987 was Pitfield Brewery's Dark Star, a strong dark ale of a type not listed by a regional or national brewery within even the longest of living memories.

Since Dark Star's triumph, Pitfield has gone on to recreate a whole raft of historic styles using recipes that are mostly more than 150 years and in one case over 200 years old. Its XXXX Stock Ale at 10 per cent alcohol is a comparative stripling, being made to a recipe dating from 1896. The London Porter recipe comes from 1850, the IPA from 1837, the Amber Ale from 1830, and the Imperial Stout from 1792. There are many frustrations in trying to re-create old recipes. The brewers of yore were often frustratingly inexact in recording quantities, boil times, and so on; and the varieties, characteristics, and methods of processing malt and hops have changed so much over the generations that we really can't know what sugars and acids the ingredients of a century ago contained, and what flavours they would yield. Nevertheless, Pitfield is only one small brewer that has looked to the past to inspire the present. The Teignworthy Brewery at Tucker's

Maltings in Newton Abbot, Devon, produces a similar range under the Edwin Tucker brand. And although its recipes are undated, beers such as Empress Russian Stout, East India Pale Ale and Old Walnut Brown root it firmly in the glory days of British brewing history. And mention must also go to Williams Brothers of Alloa, Scotland, which for many years and under several names has specialised in researching and re-creating historic Scottish beer styles remarkable for being flavoured with locally available alternatives to hops. Fraoch was the first and best-known and claims to re-create Pictish heather ale. Almost nothing is known about Pictish heather ale, as you might guess; it may even be that the heather was not actually used as a flavouring agent at all, but that its microfauna included wild yeasts and that sprigs of it were therefore used to ferment the beer. Be that as it may, Fraoch (pronounced "frook") is an excellent beer that has won friends around the world and is widely exported. Its stablemates – Alba, flavoured with spruce tips; Grozet, flavoured with gooseberries; and Ebulum, flavoured with elderberries – are not as big sellers as Fraoch but are just as... interesting.

Theming

Theming the range in this way is an enjoyable and effective way of creating a strong identity for the brewery, and of course the theme needn't be retro. In most cases it's only the beers' names that are themed, while the recipes themselves are unconnected. An outstanding example of this is Triple fff of Alton, Hampshire, whose beers are all named after classic rock singles and album tracks – hence a range that includes Witch's Promise, a 6 per cent Christmas special; Moondance, a 4.2 per cent best bitter; and Pressed Rat & Warthog, a 3.8 per cent dark mild. (Bryncelyn of Swansea, another pop-pickin' micro, has a range themed entirely on Buddy Holly.) Abbeydale of Sheffield has a huge range of beers all on a monastic theme – Absolution, Matins and so forth, although what any God-fearing abbot or prior would make of a beer called Black Mass I leave to your imagination. More common are beers named after local landmarks or local events, or even the locals themselves (not necessarily living ones, though: Wood's Local Heroes range commemorated South Shropshire worthies going all the way back to King Caractacus).

There are brewers for whom theming means more than just a common source of brand names: try any beer from Rooster's of Harrogate, for instance, and you know you're in for something hoppy. For owner Sean Franklin is a hop enthusiast, who pioneered the use of exotic American and other New World varieties. Other brewers have also looked abroad to find a common identity for their ranges. In some ways this is counterintuitive, for the genesis of the new wave brewers lies in the consumer revolt of the 1970s against keg dispense and dreadful British-brewed lagers (which in those days were often no more than very pale ales served ice-cold). In the cynical and exploitative hands of the big bad brewers of 30–40 years ago, both the "k" word and the "l"word became instantly recognisable signifiers of weak beer, brewed as cheaply as possible, and as bland as the brewer's chemists could make it. But in time, more widely travelled beer lovers – both drinkers and brewers – came to realise three important truths: that not all foreign beer is "lager"; that not all foreign beer is bad; and that if beer is brewed well, it won't be ruined by being dispensed under CO_2 pressure.

Continental Styles

These realisations have prompted many British microbrewers to experiment with continental styles, with varying degrees of authenticity and commercial success. The two don't always go hand in hand, mind: Pilgrim Brewery's excellent and extremely authentic Bavarian-style wheat beer, Springbock, is no longer brewed; while Harviestoun Brewery's lager, Schiehallion, also excellent but not nearly so authentic, wins prize after prize and goes from strength to strength. Many other microbreweries brew lagers and other German-style products with varying degrees of success, and thanks to the German influence, wheat, malted or not, has become a commonplace ingredient in British brewing. (Many regionals also brew foreign-style beers, either under their own names or under licence from foreign proprietors. Holsten, Kingfisher, Asahi, Red Stripe, Ayingerbrau – you may think they're imported, but in reality they come from exotic locations like Faversham, Tadcaster and Bedford.)

Only a handful of breweries have gone beyond producing one or two examples of foreign styles, though, and most of those that have are boutique breweries where the emphasis is as much on the style of retailing as on the beer itself. WEST brewery in Glasgow, brews a range of beers that conform with the German purity law, the Reinheitsgebot, that forbids the use of anything other than malt, hops, water and yeast in brewing. Cotswold Brewery in Foscot, Oxfordshire, only produces lager. Mash in Great Portland Street, central London, is a US-style boutique brewery with a wheat beer, a Pils, and a fruit beer on its list alongside a porter and an IPA. Meantime in Greenwich, run by

German-trained brewer Alastair Hook, produces authentic Kölsch and Altbier for Sainsbury's and a US-inspired range of specialities including chocolate and coffee-flavoured beers. West Brewing in Glasgow lists a hefeweisse or cloudy wheat beer; a Helles or everyday pale beer; a Dunkles or dark lager; and a Dunkel Hefeweisse, or dark wheat beer with yeast sediment. Finally there's Zerodregrees with branches in London, Bristol and Reading brewing German-style wheat beer, Pilsner and dark lager.

Interestingly, although many British brewers are prepared to try their hands at German beer styles, nobody I know of has dared have a go at anything Belgian yet – except, that is, fruit-flavoured concoctions whose base beer is nothing a genuine Belgian would recognise. Are Belgian beers simply too idiosyncratic and too formidable to imitate? Is the prize so far out of reach that to make a grab for it and miss is to risk a nasty fall? Whatever the reason, the field is open for a British-brewed abbey ale, strong blonde beer or oud bruin. Are you up to it?

Golden Ales

Earlier, I spoke about new brewers setting trends that regionals have followed. Would Fuller's, I wonder, have released its superb London Porter, or Marston's its authentic and stunning Old Empire IPA, if micros such as Pitfield had not delved into the brewing books of yesteryear in search of forgotten flavours? Would Badger (as Hall & Woodhouse likes to be known) have brought out its extraordinary peach-blossom beer if micros had not experimented with fruit flavours first? Would we have Young's Waggle Dance or Fuller's Organic Honeydew if Enville Ales had not pioneered the use of honey in brewing?

Regionals, though, are still pretty conservative in their attitude towards speciality beers. Mostly, given the downside of economy of scale already mentioned, their regular ranges only encompass the more popular styles; and when they do try something outrageous, it's normally as a seasonal special. But there are two trends established by microbrewers that regionals have followed enthusiastically: pretty much everybody now brews a "golden ale" of one sort or another; and regionals by and large are pursuing the green agenda as energetically as they can.

The emergence of golden ales over the past 20-odd years has been perhaps the climax of the shift away from dark beers in modern ale brewing. Porters died out in the First World War , thanks to malt rationing; and stout would have gone the same way if the government

CASE STUDY

Brewery with a Huge Range – Bartram's, Bury St Edmunds, Suffolk

When you look at the list of beers Marc Bartram produces on his five-barrel plant on the Rougham Estate in Suffolk, all you can say is: "Wow!"

The Good Beer Guide lists 22 regular brews ranging from Marld at 3.4 per cent alcohol to Comrade Bill Bartram's Egalitarian Anti-Imperialist Soviet Stout (Marc's take on the classic Russian Imperial stouts) at 6.9 per cent. Added to that is his Zodiac range, all 12 of which use the same malt grist but have different hopping regimes, and five seasonal specials.

It's not just the number of different beers that Marc brews that's impressive, though: it's the huge variations in style. His big sellers are Little Green Man, a 3.8 per cent golden bitter; The Bee's Knees, a honey-flavoured best bitter at 4.2 per cent; Cherry Stout, which is self-explanatory, at 4.8 per cent; and the aforementioned Comrade Bill's. But his other styles include the ginger and lemon-flavoured Cat's Whiskers; the gooseberry-flavoured Grozet; Catherine Bartram's, a hoppy IPA; Jester Quick One, using American hops; the straightforward Captain Bill Bartram's Best Bitter; Coal Porter; Damson Stout – need I go on?

Marc's then, is a very different approach from the strategy adopted by more cautious – and, Marc admits, more versed in marketing – brewers with industry backgrounds such as Butcombe and Black Sheep, who started by focusing on a single brand and only then establishing a wider range.

"It all comes down to your attitude to marketing," says Marc, whose career before starting Bartram's has included stints in psychiatric nursing, local government and cleaning toilets. (He also used to design ready meals for Marks & Spencer but, he says, "the tastings weren't as much fun as they are in brewing!".)

Marc first discovered that there was more to beer than just a few pints of whatever happened to be on offer at Hull Beer Festival in 1977. "I used to go out on pub crawls with the lads around Hull, and there'd be pubs owned by three or four different breweries but the beer was very much secondary," he says. But after that 1977 epiphany he started seeking out different beers in local pubs and then, as a keen home brewer, experimenting with different recipes and ingredients.

> "I was always searching out different styles and looking up ancient recipes, and diversity has been my thing since then," he says.

Marc's passion for diversity is twinned with an interest in art to create his huge and idiosyncratic range. He likes artists to come up with labels for his beers, which are mostly bottled. The Zodiac range was actually suggested by a local artist, while both the label and the name for The Cat's Whiskers came from a student of commercial art. Marc told him that the recipe included ginger and lemons, and the student came up with a painting of a ginger cat reclining on a bed of lemons. "The people who buy it often do so simply because they love cats," says Marc. "They might not know anything about beer, but the label attracts them; and if they like the beer they come back for more."

But Marc acknowledges that he could never sustain a range of anything like the size if his beers weren't mostly sold in bottle-conditioned form. He has a nucleus of eight or nine publicans who stock his beers, but most of his business is with farm shops and farmers' markets.

"For instance, I only brew a mild every two or three months, and I'll sell a few firkins to my regular accounts and bottle the rest," he says. "It's the shelf-life, you see. I couldn't sell anything like my five-barrel minimum run in cask-conditioned form before it went off; but as most of my sales are in bottle, I don't have to."

www.bartramsbrewery.co.uk

hadn't been too nervous to ration Irish brewers as well as mainland ones. Dark mild, once country's leading beer style, was a casualty – although its death has been hideously long drawn-out – of post-Second World War prosperity. Bitter or pale ale, as it used to be more widely called, was the toff's drink, expensive and aspirational, and started outselling mild in the 1950s. Lager pretty much finished it off in the 1970s, and today it is a minority style, brewed in significant volume only in Leeds and Manchester.

Golden ales as a definable beer style first emerged some 20 years ago, when Somerset's Exmoor Brewery produced Exmoor Gold and Hop Back of Salisbury came out with Summer Lightning. There had been very pale beers before then: Boddington's from Manchester was perhaps the best-known, but Wem Pale from north Shropshire was just as light-coloured, and so was the main beer from the famous Three Tuns brewpub in south Shropshire. Green's of Luton, Burt's on the Isle of Wight, and Holdens of Dudley were among the many brewers that had produced beer brands with the words "gold" or "golden" in their names in the preceding 30-odd years. But from the early 1990s onwards, ale brewers increasingly looked to much lighter-coloured and lighter-bodied beers, often with a hop aroma so pronounced as to be almost unbalanced, to challenge lager's strengthening grip on the market. (I'm proud to say that I was the first to identify golden ale as a separate beer style in print, in a 1994 issue of Brewers Guardian with a précis in the Journal of the Institute of Brewing May–June 1995.) The turning point for golden ales, though, came in 1996 when an extraordinarily hot summer plunged ale-brewing into crisis. It wasn't just that ale-drinkers turned to chilled lager as the temperature soared; it was also that the weather ruthlessly exposed weaknesses in the supply chain. Ale that had been virtually cooked in warehouses and drays arrived at the pub as an undrinkable soup; there were even tales of casks bursting in the heat.

Before then, micros had been more interested in re-creating the stronger, darker styles of the past, and the listing sections of successive Good Beer Guides are packed with stouts, porters, barley wines and old ales. The 1996 Guide, compiled in spring that year, has only 22 microbrewers producing ales with the words "gold" or "golden" in their names, and just three regionals, Belhaven, Gale's and Timothy Taylor (although Elgood's, Fuller's, and Wadworth had also brought out golden ales under different names: Barleymead, IPA and Summersault). After 1996, though, the balance began to change. Two years after that summer, the Guide listed 55 gold or golden beers from micros and 11 from regionals; the 2007 edition lists 134 and 16 respectively. Golden ales have won the Champion Beer of Britain title every year since 2001, prompting CAMRA to recognise the style with a category of its own in 2005 (since when Crouch Vale Brewers Gold has won twice in a row). So golden ale is a style pretty much invented by microbrewers that has come to a position of huge importance in the ale market.

The Range of Beers

So far we have concentrated on the huge diversity of beer styles available to you, and there are many microbrewers who have taken the fullest possible advantage of it. Bartram's of Bury St Edmunds, Suffolk, for instance, has 22 regular brews in its portfolio ranging from a 3.4 per cent alcohol dark mild to a 6.9 per cent stout and including a porter, a honey beer and a gooseberry beer. It also produces five seasonal specials and an entirely separate range under the Zodiac brand. At the other end of the alphabet, Warcop of Wentlooge, Gwent, brews an astonishing 31 regulars and six seasonals. This is rather reminiscent of some Belgian brewers who produce literally dozens of different brews, although in many cases these are actually the same beers rebadged for individual customers. Rebadging is not a practice that has caught on to any great extent in Britain, partly because there are fewer local wholesalers. But some UK brewers will either rebadge a beer as a favoured customer's house bitter, or even make a special blend of one or more of its regular beers for the same purpose. This may be a bit of a performance, but if it means guaranteeing a regular reorder it may well be worth the slight additional effort.

There is, of course, the opposite approach of brewing one superb beer and focusing all your efforts on building it up as the leading brand in its region. Black Sheep of Masham, North Yorkshire, Butcombe of Bristol, and Abbey Ales of Bath are all outstanding examples of breweries that have adopted this strategy, only starting to extend their ranges one their main brands have become firmly established. A common factor linking these three is that their founders – Paul Theakston of Black Sheep, Simon Whitmore of Butcombe, and Alan Morgan of Abbey Ales – were all extremely well-known and senior personalities in the brewing industries in their districts before they struck out on their own. They knew the free trade and the free trade knew them – and trusted them. They knew what the free trade needed, and how to tackle it. They didn't need gimmicks or novelties to help them stand out from the rest of the pack: pinning all their faith on a single beer marked them out as serious players with a serious product. They generated a confidence in themselves and their single brands that has turned out to be entirely justified. I can think of a handful of similar cases, but you have to be a very special person to succeed with the single-brand strategy; and anyway, for most small brewers the diversity is all part of the fun.

Few go the extremes that Bartram's and Warcop do, but many approach double figures in their regular ranges and produce a long list of seasonals and one-offs as well. Seasonal beers in particular have a long and respectable history – or at least, winter warmers and Christmas specials do. Generations of Londoners have eagerly awaited the arrival of winter, whose fogs and chills and long dark evenings are God's way of telling them that Young's Winter Warmer has arrived. Other regionals have also traditionally brewed strong ales for the season. Morrell's Varsity was legendary; McMullen's Stronghart was the first be lauded as "liquid Christmas pudding", although many others have hijacked the description since; and Greene King Strong Suffolk, occasionally allowed out on draught at the right time of year, is a fascinating and unique survival of the centuries-old practice of blending strong old ale with fresh pale.

Seasonal Specials

And it was the regional brewers who first extended the idea of a seasonal special to cover spring, summer and autumn as well as winter. Harvey's of Lewes claims to have been the first to launch a seasonal special – Tom Paine, a summer ale named after Lewes's most famous son, appeared in 1991. The idea quickly spread. Seasonal ales not only gave the brewers' own tenants and direct-delivered free trade accounts some variety in response to the diversity offered by the micros, it also gave the brewers a palette of novelty beers to offer the wholesale trade, which was then a new opportunity for them. Some regionals, Bateman's, for example, extended the idea beyond the merely seasonal and offered monthly specials, too. The frenzy has somewhat abated in recent years: some two-thirds of the regionals and just over half of all microbrewers still produce seasonal ranges; but in many case the "range" is a single beer, normally for Christmas.

Part of the problem is that without a definite tradition of different seasonal beers such as exists in many continental markets, brewers don't really have a clear idea of which style best suits which season. Winter ales have always tended to be strong and dark, although Young's Winter Warmer was originally the brewer's stab at a Burton-style pale ale and is only 5 per cent alcohol. The reigning champion Winter Beer of Britain at time of writing is a stout, at 4.5 per cent alcohol, Hop Backs' Entire Stout from Wiltshire. And most brewers have substantially reduced both the strength and darkness of their winter beers in recent years: they may be delicious, they may keep out the cold, but most people can only manage one pint, and where's the profit in that? Summer beers are another no-brainer that turns out on inspection to be more problematical: originally, they were in the main straightforward competitors to lager, very pale, quite aromatic and best served cold. But they failed to woo lager drinkers (by and large an incurious lot, not much given to experiment), and as golden ales became a mainstream style, they lost any claim to seasonality. As for spring and autumn, they have no seasonal styles; so with no traditional template as a guide, brewers simply produce whatever seems most suitable to them. One of the best spring beers is Harvey's Knots of May, a light mild which coincides handily with CAMRA's Mild Month campaign every May; and Shepherd Neame's Late Red is a splendid consolation for the shortening days of autumn. But no ancient tradition makes the one a spring beer and the other an autumn beer. If you are going to produce a seasonal range, therefore, you have carte blanche as to what to brew when, and no limits but your own imagination.

One-offs, especially in aid of good causes, are an altogether more straightforward business and well worth investigating. However, they tend in the main to be bottled and will therefore be considered in the relevant chapter. One exception to this is the CAMRA Festival Special. CAMRA branches like to have one, two or in odd cases even more exclusive ales brewed for the annual beer festivals. Often the chairman and other branch committee worthies will descend on the brewery or breweries lucky enough to be chosen as the provider, and "help" with the mashing. Often, too, by a mysterious and undoubtedly psychic process, the specials will emerge as medal-winners in the Beer of the Festival competition. On the whole, therefore, and despite any irritation that might be occasioned by the branch's "help", it's well worth accepting an invitation to brew a festival special. Whatever you send to the festival will sell out, and any surplus you brew is likely to be accepted by the local free trade – especially if it wins a medal.

Pursuing the Green Agenda

Not surprisingly, microbrewers have also been keen to pursue the green agenda with all its various headings: vegetarian, low-impact, organic – they're all values that can be applied to brewing.

Finding beers, particularly cask ales, that are suitable for vegans and vegetarians is surprisingly difficult. This is because the most effective substance for clearing or fining beer is isinglass, a collagen derived from various species of fish. Isinglass is a gel that, when added to the finished beer, flocculates the live yeast into a spongy mass that sinks; the beer can be strained off it. Of course, little or none of it remains in the beer – that's the whole point, after all. But vegetarians still object, correctly, that a living creature has had to die merely in order to clarify a beer. Ironically, isinglass is more rarely used in keg or bottled beer, which is filtered and therefore shouldn't need fining; and a yeasty sediment is an expected feature of bottle-conditioned beer, so that shouldn't really need fining either. But cask ale does; and while there are alternatives, they're not as effective. Filtration agents such as kieselguhr and bentonite, widely used in the wine industry, operate in a different way and can affect the character of the beer; while carragheenan or Irish moss effectively removes protein that would cause a haze but isn't so effective at dealing with yeast. Still, they are used with some success by many brewers, so if you want to appeal to vegans and vegetarians, give them a try.

Keeping it Local

There are many ways in which you can reduce the overall environmental impact of your brewery. Brewing uses a lot of energy, both to move liquids around and to heat them up and cool them down. In your brewery design you can incorporate the same sort of energy-reduction measures that might be used in any other processing plant: good insulation, short pipe-runs, solar panels and wind turbines, which will reduce your electricity bill as well as your carbon footprint. But cutting your food miles will be much harder.

One way of reducing food miles is to concentrate your sales effort on as small an area as possible. But given the structure of the trade this isn't always easy: you have to sell where you can, and if that means long road trips, either on your own truck or on the wholesaler's, there's not a lot you can do about it. Sourcing your materials locally is another possibility, but equally problematic. There's a growing trend among brewers to buy their barley from local growers – and it's not just microbrewers who are doing it, either. Shepherd Neame in Kent, Black Sheep in North Yorkshire, and St Austell Brewery in Cornwall are all keenly developing local sources of supply. But wherever you buy your barley, it all has to go to the maltster to be processed; and as there are

CAMRA, SIBA and The World Beers Awards

CAMRA, SIBA and The World Beer Awards run awards schemes. The SIBA scheme is the more straightforward. It's run on a regional basis, and member-brewers may enter any of their beers in the appropriate class. Regional winners go through to a national final held as part of SIBA's annual general meeting each spring.

CAMRA's scheme is less transparent. Local branches make Beer of the Festival awards at local events, but brewers don't enter their beers. Instead, competitors are chosen by CAMRA members by varied and mysterious means. Beer of the Festival awards are decided by equally varied methods from blind tasting to polls of festival visitors, and go through to either of the two national competitions, both of which are judged by tasting panels. The Champion Winter Beer of Britain (WBoB) is chosen at the National Winter Ales Festival in January, while the Champion Beer of Britain (CBoB) is chosen at the Great British Beer Festival (GBBF) in August. Confusingly, the winners of the WBoB competition are also eligible to win the CBoB title; but the winners of the Bottled Beer competition held at GBBF aren't.

CAMRA's categories are detailed below and note that anything unusual you brew, will be lumped in with the catch-all "speciality" category, which are increasing both in number and diversity. A gold, silver, and bronze are awarded in each, and the medallists in each category go forward to the final judging panel to win overall gold, silver, and bronze.

Judged at the Great British Beer Festival: Mild; Bitter (3.4-3.9% alcohol); Best Bitter (4-5%); Strong Bitter (5%-plus); Golden Ale (3.5-5.3%); Specialities; Bottle-Conditioned Beer. Judged at the National Winter Ales Festival: Old Ale/Strong Mild; Stout/Porter; Barley Wine.

www.camra.co.uk
www.siba.co.uk

The World Beer Awards are the global awards selecting the very best internationally recognised beer styles. This annual tasting selects, awards and promotes the 'World's Best Beers' to consumers and the trade throughout the world. Beers must be generally available and for sale in bottles or cans.

Please see page 210 for further details or go to
www.worldbeerawards.com

only four traditional floor maltings left – Tucker's in Devon, Warminster in Wiltshire, Crisp's in Norfolk, and Fawcett's in North Yorkshire – the trip there and back can be a pretty long one.

Hops, too, can have a long journey. It's all very well for brewers in Kent or Hereford & Worcester to boast about their use of local hops, but they have Britain's last hop-growers right on their doorsteps, and everybody else has to buy further afield. Having said that, hops were once grown in almost every county in Britain, including Scotland, and there's no practical reason why they shouldn't be again. The first commercial hop garden in East Anglia for many generations has just been re-established in Suffolk, and small brewers in other parts of the country, such as the Atlantic Brewery in Newquay, Cornwall, are setting about planting their own. An added advantage of planting new hop gardens is that they're likely to be pest-free, so it's certainly worth considering – especially if you know a farmer or market gardener who is keen to diversify.

Going Organic

Another part of the green agenda that you can pursue is the use of organic ingredients. Organic beers are becoming more and more popular and widespread: many brewers, even quite big ones such as Fuller's, now have at least one organic beer in their range; others have gone over to organic ingredients entirely.

The first British organic beer that I know of was brewed by the now-defunct Ross Brew House of Bristol in the late 1980s; Golden Promise, launched in 1991 by Edinburgh's Caledonian Brewery, was the first to gain national distribution. Sourcing organic ingredients was not easy: the malt was made from a strain of barley also named Golden Promise and developed for the whisky industry, while the hops had to be imported from New Zealand. The dearth of suitable ingredients meant that others were slow to follow Caledonian's example but more than 25 years on organic barley is easily available, although supplies of British-grown organic hops have not kept pace with demand, they can always be eked out with imports. There are still only seven all-organic breweries in Britain, and three or four more that are nearly there, but other brewers between them produce a couple of dozen brands at least, and several more are imported.

The demand for organic ingredients has imposed a new discipline on hop merchants and maltsters, who previously would process raw materials from different growers all mixed together. Now they have learnt to keep organic hops and barley separate from non-organic produce, a good habit that has had an equally valuable but perhaps unforeseen side-effect. For they have also learnt to process the produce of individual growers separately, so that full traceability is not only possible but commonplace these days. This means that if you only want to use malt and hops from growers in your locality, you can. Of course, they will still have had to travel to the oasts (or kilns in Hereford and Worcester) or maltings and then to your brewery, so the actual road miles involved will not have been reduced in the slightest. But discerning consumers these days put a considerable premium on traceability, and it may well be worth your while taking advantage of the fact. For although it's not easy being green, people will love you for it.

Since 1993, Britain has conformed to European Regulation 2092/91 regarding certification of organic producers. If you are brewing an all-organic beer as part of a wider range, it is not strictly necessary to be certified organic: you can get away – and many do – with simply stating on your labels that all your ingredients are from certified organic producers. However, to take full advantage of the possibilities offered by "going organic", it's probably best to get yourself certified. To contact any of the 10 certifying bodies (it's not a Soil Association monopoly!) approved by the Department for the Environment, Food, & Rural Affairs (DEFRA), visit **www.defra.gov.uk** or visit **www.organicfarmers.org.uk**

Ingredients: Malt

The malt is, of course, the heart of the beer, providing all the fermentable materials that create the alcohol (unless you're using brewing sugars as well), the residual sugars that create the mouthfeel, and many of the flavour components.

In Britain we have always tended to think of malt as coming entirely from barley, except for the handful of oats used to give some stouts a smooth, silky texture. Until recently, that is. Wheat is now a widely used grain in British brewing, so widely used, in fact, that British wheat beer ranks as a fourth category in its own right alongside Belgian, Bavarian and Berliner. Typically, the British style doesn't use nearly as much wheat as its continental cousins because the desired effect is a lighter body rather than the clove, vanilla, banana and bubblegum flavours that tend to come with higher proportions.

Malt is made, as we have discovered, by tricking barley into beginning to germinate and thereby starting the process of converting the starch in the grain into fermentable sugar. The process involves a

CASE STUDY

Renting Buildings From a Pub – Farmer's Ales, Maldon, Essex

It's a long way from the mean streets of London N1 to the rural tranquillity and rolling countryside of North Essex. But that's the journey veteran microbrewer Martin Kemp and trusty right-hand man Andy Skene made in 2006 when, 24 years after Pitfield started brewing in a former leather goods (!) warehouse in Hoxton, they upped sticks and moved to – well, to the sticks.

The brewery's first move was to Great Horkesley and an enormous but near-derelict market garden that once produced tomatoes by the bushel but was scheduled to be demolished to make way for a brand-new ecofriendly heritage centre of which Pitfield – which makes organic country wines as well as organic beers – was to have been the showpiece. Alas, the project became mired in planning red tape, and in 2008 Pitfield moved again, to a 25-acre organic farm in North Weald.

The move was a symbolic one for Pitfield, whose main outlet had always been the legendary Beer Shop in Pitfield Street. Hoxton, though, had famously become trendy and had consequently been regenerated, a process which included making parking anywhere the shop impossible. At the same time the brewery – which made history as the first micro to win CAMRA's Champion Beer of Britain competition, with Dark Star in 1987 – had gradually gone organic. Added to the virtual destruction of the shop's viability, its new green credentials made a move to the country a logical necessity.

It was in 1999 that Martin and Andy decided to try their hand at a totally natural beer, brewed without additives of any kind. At the time, organic ingredients were hard to come by but a maltster in Berwick was producing a whisky malt made from the organic Golden Promise barley. Most of it was earmarked either for distillers or for Edinburgh's Caledonian Brewery for its pioneering pale ale of the same name. But Martin managed to secure a single tonne – which he and Andy had to go to Berwick themselves to collect – and, with imported New Zealand Hallertau hops from Charles Faram of Worcester, they set about producing Eco Warrior.

At the same time Geetie Singh and Esther Boulton were starting their all-organic Singhboulton pub company, and they contacted Martin to see if he could produce a house bitter for them. Geetie suggested that Pitfield should go for full organic certification; and as Martin had happened to meet a representative of one of the government-approved certifying bodies, Organic Farmers & Growers (OF&G), at an organic show in Brighton, he approached it. It wasn't an easy process.

"Ours was a pretty straightforward case but it still took time and money," he says. "OF&G inspected the brewery and conducted a paper audit, looking through all our invoices and so on to check that all our suppliers were certified organic. That took a day. Then DEFRA, the Department of Agriculture, had to inspect the inspection, which took another day. There's also an annual fee, which depends on how much you produce – we're a five-barrel brewery, brewing two or three times a week, and our annual fee is about £600, so it's not cheap."

Pitfield also pays a premium for its supplies. Organic pale malt costs £150 a tonne more than non-organic and the coloured malts, which are used in much smaller quantities, are £400-£500 a tonne more expensive. Organic hops cost about £17-£25 per kilo compared to £5–15 for non-organic, but in the pub trade there's no organic premium on the wholesale price. That was another reason for Pitfield's move away from London. Martin still supplies the London pubs that took his beers in cask before the move, but every week he now sets up his stall at three farmers' markets where his overheads are much lower and he can charge up to £2.50 a bottle for his beers. That his bottled beers are all bottle-conditioned and made without fish-derived finings, and therefore vegan-friendly, both help establish Pitfield's green credentials. A new home in the country somehow adds to the proposition.

The switch into organic production has not been without its problems – although now he's a regular he can at least depend on supplies of ingredients – but Martin reckons it's been worth it.

"There are negatives, not least the increased costs, but on the positive side we have established ourselves in a good niche and we now have a name as an organic and vegan-friendly producer," he says. "But part of our success depends on the fact that we are small and intend to stay small. We only brew five barrels at a time, so we can get enough regular supplies to ensure production; and we can sell all we can brew."

In August 2012 Andy Skene took control of Pitfield Brewery. He rents the brewery and its brands from Martin Kemp. Martin has taken a backseat from the day to day running of the brewery, and has acquired the first Pitfield pub in over 20 years. The pub is The Reindeer in Saxon Street, near Newmarket. Andy is in the process of renewing the Pitfield brand, and aims to make Pitfield a Better known brewer of fines ale which also happen to be organic and vegan.

Andy also runs his own Dominion Brewery Company, with which he produces special hybrid and experimental beers.

Contact Andy via **www.dominionbrewerycompany.com**

Soon to be launched hopefully June 2013 **www.pitfieldbrewery.com**

lot of soaking and drying, and traditionally the drying was done by heaping the damp grain in rows on a floor for two or three weeks before drying it over a fire. The rows of grain would be turned by a man plodding up and down pulling a device like a cross between a wheelbarrow and a rake; and as is the way of things, this method of "floor-malting" has been mechanised almost out of existence. Most small brewers, though, like to get their malt from the country's four surviving floor maltings, not because the quality of the beer made from traditionally malted barley is necessarily better than beer brewed from machine-made malt, but because they say it's easier to work with. What's true is that all four floor maltings are comparatively small concerns and therefore buy in smaller batches, which they can ipso facto inspect more thoroughly, and that the man plodding up and down turning the malt is simultaneously conducting a continuous quality check. Some Scottish brewers, at the time of writing, were even approaching the few distilleries that still floor-malt their own barley to see whether they'd care to try producing ale malts alongside their whisky malts. Unfortunately most of these distilleries are owned by global corporations which still tend to seek to concentrate production on a handful of huge sites; perhaps the modest extra profits generated by making and selling ale malt on the side might keep some of these traditional little whisky maltings open for a few years more?

Types of Malt

Pale malts of various types, principally lager malt, pale ale malt and wheat malt, kilned at low temperature to preserve the golden colour of the grain, are the base on which nearly all modern beers are founded. The grists of even the blackest of beers generally have pale malts as their main constituents, the colour being supplied by surprisingly small doses of dark malt.

These are the malts that, in the late 18th and early 19th century, changed the world of beer forever. Previously, malt had been dried over peat, straw, dried furze or broom, wood or charcoal and tended to be dark, smoked, and, thanks to the difficulty of controlling the temperature of such fires, uneven in colour, flavour and moisture content. When in the middle of the 17th century the maltsters of Derbyshire started exhausting local sources of timber, they found that the locally mined hard coal produced a low-sulphur smokeless coke ideal for malting. By the end of the century malt was being dried over coke in neighbouring counties and also in parts of

Scotland (although peat remained the preferred fuel there). Coke gave an even and controllable heat and worked at lower temperatures than wood or charcoal, making it possible to kiln malts that were pale, clean-tasting and uniform, and produced crystal-clear golden or amber beers. Daniel Defoe was an early supporter of strong Derby pale ales; and Burton brewers were exporting pale ales to the Baltic before the Napoleonic War and to the East Indies after it. In the early 1840s the brewers of Pilsen in what is now the Czech Republic also discovered the potential of pale malt kilned over coke; lager as we know it was born.

Low-temperature kilning preserves the saccharifying enzymes in the malt, as well as its original colour. This means that the malt can be used with other unmalted grains such as wheat, corn and rice.

You will probably learn all you need to know about all the different types of malt and how to handle them during your training. Here, though, is a brief reminder of the wonderful variety of malts available to you, listed for convenience only by degrees lovibond, the scale used to measure the malt's colour.

- **Lager malt (2°L):** the malt that produces the world's most popular beer-style – pale gold, mid-flavoured, Pilsner derivatives – can also be used as the base malt for other beers, even dark ales. Its diastatic capability makes it ideal for use with low-enzyme speciality malts or unmalted grains
- **Wheat malt (2°L):** has a high protein content, causing wheat beer's characteristic haze. It also has good head-retention qualities: torrefied wheat is often used in small quantities for the purpose. Low in polyphenols, it is often used in modest quantities by British brewers to lighten the body of their golden ales. It has to be mashed with grains that supply a husk bed (lager malt is ideal)
- **Pale ale malt (3°L):** the malt most commonly associated with British ale is kilned at a slightly higher temperature than lager malt for a fuller flavour and darker colour. It also tends to have fewer enzymes than lager malt, although still sufficient to allow the use of adjuncts in the mash
- **Rye malt (3°L):** can be used in small quantities for its spicy flavour
- **Vienna malt (3–7°L):** produces the full-bodied amber or reddish beers that were once popular in Austria but survive mainly in Mexico, Dos Equis and Negra Modelo being noted modern examples

- **Mild ale malt (4°L):** although most surviving milds are dark, the base malt isn't, creating an amber or copper-coloured wort and producing a dry, full-flavoured beer, with brown, chocolate and/or black malts used sparingly to give the characteristic colour. Mild ale malt is also a good base for stronger ales, whether coloured or not
- **Carapils malt (7°L):** kilned at low temperature, it adds sweetness, smoothness, and body to pale ales and lagers without affecting the colour or adding caramel notes. Also aids head retention
- **Munich malt (10–20°L):** an aromatic lager malt that yields a dark reddish-orange wort and a slightly sweet caramel flavour. Munich malt comes in two grades, light and dark
- **Crystal malt (10–20°L):** produced in such a way that most of the starch is not saccharified and is then caramelised during kilning, crystal malts are very widely used in small additions in ales and lagers to give extra sweetness and flavour ranging in character from delicate light honey to rich toffee flavours. Crystal malts also thicken the mouthfeel, and create an attractive bronze colour
- **Biscuit malt (25°L):** a base malt toasted and roasted to give the beer a biscuity flavour and a deep amber colour
- **Victory malt (25°L):** a roasted malt similar to biscuit malt but with a nuttier taste. Also adds orange highlights to the colour
- **Amber malt (30°L):** similar to mild ale or Vienna malt, but with more colour and a biscuity flavour. Used mainly in old and brown ales
- **Brown malt (65°L):** traditionally kilned over a wood fire for a smoky flavour; used very sparingly in stouts and porters
- **Special Belgian malt (220°L):** a rare malt with a nutty, roasted sweetness that in small quantities enriches brown ales and porters, and in larger proportions adds a plummy, vinous quality to barley wines and strong winter beers
- **Chocolate malt (350–400°L):** its smooth roasted flavour and brownish-black colour with ruby highlights make chocolate malt irreplaceable in dark ales such as milds, stouts and porters; it can also be used in dark lagers
- **Black malt (500–600°L):** even darker than chocolate malt, with a sharp, burnt, acidic flavour that can take the sweet edge off some stronger beers

Unmalted Grains

Malting is the process of artificially stimulating the enzymes that in nature convert the grain's starch content into soluble sugar that the yeast can get at. Not all brewing grains need to be malted, though. Up to 50 per cent of the grist of most Belgian witbier is unmalted wheat, but the malted barley that makes up the other half has enough enzymes to convert the whole mash. Unmalted oats are added to some beers, especially stout, not for their negligible fermentable material but for their husky roughness, which "scrubs" microscopic particles out of the wort for a silky smooth finish. Maize and rice are almost flavourless, but are high in convertible starch. They have no enzymes of their own, but as with Belgian wheat beer, the barley malts they are mashed with have enough surplus enzymes for the whole mash. They are also very cheap, which is why they are so common in beers from the poorer Latin American countries (and in many mainstream US lagers, but that's another story). Finally, there's roasted barley, which is virtually burnt to an inky black 550°L and has an acrid charred-wood flavour. It's the signature flavour of Guinness and, to a lesser extent, Murphy's and Beamish as well.

Brewing Sugars

The use of sugar in brewing has been legal in Britain since 1847 and is now widespread throughout the world. Some brewers oppose the practice completely, because they say sugar is only used as a cheap source of fermentable material to pad out the more expensive malt; and in some cases this is true. American beers in particular use large amounts of rice or corn-based syrups, and are characteristically light-bodied (or indeed bland) as a result.

However, there are other more legitimate and indeed time-honoured uses for various types of sugar in judicious quantities. Corn-derived sugars are often used to boost the strength of the final product without significantly affecting its flavour, or to adjust from mash to mash to achieve consistent alcoholic strength. In northern Europe a derivative of beet sugar was traditionally used for the same purpose, but beet sugar created an undesirable flavour that could be avoided by "inverting" it – boiling it slowly with a little water and citric acid to produce a highly concentrated syrup, which was then crystallised to form sugar "diamonds". This sort of sugar is still used in Belgium, where it is called "candi".

Corn-derived sugars are also used in small quantities to "prime" beer before it is bottled or put into cask. Priming sugars prompt a slight secondary fermentation to give the beer a bit of fizz and create a head.

Finally, sugars of various sorts are used for their particular flavour characteristics. A dry beer can be sweetened by the use of non-fermentable lactose. Maltodextrin is a soluble starch that creates a heavier body and richer mouthfeel. Honey is an increasingly popular adjunct and retains its distinctive taste throughout the brewing process. And black treacle or molasses is sometimes added to dark beers for its density and depth of flavour.

Maltsters *(A more detailed list can be seen on page 207)*

- **Bairds Malt Ltd**
  Station Maltings, Witham, Essex, CM8 2DU
  Tel: 01376 513566; **Web:** www.bairds-malt.co.uk
- **BOCM Pauls Ltd**
  Lidum Mill, Ashby Road, Shepshed,
  Loughborough, LE12 9BS
  Tel: 0870 0500 306; **Web:** www.bocmpauls.co.uk
- **Crisp Malting Group Ltd**
  Great Ryburgh, Fakenham, Norfolk NR21 7AS
  Tel: 01328 829391; **Web:** www.crispmalt.co.uk
- **Edwin Tucker & Sons Ltd**
  The Maltings, Teign Road, Newton Abbot TQ12 4AA
  Tel: 01626 334002; **Web:** www.tuckersmaltings.co.uk
- **French and Jupps Ltd**
  Stanstead Abbotts, Ware, Herts SG12 8HG
  Tel: 01920 870015; **Web:** www.frenchandjupps.com
- **JP Simpson & Co (Alnwick) Ltd**
  Tweed Valley Maltings, Tweedside Trading Estate,
  Berwick upon Tweed TD15 2UZ
  Tel: 01289 330033
- **Muntons plc**
  Cedars Maltings, Stowmarket,
  Suffolk IP14 2AG
  Tel: 01449 618300; **Web:** www.muntons.com
- **Thomas Fawcett & Sons Ltd**
  Eastfield Lane, Castleford, W Yorks WF10 4LE
  Tel: 01977 552460; **Web:** www.fawcett-maltsters.co.uk
- **Warminster Maltings Ltd**
  39 Pound Street, Warminster, WiltsBA12 8NN
  Tel: 01985 212014; **Web:** www.warminster-malt.co.uk

Ingredients: Hops

Hops, whether used as whole flowers (or "cones"), dried pellets or extracts, contain the acids that will protect your beer from bacterial infection and provide its bitterness as well as the oils and resins that will produce much of its aroma and flavour.

A cousin of both cannabis and the common nettle, hops were used throughout the medieval period as a dyestuff and a herbal remedy for a variety of ailments (they are still used to treat insomnia). The young shoots are a spring delicacy rather like asparagus, still much eaten in Belgium, and the leaves are a good salad vegetable, while the stems yield a strong fibre suitable for both clothing and ropemaking.

Hops are first recorded as being used in brewing in Germany in the 9th century, and until recently it was thought that they didn't come to Britain until the end of the middle ages – as some versions of the rhyme have it, "turkey and carp, pickerel and beer (hopped ale) came into England all in one year" (1492). Modern research, however, suggests that even if they weren't commonly used by British brewers until the later middle ages, they were far from unknown. They may have been used by monastic brewers in Kent even before the Conquest; although if that is indeed the case it was as a purely local version of the herb grist used in different regions. Hopped beer seems to have started arriving in significant quantities in the 14th century as imports to satisfy the thirst of growing communities of Flemish, Dutch, Danish and German merchants and craftsmen for a taste of home. The first English hop garden now seems to have been planted not in Kent in the early 16th century, as used to be thought, but in Suffolk two centuries earlier. Certainly, London brewers were commonly using hops in the early 15th century and over the next 200 years their use became pretty much universal (although unhopped ale held out in the more remote parts of the Pennines until the 18th century).

Hops are now only grown on any scale in two regions, Kent and Sussex and Herefordshire and Worcestershire, although in the 19th century there were commercial hop gardens or yards in almost every county in Great Britain. And indeed there is no inherent reason why they can't be grown just about anywhere. The hedges around my home in Cambridgeshire abound with wild hops, descendants of plants that escaped from the county's hop gardens over 100 years ago; and a new commercial hop garden has just been opened in Suffolk. You might even persuade a farmer near you, wherever you live, to set aside a patch of ground for hops; the big disincentive in the past was that hops, being climbers, needed expensive poles and wirework and all the specialist skilled labour that went with them. The recent development of "dwarf" or "hedgerow" varieties such as First Gold, however, means this is no longer the case: anybody can grow hops anywhere.

One achievement of small independent brewers as a body in the last 35-odd years has been to reawaken interest in different hop varieties. Before that, hops had been prized chiefly for their bittering quality, and most of the efforts of researchers and growers had been devoted to developing high-alpha strains and finding new ways of extracting the prized bittering agents. Pioneers like Sean Franklin of Rooster's Brewery of Knaresborough and Brendan Dobbin (who briefly had his own brewery in Manchester but is best known in the industry as an installer and consultant) began working with imported varieties, mainly from the USA and New Zealand, in the 1980s. Before then, hop grists were almost universally blended from different types, with consistent performance in mind rather than the flavour and aroma characteristics of individual varieties. Today, thanks to these innovators and the many brewers who have taken their lead, single-varietal beers are, if not exactly commonplace, certainly not remarkable, and something you will doubtless want to experiment with.

Native Hop Varieties

Thanks to them, too, hop merchants like Charles Faram of Worcestershire now import weird and wonderful hop varieties from all over the world. They are too many to list here; what follows is a summary of the main native types including the hedgerow strains.

- **Admiral:** bred to complement the Target variety, Admiral produces small compact cones and has shown itself to be a good replacement for both high-alpha and dual-purpose varieties when used as a kettle hop
- **Bramling Cross:** bred in 1927 by crossing Bramling with a male seedling from a Canadian wild hop. Provides fruity blackcurrant and lemon notes and is increasingly popular in speciality beers. If used as a late hop or dry hop, the effect on the final flavour can be very interesting
- **Challenger:** bred in 1972 from Northern Brewer crossed with a downy mildew-resistant male; a cousin of Target. Gives a fruity, almost scented aroma, with some spicy overtones. A versatile kettle hop for all types of beer, it blends well with other English varieties and is sometimes used as a late hop and a dry hop
- **First Gold:** dwarf hop with a very attractive aroma, which

some liken to Goldings, but with a higher alpha content than traditional aroma hops. Suitable as a general kettle hop and also for late and dry hopping in all types of beer. First Gold has excellent aroma qualities, producing a well-balanced bitterness and a fruity, slightly spicy note in ales

- **Fuggle:** propagated by Richard Fuggle of Brenchley, Kent, in 1875, having first been noticed in about 1861 growing wild. It became the most widely grown hop in England, making up 78 per cent of all acreage in 1949, until wilt made it almost impossible to grow in Kent and Sussex. It now represents only about 9 per cent of the English crop, being grown chiefly in the West Midlands, but is also grown in the USA, mainly in Oregon, and in Slovenia, where is has changed its character and is known as Styrian Golding. Fuggle has a typical English flavour, often blended with Goldings to improve "drinkability" and add roundness and fullness to the palate. Contributes all the essential characteristics of flavour, aroma and balanced bitterness to ales, particularly as its relatively low alpha content means a high hopping rate is needed to achieve desired bitterness levels. Sometimes used as a distinctive dry hop
- **Goldings:** not one variety but a group including Cobbs, Amos's Early Bird, Eastwell Golding, Bramling, Canterbury Golding and Mathons, the strains usually being named after either a grower or the parish where they were first cultivated. Most are grown in east Kent with the exception of Mathons, which, along with some Canterbury Goldings and Early Birds, are commonly grown in Herefordshire and Worcestershire. Recognised as having the most typical English aroma, with the best flavour historically coming from east Kent, this hop is in demand for copper hopping and dry hopping of traditional ales. Goldings are also useful for late hopping lagers, when a delicate aroma is required
- **Herald:** a sister of Pioneer, and the first high-alpha dwarf hop with acceptable flavour properties. Yields large, heavy cones. In brewing trials, the flavour and bittering balance has been found very acceptable. Useful both as a bittering hop and as a dual-purpose variety in all types of beer
- **Phoenix:** developed at the same time as Admiral, this variety is a seedling of Yeoman. In brewing trials, the flavour and bittering balance has been found very acceptable. Useful as a

bittering hop and as a dual-purpose variety in all types of beer

- **Progress:** developed in 1951 as a wilt-tolerant alternative to the Fuggle. Somewhat similar to Fuggle but slightly sweeter, and providing a softer bitterness in beers of all types. With its slightly higher alpha content, it represents good value for bitterness if a recipe demands aroma hops for all the bittering elements
- **Whitbread Golding:** first cultivated in 1911 and planted extensively in the 1950s for its resistance to wilt. Provides a distinctive sweet fruit flavour similar to, but generally more pronounced and robust than, Goldings. It gives a mild, clean bitterness in traditional ales, and is sometimes used with good effect as a distinctive dry hop
- **Wye Northdown:** released for commercial production in the early 1970s, it provides cheaper alpha acid than many tradition varieties while having excellent flavour properties. A hop with a very mild, clean, neutral English flavour, it can be used in all types of beer, with no harshness of palate. Particularly with seedless Northdown, the high level of oil makes this a very distinctive dry hop for full-bodied ales
- **Wye Target:** Target, released for general production in 1972, is a cousin of Challenger and provides high yields of alpha acid with a robust aroma and is Britain's most widely used bittering hop. With strong, typical English flavour and good bittering qualities this hop is used in all types of beer. Dry hopping with traditional whole hops is also effective in some of the stronger ales, giving positive floral aromas

Getting the Most out of Hops

- **Buy** in a pack size as close to the brew length usage as you can. Aroma is lost once the pack is opened. For smaller needs buy in fresh packs or vacuum packs in double silver foil
- **Ask** the age of the hops. Buy the latest harvest hops
- **Store** in a deep freeze, refrigerator or cold room. Hop bitterness is lost faster at higher temperatures. Pellets are sometimes easier to store than whole hops
- **Learn** to evaluate hops yourself. To assess or compare hop aroma, make a hop tea, cool it and smell. Hop merchants rub and sniff hops every day, but there are two problems for the novice: hops may smell powerful when rubbed in the hand, yet may not be so powerful when in the beer; and minor

differences in the rub can mean big differences in the beer

- **Test**-brew with the new hop alone to see where it fits on the bitterness, aroma and intensity scales. Test it in a 100% pale malt beer with a standard water treatment so that the hop aroma shows through clearly. Adjust the portion quantities by the alpha content of the new hop and start with a comparison of the flavour at about 28 International Bitterness Units (similar to commercial ales)
- **Build** a hop grist by looking at the grists of beers you like and experimenting. Track down recipes in home-brewing books and on the internet. Remember, when mixing a grist, that aromas from different hops can sometimes "average out" and dilute the effect. Single-hop beers usually (but not always) have more aroma definition than mixed-varietal beers
- **Choose** your alpha hop carefully. Some alphas have strong aromas in spite of their early addition to the boil that may alter the final hop character
- **Look** closely at your yeast. The same hop can taste different when used with a different yeast. Make sure your yeast is compatible with the beer you are trying to make
- **Control** your brewing process, or other aromas might hide the hop aroma. Higher alcohols, esters, diacetyl and hydrogen sulfide will all disguise or overlay your hop aroma

All the above information and far, far more is available on the National Hop Association of England website, **www.hops.co.uk**.

Hop Merchants

- **Botanix**
  Hop Pocket Lane, Paddock Wood, Kent TN12 6DQ
  Tel: 01892 833415; **Web:** www.botanix.co.uk
- **Charles Faram**
  The Hopstore, Monksfield Lane, Newland, Worcs WR13 5BB;
  Tel: 01905 830734; **Web:** www.charlesfaram.co.uk
- **Wealden Hops**
  Congelows, Benover Road, Yalding, Kent ME18 6ET;
  Tel: 01622 817175; **Web:** www.wealden-hops.co.uk
- **Lupofresh Ltd**
  Benover Road, Yalding, Kent ME18 6ET
  Tel: 01622 815720

More hop merchants can be contacted via the National Hop Association of England (**www.britishhops.org.uk**).

Chapter Six

How to Get to Market: Draught Sales

CASE STUDIES

Realistically, where can you sell your beer? You may be based in a pub, or at least have one that acts as your brewery tap, but these sales alone won't necessarily make you a living. To get your beers out into the trade, you therefore need to understand the structure of the market. This chapter will deal with the "on-trade" – pubs, clubs and so on – and will therefore be mainly concerned with sales of draught beer.

There are two contrasting points of view here. One says that an overcrowded small brewing sector and the iron grip of the tied system mean that the traditional route to market – sales of draught beer to pubs – is dead, and that new brewers have to seek alternative outlets. The opposing view is that the long-term health of the sector as a whole demands that the tied house system be cracked open. They are not mutually exclusive. You can continue to seek draught sales to pubs at the same time as seeking new opportunities. Let us look at both, to start with, the draught beer opportunity.

There are around 180,000 on-licensed premises in Britain, not just pubs and clubs but restaurants, casinos, hotels, village halls, golf clubs, sports centre bars, theatre bars, airport bars and God knows what other varieties of bars. Few of them will have any relevance to your business. Some may provide niche opportunities, but the ones that really matter are the country's 55,000 or so pubs (but dwindling at a rate of around 30 a week during the recession) and 15,000-odd sports and social clubs. And be warned: not many of them will be buying beer from you.

As we have already seen, ownership in the pub trade has changed dramatically since the 1990 Beer Orders that followed the big Monopolies Commission enquiry into the nation's beer supply. There used to be brewery-owned tied houses – some run by managers directly employed by the brewery, others by self-employed tenants – and owner-operated free houses, many of which had accepted low-interest loans from a brewery in return for stocking only its beers, the old-fashioned "loan tie". Not any more. It's not so simple now.

The Local Free Trade

When you first decided to investigate becoming a local brewer, part of the attraction may well have been the cheery relationships you saw yourself enjoying with your local publicans, and indeed their customers, as you turned up each week with your regular beer delivery. If that's the case, disabuse yourself now. It's extremely unlikely that you'll be able to generate enough sales to sustain your business through the local free trade alone.

Perhaps 12,000 of the country's pubs are genuinely owner-operated free houses with the discretion to stock your ales. Some of them, although perhaps not as many as in the past, will be restricted by a tied brewery loan. Others buy strictly on price, or are unadventurous and favour well-known national and big regional brands, or have clienteles that are just not interested in real ale. The few that are left will be besieged by people like you. To make matters worse, breweries and pub companies are currently offering inflated prices for free houses that come on the market, thus cornering yet more of the trade.

Having said that, a guaranteed local sale, however limited, is an essential precursor to starting up: it creates awareness of your beers (and indeed of your existence) and for the first few weeks or months might be the only cash-flow you have. So even if you intend in the longer term to sell your beer much further afield, it's absolutely imperative that you research your local market thoroughly:

- How many genuine free houses are there within, say, a 20-mile radius?
- What sort of range do they stock currently – is it all from national and regional brewers, or are they apparently micro-friendly?
- How many handpumps do they have?
- Do they, or could they, stock changing guest ales?
- How many of the local free traders who blithely agree to give your beers a whirl will actually, when it comes to it, place an order – and how often will such orders be repeated?

Most parts of the country have a cadre of free trade landlords whose pubs attract diehard beer aficionados and who love to stock as widely as they can, filling six, seven, eight or even more handpumps with a never-ending rotation of exotic guest ales. (Although not everywhere: in regions where the tie has always been strong, such as Greater London and Surrey, the free trade is negligible.) Pubs like this will invariably give your beer a try, and are a useful place to be seen, but part of their unique selling point is their rapid turnover of brands, and you'll be lucky to get more than infrequent orders from them. But do take note from the case studies in Chapter Two that a single licensee in Kettering is responsible for giving a flying start to at least two local micros (actually many more); strive to meet and befriend his equivalent in your own neck of the woods.

Owner-operators

Perhaps more important is the rather more numerous breed of owner-operators who like to stock a decent if not extravagant selection that will include a local product if at all possible. It's in pubs like this that you stand your best chance of building up the steady trade you'll need; and in many cases this kind of loyal regular account was only stocking better-known regional brands before a genuinely local brewer came along and wormed his way into a regular slot. So don't be deterred if your local free house only stocks the usual suspects – Adnams, Fuller's London Pride, Marston's Pedigree, that sort of thing; it could end up as your best outlet.

You may already know most of the qualifying licensees in your area; to meet more, join the local branch of CAMRA – you can sign up online at **www.camra.org.uk** – to share local knowledge with fellow members and get personal introductions to the right retailers. Do your groundwork thoroughly, and try to establish a coterie of two or three local free houses that will have one or more of your beers permanently on tap and maybe half-a-dozen more that will order regular guests from you. But always bear in mind that only a handful of small brewers – mainly those that operate in very rural counties such as Shropshire and Devon, where the tie has never been strong – can depend on enough business in the local free trade alone to make a living. And as the number of local breweries continues to grow, the pressure on the genuine free trade increases. You will therefore, almost certainly, need to find other markets.

Brewery Tied Estates

These days, and contrary to public belief, breweries actually own remarkably few pubs. Of the four multinationals, only Scottish & Newcastle still has a retail division (actually owned by the Royal Bank of Scotland and only leased to S&N, but that's another story). Greene King and Marston's, the two "neo-national" brewers (that is, regional brewers whose estates have surpassed the old Monopolies Commission ceiling of 2,000) have about 2,500 pubs each, while the smaller regionals own another 4,500-odd between them. In these brewery-owned estates the tie is rigorously imposed: guest ales might be permitted, but only from the brewery's own list; any tenant buying outside the tie is liable to be evicted. So if there's a regional brewer's name on the pub sign, don't try and sell your beer there.

However, not all brewery tied houses belong to regionals that won't stock other brewers' brands: microbrewers themselves account for

CASE STUDY

Brewery with Large Local Free Trade – Hobson's, Cleobury Mortimer, Shropshire

When Nick Davis decided to make the switch from licensed trade property surveyor to local brewer, he had already decided on his business strategy – and that was, to remain firmly local.

That was back in 1993, when there was still a booming market in guest ales among national brewery tenants following the 1990 Beer Orders, and before the new generation of real ale wholesalers had really got into its stride. It made sound sense then to concentrate on building up a direct delivery route among strictly local outlets, and Nick determined not to canvass business further than an hour's drive from his business.

Nick's choice of location helped. The Davis family lived in the Bristol area, where Nick's parents ran a pub. But in deciding to move the 100-odd miles north to Shropshire, Nick was targeting a rural district where bigger breweries owned comparatively few tied houses and where there was therefore a strong free trade. It was dominated at the time by Banks's of Wolverhampton (now part of the neo-national group that includes Marston's of Burton-on-Trent and Jennings of Cumbria). But there were few microbreweries in the region (there are plenty now!), and Nick was certain that a locally brewed competitor to Banks's Bitter would be welcomed by many licensees and their customers.

It was a long haul, but he was proved right.

"At first it was just a question of going round letting publicans know that we were in business and making beer," he says. "It was hard work – a lot of doorknocking, and always keeping our eyes and ears open while we were out delivering."

Time on the road proved invaluable in finding out about new prospects, says Nick – often a trade customer would suggest new sales leads and just as often, pub customers would.

"You'd stop in the bar for a cup of tea after making the delivery and one of the regulars would tell you that the pub they played darts against last week desperately needed some decent beer," says Nick.

Within two years, production was up to 30 barrels a week, providing a living for Nick himself and one employee. Then in 1996 Hobson's Best was named CAMRA's West Midlands Beer of the Year, following up with a gold medal in its class in 1997's Champion Beer of Britain awards.

Competition success led to approaches from The Beer Seller and other wholesalers, says Nick, but despite being tempted he turned them down.

"The Beer Seller wanted 40 firkins straight way and the temptation to say yes was massive," he says. "The downside is that you may never see half of the casks again, which more than wipes out anything you might make on the sale."

Concentrating on the local market has allowed Hobson's to build up a strong following for a range that includes Town Crier at 4.2 per cent alcohol, Old Henry at 5.2 and more recently a 3.2 per cent traditional dark mild. In recent years its trade has come under pressure from aggressive purchasing by pubcos, which were snapping up every free house that came on the market at ridiculously inflated prices. As a result of the price inflation that they themselves have caused, the pubcos have now drawn back somewhat; and to make up for the loss of accounts Hobson's has been able to replace much better-known brands on the bars of the surviving free houses.

"The free trade cake may be smaller, but the local brewers' slice of it is much bigger," says Nick. "Brands like Banks's Bitter, Tetley Bitter and Courage Best, which used to be the big sellers, are losing popularity because people want a genuinely local beer."

Drinkers feel reassured if they have a link with the brewery, says Nick, even if it's a tenuous one. "They might know the brewer himself, but even knowing the farm where we get our hops or the mechanic who services our delivery van creates a connection," he says. "It's rather like buying your meat from a farmers' market. It's certainly not cheaper than what you'd get in a supermarket; it might not even be better. But it comes with people attached."

www.hobsons-brewery.co.uk

A Woodfordes pub

about 400 pubs. As we saw in Chapter Two, many small brewers are based in or associated with a pub, which in most cases stock a range of guest ales from other brewers too. These pubs are never going to shift huge amounts of your beer; but it's still worth trying to get the odd order out of them as a profile-raiser because they tend to attract enthusiasts who will take note of and seek out your beers in future. Similarly, some smaller breweries also have mini-estates of their own in which they usually stock not only their beers but also guests from other brewers. They don't do this out of charity: the sort of customer who seeks out their pubs is going to want a good range of different ales to try, preferably including as large a number as feasible of changing guest ales. You may very well find yourself knocking at open doors if you approach these brewers with ales of top quality; so worth a quick call, especially if they are local to you, are:

- **Banks & Taylor**, Shefford, Beds
  Tel: 01462 815080; **Web:** www.banksandtaylor.com
- **Bath Ales**, Warmley, Bristol
  Tel: 0117 947 4797; **Web:** www.bathales.com
- **Beartown Brewery**, Congleton, Cheshire
  Tel: 01260 299964; **Web:** www.beartownbrewery.co.uk
- **Burton Bridge**, Burton upon Trent
  Tel: 01283 510573; **Web:** www.burtonbridgebrewery.co.uk
- **Butcombe**, Wrington, Bristol
  Tel: 01934 863 963; **Web:** www.butcombe.com
- **Evan Evans**, Llandeilo, Carmarthen
  Tel: 01558 824455; **Web:** www.evanevansbrewery.com
- **Hop Back**, Downton, Wilts
  Tel: 01725 510986; **Web:** www.hopback.co.uk
- **Leadmill**, Denby, Derbys
  Tel: 01332 883577; **Web:** www.leadmillbrewery.co.uk

- **Milk Street**, Frome, Somerset
  Tel: 01373 467766; **Web:** www.milkstreetbrewery.co.uk
- **Milton Brewery**, Milton, Cambs
  Tel: 01223 862067; **Web:** www.miltonbrewery.co.uk
- **Oakham Ales**, Peterborough
  Tel: 01733 370500; **Web:** www.oakhamales.com
- **Ossett Brewery**, Ossett, W Yorks
  Tel: 01924 261333; **Web:** www.oakhamales.com
- **Vale Brewery**, Haddenham, Bucks
  Tel: 01884 239237; **Web:** www.valebrewery.co.uk
- **Whitstable Brewery**, Grafty Green, Kent
  Tel: 01622 851007; **Web:** www.whitstablebrewery.info
- **Wickwar Brewery**, Wickwar, Glos
  Tel: 01454292000; **Web:** www.wickwarbrewing.co.uk

Non-brewing Pub Chains

But if fewer than a quarter of Britain's pubs still fit the traditional "vertical integration" model of brewery ownership and exclusive supply, that doesn't mean the tied house system is dead, or even feeling a little unwell. The non-brewing pub companies that between them own well over 30,000 of Britain's pubs range from giants like Punch Taverns with 10,000 down to Head of Steam with just five. The smallest chains are often very receptive to local micros, but like the brewery-owned estates of old, the big chains are operated either by directly employed managers or by lessees or tenants. And like the brewery-owned estates of old, they are firmly tied; so you can't just roll up at the door and try to talk the licensee into buying your beer.

As we shall see, there are ways of getting your beer on the bar in pubs owned by the various tenanted pubcos. But you can write off almost all the directly managed town-centre "style bar" operators. Their customers would rather fill the gutters with regurgitated shots, alcopops and the like – sugary, synthetic concoctions the chain buyers can procure for less than the price of bottled water, but for which they can charge the punter as much as chateau-bottled claret – than your ale. A handful of high street chains such as All Bar One might have a solitary handpump lurking forlornly among the lager fonts, serving a musty sample of a national or leading regional brand. But by and large the high street is Budweiser territory (and I don't mean Budvar), and is barred to you and, indeed, pretty much all cask beer.

Big "family dining" managed house brands such as Harvester and Beefeater are equally unwelcoming to the craft brewer. They do, mostly, list a real ale; but inevitably it will be a cheap national brand that bears as much relation to the nectar you brew as their food does to... well, food.

JD Wetherspoon

Uniquely among big managed house chains, the JD Wetherspoon Organisation has always been a keen supporter of small brewers, but there's a limit to how many brewers' beers it can stock in its 800 pubs.

Wetherspoon's was founded in the late 1970s by Tim Martin, a barrister who decided that running pubs would be more rewarding and more socially useful than practising law. Martin was one of a handful of radical entrepreneurs such as David Bruce of the Firkin chain and Michael Cannon of Inn Leisure who set out to challenge the stranglehold then exerted over the London pub scene by Allied, Watney's, Bass, Whitbread and Courage; and his legal training was to stand him in good stead.

His chosen route was not to buy run-down pubs and transformthem, but to lease derelict shopfronts, filling stations and other high street properties, and turn them into stylish bars where good real ale was on sale at a reasonable price. It was a bold strategy: in those days magistrates refused to grant new licences where other pubs were already operating. This attitude harked back to the temperance days of 50 years previously and was intended to prevent "proliferation"; and the magistrates maintained that where there were already pubs, there was no "need" for new ones. Martin single-handedly overturned the policy, successfully arguing in court after court that the existing Big Six tenancies, run-down, restrictive and overpriced as they were, didn't answer the consumer's "need" for better pubs. His applications for new licences were invariably opposed by all vested interests – brewers and licensees afraid of the competition as well as anti-alcohol campaigners who opposed every application as a matter of principle. But he was always prepared to appeal when he was refused, and I can't think of a single case in those early days when he lost an appeal (although he has lost one or two since).

His formula worked, and in the 1980s and the first half of the 1990s Wetherspoon pubs appeared throughout London and then throughout Britain. There were even rumours of a venture in France at one stage. Wetherspoon's expansion slowed as other operators built on Martin's success at getting new licences and started competing aggressively for a finite number of suitable premises. You could even argue that Martin was personally responsible for the explosion of city-centre "drinking barns" of the last 15 years, and the company diversified out of its original music-free, TV-free, community-oriented pubs into more youthful bars (Lloyd's Number One) and even lodges. But it has always stuck to its principle of patronising local and regional brewers, sporting rows of six or more handpumps serving guest ales from all over Britain and Europe, and holding regular beer festivals.

Getting your beers into Wetherspoon's is harder now than it used to be, simply because there are more breweries competing for slots, but the company is still a great patron of local brewers. Its real ales are sourced by East-West Ales, an independent company based in Paddock Wood in Kent (Tel: 01892 725617); the people you need to impress there are Dave Aucutt and Janet Cheeseman.

Leased and Tenanted Pubcos

The biggest of the modern chains, though, run their pubs as quasi-traditional tenancies or leaseholds. These, too, are tied; but in this case it's generally the bigger chains that allow their licensees the most freedom. Ted Tuppen, founder of Enterprise Inns, said in his early days that he would be happy for his tenants to stock whatever their customers would buy, provided it all came from one supplier and at the right price. For unlike the brewers of old, these pubcos have no supply chain of their own: beers are delivered by third-party distributors such as Trade Team, Carlsberg UK and the Scottish & Newcastle-owned Waverley TBS. The range that these companies can stock and distribute is pretty big: Waverley TBS has claimed in the past to carry 2,000 draught beers from 300 breweries. However, these are mostly the large or medium concerns that can keep up with the logistics and meet the prices demanded, and few of the newer brewers have geared themselves up to operate on the scale and in the manner necessary to penetrate the pubco market successfully unaided.

The bigger pubcos have regional lists featuring beers from local brewers as well as seasonal ales drawn from a wider supply base and they sometimes host promotions that they call beer festivals, during which they temporarily allow their tenants to stock from a wider list still. But generally speaking your best way of getting your beers into their pubs is via the Society of Independent Brewers' Direct Delivery Scheme.

That accounts for the bigger pubcos. But there are many smaller ones as well. Some are run as property companies that squeeze as

SIBA Direct Delivery System (DDS)

Launched in December 2003, the Society of Independent Brewers' Direct Delivery Scheme followed a long campaign aimed at winning access for local brewers to pubco-owned tied estates.

Previously, small brewers had been frozen out by the big pubcos' highly centralised ordering and logistical systems, which favoured the national and bigger regional companies. Protests by small brewers had failed to prize open the crucial pubco market, so SIBA's Nick Stafford, who runs the Hambleton Brewery in North Yorkshire, decided on a co-operative approach instead and set out to find a way of enabling small brewers to meet big pubcos' trading requirements. The system he worked out, SIBA DDS, has at the time of writing been taken up by retailers including Admiral Taverns, Punch Taverns, New Century Inns and Enterprise Inns among pubcos and, in the take-home trade, Edinburgh Woollen Mills and Asda; the list continues to grow.

The aim of SIBA DDS is to allow small brewers to sell their beer to pubco-owned pubs in their neighbourhoods directly, without having to go through the pubco's central listing, ordering, stocking and delivery systems. It allows participating brewers to make their sales pitch direct to participating licensees, deliver the beer to the pub itself rather than to a regional depot, and collect their empties themselves rather than have to wait months for them to trickle back through the pubco's logistical chain. As well as saving time and effort, it allows small brewers to work in close partnership with publicans and increase trade to the benefit of all parties. Since its launch, SIBA DDS has proved that when pubco lessees have the chance to stock local beers from local brewers, they can enhance their businesses and increase their cask beer sales by offering their patrons a wider and more attractive range. Another attraction for the pubcos is that creating a single point of contact between their head offices and an organisation representing literally dozens of small brewers minimises their administrative burden and costs.

So how does it work? Well, it does depend on its members being relatively IT-savvy and very conscientious about keeping their admin up to date. Essentially, once the member brewer has struck up a trading relationship with a pubco tenant, the licensee orders his beer from the pubco's telesales department in the normal way. The pubco then places the order with SIBA, which promptly issues a delivery note to the brewer. It may seem a trifle circuitous, but because it's all done on the internet the order can be with the brewer within hours of the licensee placing it.

And because the order is added to the licensee's regular statement from the pubco, payment should be equally quick – no more wasting hours and even days chasing an ancient invoice for a single firkin!

It all sounds too good to be true and of course there are drawbacks. For a start, not every tenant of every participating pubco is allowed to take advantage of it: in most cases, before he can stock beer from the local micro, the tenant has to get his regional manager's approval, which for many reasons may not always be forthcoming.

And SIBA DDS is not without its administrative burden. It requires standardised e-forms that the member brewers have to learn to use. They have to have the right software (specially created for DDS) too, and hardware that can read it. They have to check their e-mails frequently and respond promptly and correctly, whether they're exhausted at the end of a long day or not. They have to pay a £250 registration fee (plus VAT, of course). And they have to accept a fixed price, agreed individually between SIBA DDS and each participating retailer, that may not be as high as what they would usually charge. This is a big sticking point for some brewers, who have decided not to join SIBA DDS on the strength of it. But the effectively discounted price has to be offset against greatly improved cash-flow and vastly reduced exposure to bad debt.

According to SIBA, DDS is a system that benefits everyone – pubcos, their main national distributors and brewers alike. To find out more, visit **www.siba.co.uk** and follow the link to "SIBA Direct Delivery Scheme".

much rent as possible out of their tenants while buying the cheapest beer they can find. Not much opportunity for you in these cases – but on the other hand some of the smaller locally based chains are extremely friendly to local brewers and indeed specialise in stocking their beers. Tynemill in the East Midlands, David Bruce's Capital Pub Company in London, Market Town Taverns in West Yorkshire, Brunning & Price in Cheshire and North Wales, and English Inns in Bedfordshire and Hertfordshire are all good examples. But they probably own fewer than 300 pubs between them, and in trying to sell to them you are joining a very long queue.

Listed below is a selection of micro-friendly pub companies. There may well be other small chains of a similar bent in your area. If you know of any pubs that seem to serve a wider than usual range of ales and don't obviously belong to a brewery, ask the licensee who the owner is. It may be one of the national giant pubcos but it may well be a smaller local outfit that can be persuaded to stock your beer.

- **Brunning & Price**
  Yew Tree Farm Buildings, Saighton, Chester CH3 6EG
  Tel: 01244 333100; **Web:** www.brunningandprice.co.uk
  15 pubs in the north-west and north Wales.
- **Capital Pub Co**
  1 Relton Mews, London SW7 1ET
  Tel: 0207 589 4888; **Web:** www.capitalpubcompany.com
  Founded by Firkin brewpub entrepreneur David Bruce.
  26 pubs in London.
- **Cascade Public House Management**
  5 Merlin Way, Bowerhill Trading Estate, Melksham,
  Wiltshire SN12 6TJ
  Tel: 01225 704734
  14 pubs in the South-west. Sister company of Moles Brewery.
- **Chapman Group**
  Syon House, High Street, Angmering, W. Sussex BN16 4AG.
  Tel: 01903 856744
  44 pubs in the South-east and Gloucester.
- **Sir John Fitzgerald**
  Café Royal Buildings, 8 Nelson St, Newcastle on Tyne
  NE1 5AW
  Tel: 0191 232 0664; **Web:** www.sjf.co.uk
  28 pubs, mainly in the North-east.
- **Head of Steam**

Manesty, Leazes Lane, Hexham, Northumberland NE46 3AE
Tel: 07803124508; **Web:** www.theheadofsteam.co.uk
Four railway-themed managed pubs in the North of England.

- **Market Town Taverns**
  6 Green Dragon Yard, Knaresborough,
  North Yorkshire HG5 8AU
  Tel: 01423 86610; **Web:** www.markettowntaverns.co.uk
  Eight pubs in Yorkshire.
- **Castle Rock Ltd**
  Queen's Bridge Rd Nottingham, NG2 1NB
  Tel: 0115 985 1615; **Web:** www.castlerockbrewery.co.uk
  17 pubs in East Midlands; founded by former CAMRA National Chairman Chris Holmes.

The Wholesale Trade

Not all that long ago, cask ale brewers were deeply suspicious of the wholesale trade. Brewers were understandably paranoid about the way their beers – living, breathing products, after all – were being handled. How long did they take to work through the distribution chain? How many drops and pick-ups were there between the brewery and the beer's final destination, the pub? What temperature control, if any, was there in the supply chain? What state would the casks return in, and when? The wholesale trade did little to allay their suspicions.

But the way the market was changing left small brewers, and indeed many regional brewers, with little alternative but to trust the wholesale trade. The guest ale market that opened up in the 1990s was a truly national market: ales from Kent were called for in Cumbria and vice versa, and only the independent wholesale trade could handle the traffic.

It was at this time that independent wholesalers specialising in real ale such as The Beer Seller, Flying Firkin, East-West Ales, Small Beer, Little Ale Cart and many others set up in business. Some disappeared; others prospered – The Beer Seller, in particular, became a major national player, supplying not only small free traders but also the big pubcos. It has now been snapped up by Scottish & Newcastle and renamed Waverley TBS, Waverley Vintners having been S&N's existing wholesale division. That leaves Flying Firkin – northern-based, but now with a much longer reach having acquired the non-Wetherspoon business of East-West Ales and its southern customer base – as by far the leading independent.

The new generation of specialist wholesalers did their best to speed up turnaround times so that the beer arrived at the retailer reasonably fresh and the casks got back to the brewery reasonably quickly. They also sought, mostly although by no means universally, to introduce refrigerated storage and transport so that the problem of beer-barrels exploding on wholesalers' lorries, which had actually happened during the very hot summer of 1996, occurred no more.

These new wave distributors existed alongside a big national network of well-established locally rooted general licensed trade wholesalers, some of them family firms several generations old, who dealt in beer and lager as well as wines, spirits, soft drinks, bag-snacks and sundries. Many of these older companies still exist, and 26 of them with 53 depots in England and Wales between them are now members of a consortium called National Drink Distributors (NDD). They deal in a wider range of beers than they used to, running guest and seasonal ale programmes for their 25,000 regular accounts, and in many cases trying to stock at least some local products; so if they haven't traditionally been big players in the microbrewing world, they are well worth cultivating now.

Before tackling the wholesale trade, though, you have to ask what it will require of you. The answer, in a nutshell, is reliability. That's why most wholesalers won't handle your beers until you've been in business for a while: they don't want to go to all the effort and expense of offering their customers a beer that may very well not be available tomorrow, so they'll want to know that you're sound and solvent. And they want reliable quality as well as reliable supply: nothing upsets them more, as you can imagine, than getting a load of returns from an angry licensee because the beer tastes like vinegar. So even if, as often happens, your very first brew is enthusiastically greeted with a clutch of CAMRA Beer of the Festival awards, most wholesalers will want to be sure you're thoroughly bedded in before they commit resources and energy to pushing your beers.

Selling through wholesalers does have its drawbacks. You have no control over the quality of your beer as it passes through the supply chain. You have to share your margin, which may be slim enough already, with a middleman. You may have to wait quite a while to get your casks back. And you will almost certainly have to wait to get paid, especially if you're dealing with some of the big national chains. They might claim to have a credit period of 60 days, but a glitch – carelessness, laziness or simply the inevitable cases of human error – can wreak havoc with your cash-flow. Marc Bartram tells a story of chasing his payments from the very biggest chain for month after month, only to find out entirely by accident that his cheques were being sent to a very bemused Buffy's Brewery, which, like him, could get no sense at all out of the bureaucrats in the wholesaler's accounts department. These kind of glitches can take a lot of time and effort on your part to sort out; Marc is not the only brewer who swears he will never sell to a wholesaler again unless they pay upfront – which, of course, they don't.

But there's a limit to how much effort you can put into sales yourself;,and as your business grows you may well find the services of a wholesaler indispensable. To meet reliable wholesalers, join SIBA, it has many of them among its associate membership, and Flying Firkin in particular is commendably eager to give beers from new breweries a go.

A last word on the wholesale trade concerns brewery swaps. Many small breweries try to vary the repertoire of beers they can offer their customers and, at the same time, to get their beers more widely known, by swapping beers with other breweries. This isn't quite the same thing as straightforward wholesale distribution, and it doesn't in itself increase your sales volume. But it's a trading model that has worked very well for very many small brewers for very many years; a little further down the line, perhaps when you're more established, it could prove a handy boost for you too.

Niche Opportunities

One of the very first new breweries to be established, way back in the early 1970s, was the Miners Arms at Priddy, Somerset. It wasn't a pub, though; despite the name, it was a restaurant. And you'd think, wouldn't you, that hotels and restaurants would be keen to offer tourists and diners craft-made local products? Well, in 99 per cent of cases, they're not, and on the rare occasions when they are, they tend to be more interested in bottled beer than draught, so perhaps we'll deal with them in the next chapter alongside the tourist gift shops and other miscellaneous outlets.

The Club Trade

One large part of the on-trade that hasn't really been penetrated by local brewers, though, is the club trade. Not nightclubs, but Britain's 15,000-odd traditional sports and social clubs.

In theory, your local traditional workingmen's club ought to be the ideal home for your ales. You're a local outfit, maybe employing local

CASE STUDY

Using Wholesalers, but Cautiously – Mordue Brewery, North Shields, Tyne & Wear

Things went a bit mad for Garry and Matt Fawson of Mordue Brewery when their best bitter, Workie Ticket, won CAMRA's Champion Beer of Britain competition in 1997. The brewery had only been open for a couple of years, and the brothers simply weren't ready for the demand the award generated.

It was a mad scramble for a while, and even when the rush had settled down it was still obvious that the original site wasn't big enough. In 1998 Mordue moved to bigger premises, and Garry and Matt were able to concentrate on building up sales in a more measured way. So In 2005 they had to move again, to a new 12,000 sq ft, £1.5 million development.

In those early days, anyone who called themselves a wholesaler and owned a van was knocking at Mordue's door demanding to be allowed to buy its beer. Producing enough to satisfy them all was a nightmare. A nice problem to have, you might think. But the trouble with an award like CBoB is the weight of expectations it carries with it. Yours is officially the best beer in the country, so if it's less than perfect the customers aren't just going to shrug and try something different – they're going to feel cheated. And, says Garry: "They're not going to blame the landlord; they're not going to blame the wholesaler – they're going to blame the brewery."

Early on, then, the Fawsons learnt not to trust the wholesale trade – not blindly, at least. They wanted to be sure their beers were being treated as reverently as the reigning CBoB deserved, so customers always got the quality pint they were expecting. And while it was beyond their power to check every pub where Workie Ticket was sold, they could at least check all the wholesalers who delivered it. What they wanted was cold storage, proper stock control and the reasonably speedy return of their precious empties; and to make sure they got it they insisted on vetting new wholesale customers before agreeing to supply them.

Their experience was that many smaller wholesalers simply didn't have the facilities to handle a living product like cask beer properly, so they stopped dealing with them. These days, around half of Mordue's 8,000-barrel a year output is delivered through third parties, but they're third parties the brothers can trust. Something like 30 per cent of the output is carried by hauliers to the depots of big pubcos such as Punch Taverns and JD Wetherspoon, but the Fawsons won't use groupage hauliers who deliver to and collect from central hubs. Instead, they only use hauliers they know. Everything goes on overnight deliveries direct to the customers' depots, a bit of fastidiousness that means every pallet of

18–24 firkins costs around £40 to send out rather than the £20 it would cost on economy.

Wholesalers buy about 20 per cent of the brewery's output. Mordue sells to the biggest of them, the Scottish & Newcastle-owned Waverley TBS, and Garry says standards at the company have increased dramatically in recent years. "They've got temperature-controlled vehicles as well as depots and they promote our beer properly so that it turns over quickly and we don't get returns that are past their expiry dates," he says.

Flying Firkin, the Lancashire-based specialist in microbrewery beers, is another favoured customer, as is the National Drinks Distributors network of local independent wholesalers. "They're very strict on quality, and very good," says Garry. Mordue also sells to a handful of small local outfits such as Small Beer of Lincoln, but only because the brothers have personally inspected their facilities and are satisfied with their standards.

"My advice to anyone starting out in the business is not to be tempted by wholesalers you don't know and haven't vetted," says Garry. "It may look like a way of generating revenue quickly, but it's not a long-term strategy because quality is so important. Some of them are pretty good, but you have to be sure."

Another drawback of distributing so widely through third parties is the demands it places on your cask population. Mordue, although not a huge brewery, has more than £500,000 tied up in its stock of around 10,000 firkins.

"It's quite possible for your cask population to be worth more than your actual brewery," says Garry. "It takes six to eight weeks to get your casks back if you're selling through the larger wholesalers, longer if you've sold them to smaller companies. That's something that many people setting up in the business don't realise – but it's all got to be built into your costings and your cash-flow forecasts."

www.morduebrewery.com

people, which should appeal to the committee that decides what beers to stock. You should be able to offer an attractive price: if you brew less than 3,000 barrels you qualify for 50 per cent duty relief, which you can share with a big, steady customer such as a club. And the club is big, and usually busy, and full of thirsty working-class drinkers who like their ale; so your beer will turn over quickly enough to ensure that it's always in excellent condition.

Workingmen's clubs, though, have remained stubbornly resistant to the advances of local brewers. Many are still loan-tied, for a start; even if they're not, committees tend to remain loyal to traditional suppliers who often offer not inconsiderable inducements. An all-expenses-paid annual day out at the races, complete with champagne buffet and executive marquee, for the entire club committee including spouses may not actually qualify as bribery, but it's not the sort of jaunt the average working family would indulge in if the brewery wasn't paying. The wholesale price clubs have to pay is one you would find hard to match, even with your duty relief. And finally, they may be big, busy and full of good, steady drinkers, but the drink they drink these days tends to be lager rather than ale. And the ale they do sell tends to be keg rather than cask, for ease of handling.

Even at the traditional end of the club trade, though, this resistance is beginning to crack. The truth is that membership of workingmen's clubs, Royal British Legions and the like, is tumbling, despite the cheap beer, because their offering is simply not up-to-date. Leatherette banquettes, scruffy lino floors, keg ale, bad singers with obvious comb-overs, and bingo are just not attractive any more, and clubs are having to raise their game to stay alive. Some of them are now offering equal rights to women. There is no reason why real ale shouldn't be part of the equation, especially if it's locally brewed. So it's up to you to go to see your local club steward and secretary, demonstrate that stocking real ale is perfectly possible (even if there isn't a proper cellar, there's bound to be a cold room), and persuade them that well-priced and properly promoted cask beer will attract members. They'll give you a hearing, and the worst they can say is no. Then you go back next year and see if they've changed their minds yet.

You will probably have less trouble getting a sympathetic hearing at sports clubs. There are two sorts here, though: those whose bars are only busy – or in some cases only open at all – on match days aren't going to stock cask ale, even from a local brewer.

Our Complete Solution For All Your Stainless Steel Tanks and Vessels
Looking for new Brewery equipment?
Call Kerry 01777 712110
• Brew Kettle
• Mash/Lauter Tun
• Whirlpool
• Hot/Cold Liquor Tanks
• Fermentation Tanks
• Conditioning Tanks
PART OF THE MOODY GROUP OF COMPANIES
Moody Systems Limited, West Carr Road, Retford, Nottinghamshire, DN22 7SN, England
Tel: +44 (0) 1777 701141 Fax: +44 (0) 1777 709086
Email: info@moodysystems.co.uk
Web: www.moodysystems.co.uk
MOODY Systems

But the larger, busier and – though I hardly dare utter the words – more middle-class establishments such as golf clubs are proving fruitful territory for those cask ale brewers who actually try to tackle them. Middle-class men tend to drink real ale rather than lager, it's a fact. They also tend to be more appreciative of lesser-known and local brewers whose products have the cachet of, if not exclusivity, certainly rarity.

So, don't ignore the club trade. Clubs may not be easy accounts to crack, and they may not pay top dollar, but they are in general a growth area for real ale, and listing one or two local sporting or social clubs among your regular outlets will pay off in terms of long-term, reliable, regular business.

Outside bars

Running outside bars at events private and public is an opportunity that more and more small breweries are learning to exploit. It's an area that offers two main advantages: the overheads are comparatively low and the margins are comparatively high. It also has its drawbacks, as we shall see.

The half-full glass first. Even at public events such as village shows, your overheads are likely to be minimal. In capital terms, you need a strong trestle table, an awning of some sort, a stillage system on which to mount your casks, some cooling jackets and a biscuit-tin for the cash. If you're selling ale racked off bright into polypins rather than from the cask you can even dispense with the stillage, although a reliable cooling system is still indispensable. These will all involve in you a modest one-off outlay. You might also need some assistants, who will want paying. To offset the expenses you get the full retail price of the beer – £180 a firkin rather than the £50–60 you get wholesale – and if you can't make a nice margin on that, you're doing something wrong.

There are drawbacks. You will need a personal licence (see Chapter Eight), but then you should get one of those anyway. It takes time to set up and take down. There will be the occasional disaster when only half the predicted numbers turn up; and if you've racked beer off specially then the unsold surplus will, unless you have another outside bar the very next day, be spoiled (you can get duty drawback on it, but that's not much help). (I have heard of brewers sneaking very small quantities of freshly racked ale back into the system, but it's risky. You never know what infections are lurking in the sort of barns where ceilidhs are often held, and you certainly wouldn't want them in your brewery. If you really can't overcome the temptation to spread unsold bright beer among a few casks, make sure it's from unbroached containers only. It's safer just to stand the loss, though.)

As well as the cash they generate, though, outside bars have a promotional value. In brewing as in life, you can never have too many friends; outside bars spread your name around in the neighbourhood. So: hard work; chancy; but a good sales opportunity and an equally good profile raiser.

Marketing

Naturally, your first efforts in marketing are going to be concentrated on finding outlets for your beers. But equally important, and in some ways more difficult, is making an impression on the beer-drinking public. Getting repeat orders from your trade customers will depend on how much of your beer they can sell to their patrons and although you might say that they're the retailers and it's up to them to promote themselves, there's a great deal you can do to help.

Your marketing effort starts at the simplest possible level – with the names of your beers, and the design of your pump-clips. In most of the pubs where your beer is on sale, your pump-clip is likely to be the only way you have of communicating directly with the public and the only chance you will get to persuade people who have never heard of you to buy your beer. What sort of people are they likely to be, and what sort of message are they most likely to respond to?

Branding

In the early days, many microbrewers chose branding that was blokeish, jocular and often rustic. Beers had names like The Dog's Bollocks, Piston Bitter (a variation on pissed 'n' broke) and Old Fart, and the pump-clips were often not much more than photocopies of the head brewer's late-night attempts at postcard art. Often it seemed all too crass and amateurish but it struck a chord with an ale-drinking public that was itself, in many cases, blokeish and jocular. Real ale in those days belonged in the same cultural cage as rugby club dinners and dwyle flonking; today, with beer-drinking generally in decline, reading and catching a market is a much more subtle and demanding business.

This is no place to lecture you on design, on what colours to choose, on what note to strike. Suffice it to say that unless the former life you're going into brewing to escape was in graphic design, you should hire a professional. It does make sense, though – and it's surprising how

An example of a Tunnel Brewery label

many brewers have let this obvious point fly past them – to maintain uniformity of branding across your range. In a crowded bar on a busy night, the size of the type identifying the pump-clip as yours will defy the cursory inspection that may be all that is possible. Unless all your pump-clips have the same shape and colours, it will not be apparent that the best bitter on tap is one of yours; anybody who tried and liked your cooking bitter and would try more of your beers if given the chance will remain in ignorance. So establishing a clearly recognisable identity via a uniform pump-clip design will be a great help to drinkers who want to sample your range and to publicans who want to sell it. Some breweries take it a step further and give their whole range a common theme: Cotleigh, with its birds of prey, is perhaps the supreme example. That may be a little excessive – it seems to me a little hasty to narrow your possibilities without good reason – but for the perfect example of unobtrusive yet clearly identifiable branding across the range, check out Tunnel Brewery, see **www.tunnelbrewery.co.uk**.

As for names, jokes for blokes are as much of a turn-off as a come-on these days. Craft-brewed beer is trying to get serious and to press a whole range of contemporary buttons: it's foodie-friendly; it's got good green credentials; it's the very antithesis of all that's corporate and globalised. And perhaps because so many of these issues resonate so strongly with women, more and more of them are drinking real ale. You are not going to appeal to them by christening your beer Maiden's Ruin, Willie Warmer or Old Horizontal (all genuine examples).

Working with CAMRA

Having created a range with names and branding calculated to appeal to the customers you have in mind, you need to find cost-effective ways of getting your name around. I have already mentioned

CAMRA membership as a way of making contact with local micro-friendly licensees but CAMRA can do more for you than that.

The Brewery Liason Office (BLO)

The first thing your local CAMRA branch should do for you is appoint a brewery liaison officer, or (BLO). The origins of this office lie far back in the mists of the Campaign's early days, when new breweries were rare and tender shoots that CAMRA branches felt they had to nurture and fuss over. The BLO's job was to check in with the brewer from time to time, make sure that everything was all right, and report back to the branch with news of new brews or looming problems. In the best of cases, the relationship between brewery and BLO would develop much further, with the BLO taking responsibility – in return for free pints and being made to feel special – for publicising and promoting the brewery. In the worst of cases the very opposite happened, and the BLO turned out to be a curmudgeon for whom everything the brewery did was a mockery. In these, thankfully few, cases, the BLO was very soon made persona non grata at the brewery. More often, the BLO turned into a valued and indeed valuable friend who even, on occasion, ended up as a partner or employee. So provided your BLO is bearable, cultivate them. Give them things to do like writing press releases or delivering leaflets to your retail customers. They will not resent it or regard it as imposition; they will be genuinely thrilled to help.

Every CAMRA branch also organises an annual beer festival – and some of them run a winter ales festival as well. Your local branch will undoubtedly order one or more of your range for its festival. In volume terms this is insignificant – three or four firkins at most, unless you're in one of these places like Peterborough, Cambridge, Worcester where the local beer festival is huge. So although there's not likely to be a huge amount of money involved, a presence at local festivals is great for public exposure. The local media – broadcast as well as print – will often seize on a new local brewery as the news "hook" on which to hang their festival report, especially if you have an unusual or catchy story to tell. More important still, even quite small festivals attract CAMRA members from surprisingly far afield, who are always on the look-out for new brews and will spread the news of your arrival – and their opinion of your ales – to other CAMRA members, beer festival organisers and micro-friendly licensees back home. Winning the Beer of the Festival award is better publicity still, and it's surprising how often the title goes to a new local brewery. These things are supposed to be unfixable, and in the most obvious sense they are (waving brown envelopes is likely to be counter-productive, I'm happy to affirm). But sentiment and local loyalty often play their part and who's going to complain about that?

Public Relations

Using the media is something local brewers are often very bad at. There are only 24 hours in the day, and at least two or three of them should be spent asleep; media relations tend to come far down the busy brewer's list of priorities. It shouldn't be that way, though: building strong local awareness is one way you can help your trade accounts shift your beer, which will help you win new accounts as well.

There's no publicity like free publicity, and putting out a constant stream of news stories is one good way of getting it. Some local breweries even hire public relations agencies to do it for them, and to good effect: Wood's of Wistanstow in Shropshire has used local PR agency Seabury Salmon for years and has become one of the best-known brewers in the West Midlands as a result.

If you're doing it yourself, you need to get to know your media. Start with the CAMRA branches in your trading area: they almost all publish local newsletters – the larger branches print thousands of them – which are distributed around ale-friendly pubs for customers to browse in odd moments. Find out who the editor of your local newsletter is and keep him or filling these publications can be difficult and the editor will be glad of news of new brews, new members of staff, funny stories and the like. You may even get your brewery featured in a profile article.

You (or even your BLO, if he or she is canny, aware and willing) should also try to develop a relationship with a favoured reporter at each of your local newspapers and radio stations. This is, remember, a two-way street: the reporters will help you by getting you free publicity, and you will help the reporters by establishing their reputation as a keen and productive newshound.

As for what kind of stories local media are looking for – well, buy your local paper and see for yourself. New beers, especially charity brews (Wood's, that PR-canny outfit, is one of many that regularly creates special brews with part of the proceeds going to newsworthy local causes), new recruits, the brewery cat's mouse-hunting prowess – it's all grist to the local reporter's mill.

Ask yourself, also, what format your local media want to get their information in. A story is more likely to make it into the paper or onto

CASE STUDY

WEST Brewery. From Micro to Macro

WEST situated in Glasgow was started by Petra Wetzel in 2006 and has gone from being a microbrewery producing about 10 kegs a week to now producing over 250,000 litres a year. There are plans for a new £6.6m brewery to be built nearby which will increase production by twenty times and is due to open in Spring 2014.

Petra first visited Scotland aged 13 and later went to Glasgow University in 1994 to study Politics, History of Art and Management. After university she went to work at the Glasgow Tourist Board as a roving ambassador and attended trade fairs all over the world. After a chance meeting with a lawyer she changed tack and started a legal career at Maclay Murray Spens and worked there for five years. When her father complained regularly on his visitsto Glasgow that he was unable to find good German beers an idea started to form about starting a brewery. In 2006, with the backing of her father, Petra started the business WEST and today the company has a turnover of more than £1m and employs over 40 staff.

WEST is located in part of the old Templeton carpet factory (on Glasgow Green) which was completed in 1892 and designed by William Leiper. It is a stunning building and the design is directly inspired by the medieval Palazzo Ducale in Venice. The building was used as a production carpet factory until the 1950s. The building is unique in terms of its design and construction, and is now one of Glasgow's most protected sites. The Winding House annex was completed in 1904, and housed all the winding equipment for carpet production. Templetons was a hugely successful

company, producing carpet for the West Wing of the Whitehouse, the Taj Mahal and the Houses of Parliament.

The brewery is located downstairs and you can see from the bar the two massive copper brewing kettles. The brewery now houses 12 fermenting vessels and a very unusual mash filter which enables the brewery to produce a wheat beer which contains 80% wheat malt, the only brewery in the UK to be able to do this. The brewery uses Scottish and German malted barley and employs two full time brewers and a number of other staff who are in charge of deliveries and logistics.

WEST now produce a range of beers all of which are brewed in accordance with The Reinheitsgebot known as the German Purity Law 1516. This means that they only use the four core ingredients, water, malt, hops and yeast in brewing all of their beers, adding no artificial additives, colouring or preservatives. These include St Mungo lager ABV 4.9% named after Glasgow's patron saint, a WEST dunkel, a dark lager at 4.9% and at Christmas they produce a special beer called Weihnachtsbier.

With the success of the brewery has come other rewards as in 2011 WEST was voted Scotlands best Family restaurant of the year. Petra's aim has always been to make the restaurant and bar family friendly with the emphasis on welcoming children and dogs and they run brewery tours every weekend.

www.westbeer.com

the airwaves if it's presented in a way that needs minimum processing – as a proper press release (see panel piece on page 134). In the old days these would be hard copies posted by snail mail to the newsdesk; these days, a Word document attached to an email (not a cut and paste job, preferably) addressed personally to your pet reporter is probably your best bet.

Advertising

Advertising is a trickier business altogether. You're not Carlsberg, spending a fortune to establish and maintain a national brand; and you're not a scaled-down version of Carlsberg, either, trying to achieve locally what it does nationally. Brand advertising doesn't work like that: it has to make your brand stand out from the others, and if it's not ubiquitous and relentless, it's a waste of money. It also only works if your brand already has excellent distribution – there's no point advertising a product no one can buy! So don't be tempted to take out big ads for your brands in the local paper, or even in specialist media such as What's Brewing. That's for the future, perhaps, but for now it's simply not effective.

That doesn't mean you shouldn't advertise at all; but to repay the investment, your advertising needs to be highly market-specific and well planned.

- **First**, make a forward planner highlighting events such as your open day/beer festival, the launch dates of special or seasonal beers, important local dates where your beers can be highlighted, and joint promotions with stockists
- **Second**, set a budget. This will be based on the advertising rates of the local media you intend to use, but remember – nobody but a mug pays the ratecard price. Discounts are always available, based on repeat or regular advertising. Once you've made your budget, stick to it. You might want a small contingency fund to exploit unforseen events, but don't be tempted by slick sales patter into spending more than you have allocated
- **Third**, choose your medium, basing your decision on "bangs per buck" – not the overall readership or audience of the medium, but what proportion is likely to buy your beer. Always remember: a small ad strategically placed is more effective than a big ad in a vacuum. And one sure way of

testing the effectiveness of each medium is to include a money-off voucher in some of your adverts (but not all of them – promotional discounts are part of your advertising budget, which is unlikely to be huge) and analyse the redemption rate

Local CAMRA branch newsletters, distributed in hundreds to local pubs to be picked up and put down by drinkers whiling away a few idle moments, are ideal vehicles for your advertising because everybody who reads them is ipso facto in a pub and willing to buy. Local A5 newsletters of the type that are appearing in villages all over the country are also cheap and well-read; but here, as with local newspapers and local commercial radio, you need to tie your advertising to a particular event. Is the local CAMRA beer festival being covered? Is a pub where your beers are on sale being profiled? Is a special feature being published or broadcast on drinking and dining out in the district? Supporting this sort of coverage with a well-designed ad is effective both in raising your public profile and earning brownie points with your retail partners. Then there are specialist national media: not only beer-related titles such as What's Brewing and Beers of the World, but all sorts of hobby and special interest media too. Do your beers, for instance, have a steam theme, or are you involved in a promotion with a local preserved railway? Are you donating part of the proceeds of a special ale to a good cause? Invariably there is a magazine or newsletter that will give your efforts exposure to relevant consumers, and normally at a reasonable rate too.

Even where there's no editorial involved, working with your stockists makes advertising more effective. One of your best accounts may be advertising its quiz nights, or a forthcoming beer festival, or a special menu for Valentine's Day or Father's Day. Share the cost and get your name in the ad, and you both save money and share the benefit.

One design tip: keep your ad simple. When you're paying by the column centimetre, the urge to cram in as much information as possible is almost irresistible. But do resist. Save the lingering descriptions of your hop varieties, your malt grist, your pure well-water and the rest for your website, your brochures and your back-label. Your ad should be clear, uncluttered and eye-catching. You only have a split second of your customer's attention, so use it wisely! And if in doubt, use a professional graphic designer – traditionally a service provided free or at little extra cost by local newspapers.

Promotions

As with advertising, organising promotions jointly with your retail partners (and do try to think of them as partners) is often more rewarding than going solo, spreading the work and the cost and sharing the benefit. You might provide prizes – free pints, for instance, or merchandise, or brewery tours – for pub promotions such as quiz nights or charity raffles. But don't go mad with money-off or free-pint offers: they can turn out to be more popular than you anticipated and end up blowing a hole in your promotional budget.

If one of your stockists is running a beer festival, you can be of practical help by pulling pints (staff are always short on occasions like this) and providing technical support. Naturally, your beers will be on sale, and you might also be allowed to leave your dray prominently sited in the pub car park for the duration, and to put up a few posters or distribute your leaflets. If you want to run a beer festival or open day at the brewery but don't have a liquor licence of your own, a friendly publican might act as licensee for the event and also help out with cellar and dispense equipment in return for a cut of the profits and some publicity for the pub. Involving a local charity is also a good idea: it guarantees publicity, and volunteer fundraisers will both provide a workforce and drag their friends and family along.

It's all about mutuality, building relationships, creating bonds that will last. There's only a certain amount you can do alone, especially if you're a one-man band, so work with your stockists, your CAMRA branch, and local voluntary groups, share both the burdens and the benefits – and make every casual customer a regular.

Electronic Marketing and Promotion

Marketing and promotion doesn't depend entirely on the old-fashioned print and broadcast media these days. More and more, they make use of websites, blogs and emails.

A website of your own is pretty much a must these days. It's a low-cost, low-maintenance shop window, and one that the entire world can mooch past and gaze into. And not only does it show the world what you have to offer, it can also invite them in to buy. We'll deal with e-commerce later: if you decide that it's one of your main routes to market, you'll soon find that it has problems and peculiarities of its own. But as a complement to your core business, it can be a very useful way of generating sales and awareness. It's also one way of announcing new launches and other promotions and developments.

However a website is not cost-free: designers charge, and a webmaster will also charge to keep the site up to date. You can pay anything from £200 to £2,000 or even more, and in web design as in life, you get what you pay for. A cheap designer will produce a cheap-looking site that will probably break down. The best advice is to shop around and get plenty of quotes; fortunately, web designers are very easy to find. You can either Google "web designers", and literally hundreds will come up, or you can be a bit more selective and revisit a few sites you've been impressed by in the past. They'll normally have the designer's name and contact details on them somewhere (it would be odd if they didn't!) and the designers' own sites should have links to other sites they've designed.

Website Content

But before you rush in to getting quotes from designers, think of what you want your site to do. First, you want it to say who you are and what you do. A nice, chatty site with plenty of background detail is more effective than a dry, terse, and straightforwardly factual one. You want potential customers to get involved with you and your venture on a personal level – it makes the beer taste better and seem cheaper! Having engaged the public's attention, give them the facts about your beers and where people can actually buy them; keep them abreast of new and forthcoming beers, new outlets, events of your own such as brewery open days and third-party events such as beer festivals where your ales will be on sale. Using the site for e-commerce we have already touched on; but you can also use it for trade telesales, with a password-controlled page where customers can not only place orders but also pay for them.

Another way of using new media to establish a link with your trade customers is to set up an email newsletter. This is free, takes hardly any time at all, and communicates with all your stockists at once at the click of a mouse. You simply tap all their email addresses into your system, and every time you have a new product or promotional price or any other information you need to communicate to them, just write a little sales letter and click on "send". It couldn't be simpler, it couldn't be more tightly targeted, it couldn't be more effort-efficient. And it's not strictly speaking spam, either, since your customers will have given you their email addresses voluntarily.

Chapter Seven

How to Get to Market: Bottled Sales

Local brewers have always been associated in the public mind with cask-conditioned beers or, to use the vernacular, real ale. But bottled beer sales are becoming more and more important both to the brewing industry as a whole and for smaller brewers in particular.

Ten years ago I sat down with a set of statistics detailing the trend in draught versus packaged beer sales volumes. At that time packaged sales (bottled and canned) represented fractionally over a third of all volume, but the trend was gradually upwards. I worked out that on a straight-line graph, packaged volume would overtake draught in 2007. I haven't seen the most recent statistics; but by 2002, according to the British Beer & Pubs Association, packaged volume had reached 41.7 per cent of all sales, so my prediction was on track at the halfway stage.

For not only have take-home sales continued to grow at the expense of pub sales (not least because beer is so much cheaper in supermarkets than in pubs), but in a large section of the on-trade – the town-centre drinking barns – bottled beer sales have been growing at the expense of draught. I cannot myself see the advantage of necking a 330ml bottle of Budweiser when a whole pint of Stella costs not much more, and proportionately rather less, but apparently I am in a minority.

Bottling some or all of your output is a second, and increasingly important, string to your commercial bow, and does have several advantages over draught. Bottling extends a beer's life from a few days to a few months; it gives you something to stock your own shop with, if you have one; and some shops in your area – small independent grocers, especially – can usually be persuaded to stock a case or two as a showcase for (or gesture towards) local produce. Bottled versions of your range make great trade samples when you're out on the road trying to drum up new business. And it's relatively efficient in that you don't have to worry about your empties as you will, constantly, with casks.

But as with the pub trade, so with the bottled trade. Most of it, quite simply, will remain firmly closed to you. The Competition Commission adamantly refuses to accept that the "multiple grocers" warrant investigation, even though the Big Five supermarkets control as large a slice of the market as the Big Six brewers did back in the late 1980s, when they were found to be a complex monopoly operating against the public interest. To pursue the analogy, the independent sector – effectively, Thresher's and second-string supermarket chains such as Budgen's, Somerfield and the Co-op – is

CASE STUDY

Flipside Brewery

After years of working in the IT industry, and often working away from home, Andrew Dunkin could see his role was coming to an end. Combining a desire to spend more time at home for family reasons with a desire to get out of IT but still run his own business, Andrew decided to follow his passion for real ale.

Andrew and wife Maggie used their kitchen (and, indeed – patio) to do test brewing for about eight months and armed with some potential recipes, opened a new brewery.

Flipside Brewery is based in Colwick, Nottingham and they supply many of the local pubs with their beers as well as the general public. After a lot of thought about a name for a brewery they came up with Flipside, a reference to the coin collecting hobby of the head brewer. This also provides the theme for the names of the beers which are loosely based around coins and currency, apart from when they can't resist using the word "flipping"!

Setting the business up was fraught with problems – getting planning permission, effluent permission from Severn Trent, organising builders and finance along with lead times like nine months for installation of the brewing equipment and three months for the casks made it a logistical headache. However, it all came together eventually and the installation and first brew went ahead on the 4th May 2010.

Flipside Brewery's first two beers, Sterling Pale and Flipping Best were seen for the first time at the Newark Beer Festival at the end of May 2010. They then held a launch event at the Cross Keys on Byard Lane which was attended by CAMRA members and some local publicans.

Since then they have brought out two more beers - Copper Penny and Dusty Penny. They have received excellent feedback on all the beers and they are proving to be very popular with the discerning drinker.

Andrew Dunkin says "We've had a great time since opening the brewery. Everybody that we talk to is so enthusiastic: the suppliers, the publicans, the other brewers and CAMRA are so pleased and fall over themselves to help you, it's amazing, We've never known an industry like it, everybody is so helpful and friendly."

Maggie, who works full-time as a computer trainer and their son Luke help out on brew days and with racking up. Luke is very keen and can be bribed to clean out the mash tun and copper.

www.flipsidebrewery.co.uk

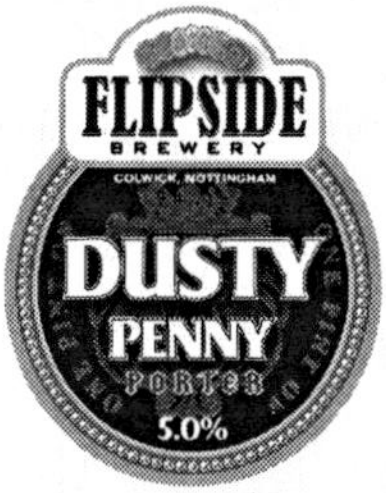

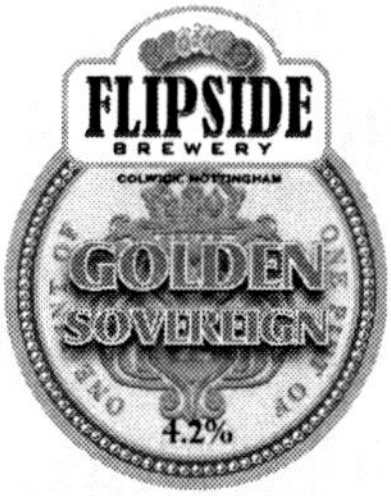

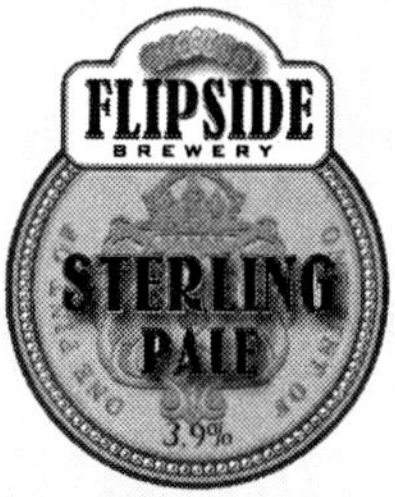

just as restrictive as the regional brewery tied estates of the 1980s were. And the take-home equivalent of the free trade – small family-run convenience stores and corner shops – is largely supplied by cash and carries that will be closed to you as firmly as Tesco or Morrisons.

Nevertheless, there are opportunities out there that will almost certainly make bottling some or all of your beers – either yourself, or by a contractor – worth investigating. But before we explore these opportunities, let's take a look at the practicalities.

Bottling Your Beers

Bottling on a small scale is time-consuming and monotonous, and unless it's done carefully the beer is very prone to infection and oxidisation. On the other hand, starter-level bottling and corking machines are almost ridiculously cheap: an initial outlay of £600 will provide you with all the equipment you really need.

The most primitive four-headed hand-bottler will cost you about £320 + VAT and stands quite happily on a trestle table, leaving plenty of room for a simple crown capper (about £165 + VAT) to stand next to it. The bottle filler is connected directly to a beer barrel or conditioning tank; the operative simply holds the bottles to the spouts, two at a time, and pays enough attention to make sure they don't overfill. Two operatives, one on the filler, one on the capper, can easily fill 1,500 bottles (more than an entire pallet) a day on this simplest of items; or, if you're a one-person band, you can do the whole job yourself and fill 750, which is the equivalent of nearly 10 firkins. If you do the whole thing yourself, though, remember to swap from filler to capper frequently: not only does it make the job marginally less boring, but the longer the bottles stand around without caps, the likelier they are to pick up an infection. And however bored you get, remember that if you're retailing the beer yourself, either through your own shop or at farmers' markets (see below), your day equates to sales of £1,500 or more, depending on the strength of the beer.

Of course, you'll need slightly more equipment than that, but none of it is expensive. If you're filling bottles direct from the shrink-wrapped pallet on which they were delivered they shouldn't even need sterilising. But if you're only bottling a few cases at a time, and the bottles are likely to stand around unprotected for any period before use, a simple rinser using an odourless peracetic fluid costs £30 and a plastic draining tree £24.

Consumables

Then there are the consumables, which unfortunately are more expensive than you might think. If you're buying in bulk – and by bulk I mean a container-load of 50-plus pallets of 1,200 bottles apiece – you can probably get your bottles for 15–16p each, with delivery thrown in. If, though, you can only afford and store a pallet or two at a time you'll have to pay top dollar, with £45–50 a pallet delivery on top, so you may be paying more than 20p for a simple brown 500ml bottle. (Always use brown glass, by the way: beer kept in clear or even green bottles is affected by ultraviolet light, producing an off-flavour known as "light-struck".) Crown corks cost a penny apiece in quantities of 1,000 or less; you could pay less than half that if you could afford and store a minimum order of 10,000. (From this I think you're beginning to see that you could get your consumables much more cheaply if you joined forces with two or three other local breweries.)

Labels

Labels are more problematic still. In bulk, they're quite cheap – but bulk, in this case, does mean quite improbable quantities. There are several firms that specialise in typesetting and printing reel upon reel of bottle labels, but the origination can be expensive. Designs stored on your personal computer (and do get a professional designer: amateur labels are rarely good enough in a market where presentation is much more important than many small brewers think) can be emailed as PDFs or taken on a disk to a local commercial printer and turned into self-adhesive labels at 8–9p each. The drawback is that these have to be applied by hand (the bottles carry a line to help you position the labels accurately), which you may well find repetitive and time-consuming enough to make you rethink your new career entirely. However, an entry-level automatic labelling machine can set you back £1,800, so perhaps it's something for later on. And it's worth reflecting as you toil that the bottle, label, and cork for a package that can retail at £2.50 should only cost about 30p.

A final point on hand-bottling: the expiry date. There is no legal formula for deciding this: most breweries opt for 12 months, or 18 months for beers of 6 per cent alcohol or more. But it's up to you. Set too short an expiry date and you can have trouble selling the beer to wholesale and retail accounts who are especially picky; too long and you risk beer being on the shelf when it really shouldn't, which can harm your reputation.

Which brings us neatly to bottle-conditioned ales (BCAs): in the case of the strongest BCAs, expiry dates are a complete irrelevance, since the beer is still gently working and often improves over 10 or even 20 years.

Bottle-conditioned Ales (BCAs)

Bottle-conditioned ales are those that are unfiltered, unpasteurised, and uncarbonated, and bottled with a certain amount of live yeast – either the original yeast with which they were fermented, or a special flocculating culture that will sink to the bottle of the bottle and form a sediment. Sometimes they are also primed with a tiny dose of sugar to help the yeast continue to work.

Some customers don't like the sediment, but if the beer is poured carefully the yeast should stay put in the shoulder of the bottle. Actually, though, it's packed with Vitamin B and is very good for you; its only deleterious effect is that it makes the beer cloudy in the glass. An old trick is to pour the beer carefully and indeed theatrically, expertly leaving all the sediment behind – then to swirl the bottle with a triumphant flourish and tip the yeast into your otherwise brilliantly clear ale.

Bottle-conditioned ales are promoted by CAMRA as "Real Ale in a Bottle" or "RAIB" (although what's wrong with the technical expression is beyond me: RAIB is a singularly graceless set of initials), and are now quite common, but until a few years ago they were in danger of dying out. When CAMRA was founded in 1972 the only BCAs (a far less clumsy expression than RAIBs, so I shall stick with it) on the market were Guinness, Gale's Prize Old Ale, Courage Russian Imperial Stout, Eldridge Pope Thomas Hardy's Ale and Worthington White Shield. Bottled Guinness is now pasteurised and carbonated, and has been for over a decade. Gale's has been bought by Fuller's, which seems disinclined to produce any more of the marvellously rich and vinous 9 per cent alcohol Prize Old. The last vintage of Imperial Russian Stout was brewed in 1993, although there have been periodic rumours of another brew. Eldridge Pope of Dorchester first brewed Thomas Hardy's Ale in 1968 to commemorate the centenary of the novelist's birth but shut down long ago; the beer is now brewed across the border in Devon by O'Hanlon's. Finally, Worthington White Shield is a classic Burton pale ale whose production has been shunted from brewery to brewery over the last decade. It was briefly contract-brewed by King & Barnes in Sussex, but when King & Barnes was closed White Shield went home to Burton, where it is now produced at the small demonstration brewery (part of the Coors Brewing Museum) that bears its name.

That was then. Thanks almost entirely to microbrewers, bottle-conditioning is now popular again, and thank goodness for that. For BCAs are usually much more complex and less gassy than pasteurised and carbonated beers, and some of the stronger ones can be laid down like a claret and will improve for years. CAMRA's Good Bottled Beer Guide listed over 600 examples from 143 breweries and the number continues to grow (and most CAMRA festivals these days have a bottled beer bar that stocks only BCAs, which means a few more sales and a bit more exposure for you). Some brewers won't touch BCAs because they think they're tricky to bottle and are easily spoilt, but keep everything spotless and your BCAs should be fine. Opting to bottle-condition your beers will also save you the not inconsiderable expense of installing filtration, pasteurisation and carbonation equipment.

So you've bottled your beer. Now it's time to sell it.

Supermarkets

Supermarkets generally are not regarded with much fondness by microbrewers, who often claim they are either exploited or frozen out altogether. To be fair to the big multiples, though, there is only a limited amount of shelf space they can devote to premium bottled ales (PBAs), although the fact that their wine aisles are several times longer than their beer and cider aisles seems bitterly unfair. And it's logical that most of the space allotted to PBAs (a fraction of the beer aisle, most of which is filled with "slab packs" of cut-price canned lager) should go to beers from the regionals, which are cheaper for the supermarkets to buy and just as attractive to an undiscerning public.

Most of the bigger chains are beginning to shift, though, if only at the pace of a turning supertanker. Many of them now have "regional lists", and Asda has even joined SIBA-DDS. Tesco and Asda also run competitions whose winners are assured a six-month national listing, which often leads on to listings with other chains and a good steady national sale. But there are some 600 brewers banging at the buyers' doors, and without a seismic and frankly unforeseeable change in their attitude to beer retailing, they are never going to find shelf space for more than a couple of dozen of them.

And frankly, even if you do get on a supermarket's regional list, it can be a mixed blessing. The surge in demand can be hard to satisfy and, unless you get a batch or two of beer expensively contract-bottled, may mean letting down your regular customers – always something to be wary of. As with the national wholesale giants, getting paid can be a nightmare: there's a huge bureaucracy involved, in which it's all too easy for your invoice to get lost – more so if you're not a regular supplier – and mistakes once made are hard to get straightened out. Then again, it may boost your sales volumes if your beer is put on a four-for-five promotion or even declared a BOGOF, but don't think the retailer is absorbing the discount. It gets passed down to you, and even the slim margin the supermarket allows you can easily be swallowed up in this way.

Having said that, the supermarket buyers I know say they don't deal with regionally listed micros quite as harshly, if that's the right word, as they do with bigger, more efficient, brewers. They say that once they've decided that stocking a range of local products is in their commercial interest, they'd be stupid to put their local suppliers out of business. Hence the margins should be somewhat better for small local brewers, and their beers are unlikely to be put on promotion.

And of course, some independent brewers do very well out of the supermarket trade. Refresh UK, which supplies the Brakspear, Wychwood and Duchy Original brands and is perhaps not so small, is hugely successful in this field; last time I met any of its directors, they weren't looking too hard up. Hop Back Summer Lightning, Arran Blonde and Coniston Bluebird are also pretty ubiquitous. What these brewers have in common is the efficiency to work on slim margins, promotional discounts and long credit periods. In short, they're geared up for it.

But suspicion remains: for every supermarket-friendly small brewer there are a dozen who say from experience that no spoon is long enough to sup with this particular devil. Anyway, don't build the supermarket trade into your business plan: it'll be a year or two before you're ready to start submitting samples to them, if then.

Local Stockists

Far more important to the start-up small brewer – and, indeed, the established small brewer – is to build up a strong local trade. Looking round the retail scene in your trading area, this may not at first seem too promising a strategy: you can get canned Carling anywhere, but how many off-licences stock a local beer?

That's in large part due to the fact that the big cash and carries used by many independent retailers are even less micro-friendly than the supermarkets. But in East Anglia there's a dedicated band of small brewers who have joined forces to create an alternative market for their bottled beers, under the radar of the supermarkets and cash and carries, and perhaps everybody could learn a few dodges and wrinkles from the East Anglian Brewers' Co-operative (EAB).

EAB was founded in 2002, principally as a buying consortium, and now has around 40 brewery members. It sources all kinds of equipment, services and ingredients for its members, including barley from one of the region's top growers, which is specially (and separately) processed by Crisp's, a traditional floor-maltings at Great Ryburgh, Norfolk, thus guaranteeing traceability. EAB soon evolved from a buying consortium into a marketing operation helping to build collective sales of its members' bottled output, although their cask beers still compete freely in the pub trade. The philosophy is simple. While a pub may have only three or four handpumps and can only stock beers from one or two brewers at a time, beer shops and off-licences benefit from having as wide a variety of beers on the shelf as space allows. Therefore, while the EAB's members fight merrily for pub accounts, they see the wisdom of collaborating in the take-home trade. Being a "secondary co-operative", or one made up of independent members, EAB also qualifies for grants and other help from the region's Social Enterprise Board. It's been a highly successful experiment, albeit one that small brewers in other regions haven't copied.

When EAB was founded, the take-home trade in the region was as hard to penetrate as anywhere else. The group therefore set about targeting a new area of the trade: farm shops, garden centres, gift shops, post offices, delicatessens, camp-site shops and bars, traditional butcher's shops – any retailer, in fact, with a demand for genuine craft-made local produce. Many of these outlets aren't licensed: EAB therefore pays for them to get licences. The £400 it costs can quickly be recouped, and a whole new layer of retail opportunities is gradually being created.

Even if you're not a member of a co-operative like EAB, it can still be worth your while paying a shop of this kind to get a licence, and if you're paying, you can then institute an exclusive supply arrangement – a sort of tie, in fact. You might even share the cost with a local craft cidermaker or vineyard whose products will complement yours rather than compete with them. And if the shopkeeper is still a little wary, there's another step you can take: instead of selling them the beer in the normal way, you stock their shelves yourself and pay them commission on everything they sell. That way, the shopkeeper doesn't even have to pay upfront for stock and is taking no risk whatever. An inexpensive advertisement or two in the local CAMRA branch newsletter, with the cost perhaps shared by the shopkeeper, should generate enough custom to make the venture worth everybody's investment.

Farmers' Markets

Farmers' markets grew up as an evolution of farm shops, and in response to the difficulties farmers encounter in getting a decent price for their produce from the country's high street retailers. There are now around 500 regular markets throughout the UK, half of which are certified by the National Farmers' Retail & Markets Association (FARMA - www.farma.org.uk).

In their purest form, farmers' markets such as those certified by FARMA insist that stalls are only available to bona fide growers and producers, and there is sometimes an additional requirement that stallholders must be genuinely local. This is a source of much contention, often reaching Byzantine levels of intricacy. The purists say that shoppers who are led to believe that they're buying only local produce need to be protected from carpetbaggers who aren't really producers and aren't really local. But many small independent traders say that it only needs to be clear that all the stallholders are small and independent, and complain that they are often excluded from farmers' markets on mere technicalities. You may well think that the ideology of farmers' markets obscures their real point – that they're not Tesco; that shoppers are guaranteed a range of interesting, unusual and top-quality products; and that producers frozen out of the mainstream can find an outlet for their products. Be that as it may, farmers' markets are proving a useful outlet for more and more small, independent brewers.

The advantages are many. The overheads are low, for a start. Your stall can be as simple or as elaborate as you wish, but even the most ornate is only a comparatively modest one-off cost, and your pitch fee is likely to be £50 or less – much less, in some cases. Your margin will be good: one small brewer I know charges £2.50 for a 500ml bottle of a 5 per cent ABV beer, which is 70p more than a supermarket would charge for a similar product; and of course without the middleman, all the profit margin is his to keep.

There are drawbacks, though, says this same brewer. Markets are quite expensive in terms of time, he warns. Don't forget that you have to put up and take down your stall, which can add many hours to the proceedings. You also want to be the only brewer at that particular market: competition from a local cidermaker or vineyard is one thing; but there's only so much people can carry away from the market, and if they're buying beer you want to be sure it's your beer. Then again, you are very much in the hands of the weather and the organiser. A run of rainy market-days can see takings plunge, and if the organiser is no good at promotion you can stand there for an awfully long time without any customers. My friendly brewer's advice, though, is to stick it out for a while – if your first stall at a particular market doesn't work, give it two or three more attempts. Your first bad day may have been a one-off, so persist for long enough to see whether it's usually a busy event or not.

In licensing terms, you'll need a personal licence to run your stall (see page 161). What sort of licence the promoter needs is a much more complicated question. It could be argued that the marketplace needs a premises license, which is a virtual impossibility. It is theoretically possible to licence an open space, but the owner of the marketplace (normally the district council, which also happens to be the licensing authority) is unlikely to go to the trouble. On the other hand, it might be permissible to hold a farmers' market with only a temporary event notice (TEN), and indeed many are held on precisely that basis. However, a TEN only allows for 500 people to attend at a time. Either way, there is a "due diligence" defence for the stallholder, and provided you've done your absolute best to ensure you're operating legally, you'd be very unlucky to be prosecuted. As to whether the licensing authority would prosecute the promoter or the landowner (which is usually, as we've seen, the licensing authority) – well, Kafka must be chuckling in his grave. Meanwhile we await either a test case or, preferably, new guidance from the Department for Culture, Media and Sport (DCMS), which inexplicably is in charge of liquor licensing.

Assuming the current state of uncertainty in the licensing regime hasn't put you off, you need to find your nearest farmers' market. The FARMA website mentioned above is the best place to start, but it's an unregulated and largely unorganised (as opposed to disorganised) activity, and simply Googling "farmers' markets" will throw up many useful sites such as www.lfm.org.uk (farmers' markets in London) and www.scottishfarmersmarkets.co.uk (farmers' markets in Scotland). Other than that, it's largely a matter of local knowledge. Is there a farmers' market in your area? Has it got a brewery stall? If the answers are (a) yes and (b) no, then pounce.

CASE STUDY

Norfolk Brewhouse

The Norfolk Brewhouse is owned and run by Rachel and David Holliday – a couple with a passion for crafting distinctive, high-quality beers.

Founded in May 2011 the brewery has breathed new life into a redundant old barn, on a family farm, in North Norfolk right in the heart of the country's finest malting barley region

They make the most of all that is around them - using chalk-filtered water from the brewery's own well and using Norfolk's finest malting barley.

Even the names of the beers – Moon Gazer Ale and StubbleStag premium lager – were inspired by the wildlife around them, reflecting as they do the stunning sight of wild hares transfixed by the moon.

Taken from their excellent blog on their website **www.norfolkbrewhouse.co.uk** there follows a stage by stage account of how the microbrewery got up and running.

May/June 2011

Work starts and very importantly up goes the sign board announcing the arrival of The Norfolk Brewhouse.

June 2011

The barn was not big enough so, it was time to knock down one of the old walls, allowing us to extend out. By now we were working with AB UK brewery equipment suppliers.

Grateful for their help and support we gave them a big cheque to send to DME in Canada who would be fabricating our purpose built Norfolk Brewhouse.

July 2011

The availability of pure, naturally chalk-filtered water is one of the key advantages of locating The Norfolk Brewhouse at the family farm. Down the piling went to about 190 ft before crystal-clear and very cold water flowed up from the underground source. Not only was it fantastically clear it was also very cold – a feature which will allow us to cool down the wort after boiling – saving a significant amount of water and energy.

This chalk-filtered water is the starting point of all of Norfolk Brewhouse's real ales and lagers. Norfolk ale, hand-crafted using Norfolk water, now that can't be bad now can it?

August 2011

The main structure of The Norfolk Brewhouse has taken shape, and builders have moved inside to start the internal works which would transform the old barn into a modern brewhouse.

Passers-by must be wondering just why we have what looks like a submarine in the grounds. Well, with brewing beer comes lots of water. On average you can use six pints of water to make one pint of beer – maybe a bit more, or a bit less, depending on how efficiently you plan and run your brewhouse. So the submarine needs to trap all of that water.

Our brewhouse is in a very rural part of the county and feeds into a fairly old, small-scale waste water works at Langham. This means that, in order to meet with Anglian Water's waste water consent, we are obliged to fit a very large settlement tank.

This allows us to catch, monitor and control the water before it enters the Langham plant.

December 2011

ABUK are here along with Tim from Greens – the equipment fitting company – closely followed by a very large truck with a mash tun, copper, hot liquor tank and auger and more equipment arriving later in the month.

February 2012

What a day – it's brew day. There were times when we thought it would never happen but today was the day. An early start – Scott Barnet, our brewing 'commissioner' from Quartz Brewery, got us started at 7am and by 11am the steam was bellowing into the clear blue sky, and the aroma of malt and hops was a true delight but come six in the evening we cracked open a beer and toasted the arrival of The Norfolk Brewhouse.

June 2012

After three weeks fermenting and three months conditioning the wait was finally over. Today is the day when StubbleStag Norfolk lager was served on draught for the first time.

October 2012

The Norfolk Brewhouse debut at the Norwich Beer Festival. Our beers on offer included; Moon Gazer Amber ale, Moon Gazer Ruby ale and StubbleStag cask lager.

March 2013

A new Wetherspoon pub in Fakenham will be serving a new and exclusive locally brewed beer from The Norfolk Brewhouse

May 2013

Members from the CAMRA Norfolk & Norwich Branch raise a glass at the first official brewery tour for The Norfolk Brewhouse.

www.norfolkbrewhouse.co.uk

Specialist Shops

Dotted around the country are 100 or so specialist beer shops, ranging from the enormous Beers of Europe, www.beersofeurope.co.uk in its vast complex of former granaries at Setchey just outside King's Lynn to the tiny Beer Essentials squeezed into a pedestrianised street in Horsham town centre.

Their number is slowly growing, but the amount of beer most of them can realistically be expected to order is measured in cases rather than pallets, and getting a few cases to an outlet on the other side of the country is prohibitively expensive. There are a handful of wholesalers that specialise in bottles, such as Beer Paradise, www.beerparadise.co.uk, also the parent company of the Yorkshire-based BierRitz off-licence chain (mailto: sales@beerparadise.co.uk). So far, though, no one seems to have investigated the same sort of beer swaps for bottled beers that Crouch Vale and others do for cask beers; perhaps it's time someone tried. In the meantime you might consider pooling deliveries with one or more of your small brewing neighbours to target specialist shops in other parts of the country. If you can jointly rustle up an order of 60–70 cases (and in some cases, small shops will "share" orders with others) it's worth making up a pallet, which can be delivered anywhere in the country by a groupage distributor such as Palletways or Palletline for less than £50.

For complete details of specialist off-licences refer to the Good Bottled Beer Guide by Jeff Evans published by CAMRA.

Hotels and Restaurants

Brewers large and small have long bemoaned the fact that however good their beers are, the foodie end of the on-trade simply won't take them seriously. Wine lovers are well catered for: no restaurant would be caught dead with Piat D'Or or Blue Nun on its list. But ask for a beer in a restaurant with a wine list that starts at £20 and you're likely to be offered a bottled international lager, take it or leave it. Chefs and sommeliers are, by and large, horribly ignorant about beer: I was consulted on what real ales Gordon Ramsay Holdings should stock in its new pub in north London, the Warrington, and Ramsay's executive chef suggested Caffrey's! I also once, many years ago, had the temerity to complain that there were no American beers on sale in the American Bar at the Savoy; I might at least have expected Budweiser or Miller, although I suppose Sierra Nevada or Rogue would have been too much to hope for. I got a very haughty reply saying that these beers were too expensive. So does the Savoy buy its wines on price?

National newspapers are willing conspirators in this neglect of beer: all the broadsheets have wine writers; not one has a beer writer. And if, grudgingly, the food and drink editor condescends once in a while to include a piece about beer, the writing is generally deputed to the wine writer. This doesn't happen in Belgium, where top chefs as well as humble proles place craft-brewed beer among the nation's glories and regularly use it in cooking, as well as stocking it in their restaurants. In desperation, the British brewing industry declared 2004 as "The Year of Beer and Food", laying on food and beer banquets, dishing out recipe and beer-matching cards, and promoting food-friendly beers alongside the fine and not-so-fine wines in the dining rooms of their managed houses. It was a promotion that frankly sank without trace, not through lack of effort or lack of merit, but through the total indifference of the media and the catering trade. True, a couple of high-profile restaurateurs sourced pretty skimpy beer lists almost as a token gesture. But the fact that Michel Roux jnr turned to InBev for all the beers on the highly publicised list he introduced at Le Gavroche demonstrates the scale of ignorance about beer at the top end of the catering trade.

Bottled beers ought to be the ideal way for hotels and restaurants to introduce foreign visitors and native gastronauts alike to the "wine of the country". Bottles are easy to store and dispense; they have a long shelf-life, so they won't go off; and they can be bought tentatively, a case or two at a time, so the financial risk is no more than the chef's teenage daughter's pocket-money. And for a restaurateur, a beer list alongside the wine-list creates consumer interest and is a good point of difference.

These are strong plus points with which enterprising small brewers can tackle the more enlightened hoteliers and restaurateurs in the district. But there's more: your beer's local provenance ought to be another attraction in these days of eco-consciousness; and if that isn't enough to make the sale, you could always offer the same deal you've offered the farm shop: no money upfront, but a commission on sales. But if this is a market you've decided to try, you might have to work harder still to make your sales. Offer to create a beer list tailored to the individual restaurant's cuisine, with full tasting notes and a description of each beer (not too gushing, though – adjectives exist to shore up a bad product, and a good product shouldn't need them). If it doesn't work, you've wasted nothing but time; if it does, you've got your beer into a high-profile outlet that will benefit your reputation as well as your balance sheet.

Mail Order and Web Sales

Mail order has always been a classic route to market for small manufacturers and traders in all sectors, including food and drink, who for one reason or another have found it difficult to break into the high street. And since the arrival of the internet mail order, or more properly e-commerce, has become an even more important alternative sales platform.

For obvious reasons, though, bottled products and couriers tend not to mix. Even those who claim to be specialists in the field can be more than a little heavy-handed. This has not deterred wine merchants, many of whom have long been enthusiastic exponents of mail order. But there's a good reason for their enthusiasm. Packaging bottles to withstand the rough handling they will undoubtedly suffer is an expensive and time-consuming business. Couriers aren't cheap, either. But when you're selling high-value merchandise such as a case of 12 bottles of good-quality wine at £6 or £7 the bottle, the additional cost of packaging and delivery is a fairly, or at any rate acceptably, small addition to the price and the consumer is happy to pay it. However, the costs are exactly the same for a case of 12 bottles of beer at £2 or £3 the bottle, and suddenly the value doesn't look quite so good.

The cost of sending cases of beer around the country, and more particularly its effect on the consumer's perception of value, is one factor that deters many small brewers from getting involved in e-commerce. Dealing with couriers is an even greater one. It can take a long time to find a courier who understands that glass is fragile, according to Stuart Evans of Uncle Stuart's Brewery of Norfolk. And trying to claim insurance for the inevitable breakages is such a tangle, he says, that given the small sums involved it's not worth the effort. In fact Uncle Stuart's is getting out of ecommerce altogether, having tried a long list of courier companies and every variety of protective packaging under the sun.

Not all small brewers have had such an unhappy experience, though. Pitfield Brewery has had a profitable sideline in mail order sales for a quarter of century, and brewer Andy Skene says that after trying many couriers, the proportion of breakages is now well within acceptable limits. The couriers' most annoying habit, he says, is not informing the brewery of breakages. Either they're not delivered at all and are left in a corner of the depot, with brewer and customer wondering what has happened; or they're delivered, rejected, and returned to the depot as above, with the brewer not knowing anything is wrong and the customer fuming for a replacement or refund.

Having said that, there are still one or two highly successful ecommerce beer sellers, probably the biggest being Beers of Europe (BoE) of Setchey in Norfolk. It did take a lot of chopping and changing before the company found a courier it could trust (it won't divulge the company's name, though); and BoE's packaging is second to none.

Not only is BoE a big company itself (by the standards of the sector, at any rate), it's also part of a much bigger one. That means there's a lot of waste cardboard on its vast site – enough to justify investing in a shredder that cost "several thousand pounds". BoE now delivers its beers in two boxes, the outer being of chemically toughened cardboard, and the whole parcel is stuffed with shredded card which not only cushions the contents against the courier's buffeting but also efficiently soaks up the spillage when a bottle does get broken. This has proved so successful that BoE now collects and shreds waste cardboard from other factories and warehouses in the area. It's not without cost, though: BoE charges its customers a standard £7.49 per delivery, of whatever size. In theory the smaller orders subsidise the larger ones; actually the charges never quite cover the expenditure. On the plus side, the standard charge encourages customers to place bigger orders, so it's six of one and half-a-dozen of the other.

In conclusion, is ecommerce worth your while? Having spoken to two or three mail order/e-commerce specialists, I think Uncle Stuart's has been exceptionally unlucky; perhaps the way to find out whether your luck is equally bad is to give it a try. Certainly there are enough brewers and retailers engaged in selling beer around the country by the case to suggest it's worth their while. And thanks to the world wide web, you can now dip a toe into the mail order business without too much financial risk. In the past you had to advertise widely, persistently, and expensively to get a worthwhile response, but these days your website should generate enough orders to give you sufficient experience to base a longer-term decision on.

A final word on mail order, and perhaps one that might sway you one way or the other: it's specifically mentioned as a licensable activity in the latest guidance notes from the DCMS, and that means you now need a premises licence to do it. Having said that, getting a premises licence for a mail order operation where the public don't actually come on site should (according to my local council's licensing officer) be a pretty simple procedure, and you're probably getting a premises licence anyway to enable you to hold open days and beer festivals.

Chapter Eight
A Place of Your Own

CASE STUDIES

There's a widely held view in small brewing these days that anyone who opens a brewery without a pub is mad. That's because the sector has become so competitive that custom is increasingly hard to come by and wholesale beer prices have actually fallen to £50 a firkin or even less.

As we saw in Chapter Two, though, owning or leasing a pub of your own isn't necessarily the key to success; in fact the opposite can be true. For a pub is a huge investment – bigger, almost certainly, than the brewery itself; and if it turns out to be a dud, or if you're unlucky enough to appoint an unreliable manager, it can drag the whole enterprise down. The truth is that running a brewery and running a pub are quite different disciplines; despite the superficial appeal of guaranteed retail, most new brewers prefer to concentrate on what they're good at, for the simple reason that they're good at it.

Nevertheless, local brewers are either based at or very closely associated with about 130 pubs. In many of these cases, a publican has decided to turn brewer. Or, as we saw earlier, a brewer has found that a pub's outbuildings make a convenient home. And then there's the handful of local breweries including Beartown, Bath Ales, Ringwood, Butcombe and Hop Back that have made conscious decisions to go into retail wholeheartedly and have set about building up miniature tied estates of their own, although few own more than half-a-dozen or so pubs. But between them, small brewers only own a combined total of about 350 pubs, or 0.7% of the national stock; so they haven't exactly been falling over themselves to invest in licensed real estate.

Still, and even if the hundreds of thousands of pounds it takes to buy a viable or potentially viable pub is a bit rich for your blood, the fact remains that building up a decent delivery route gets harder and harder, making it a good idea to take as much control as you can over the retail of your beer. A pub of your own, though, isn't the only way to achieve it. But the whole process starts with getting a licence.

The 2003 Licensing Act

The Licensing Act 2003 which came into force in November 2005 brought fundamental changes to the system. Its most highly publicised provision, and the original idea that underlay its long and tortuous journey from green paper to enactment, was the abolition of the century-old concept of "permitted hours". Pubs can now, in theory, open whenever they choose – provided, of course, the licensing authority agrees.

And that was the Act's second main provision: it took the licensing function away from the magistrates and awarded it instead to district and borough councils. What this means, in practice, is that there is much wider and more detailed consultation than in the old days. The Act not only demands that licensees understand and uphold public policy on alcohol-related disorder, protection of minors, and so on; it also ensures that the police and fire services as well as the local community are closely involved in granting of liquor licences and imposing conditions.

In one sense this is a headache for licensees. Getting a new licence, or having an existing one varied, is far more costly and time-consuming than it used to be. On the other hand, you only have to jump through the council's hoops once; when you've done it, you're on a much more secure footing than you used to be. Provided you respect the terms of your licence, it's hard for the police or neighbours to get it reviewed or even revoked altogether (although the Act does give the police powers to close "trouble pubs" on the spot). You've filed your flight-plan, as it were, in exhaustive detail and everyone concerned has had a chance to suggest alterations and restrictions. Everything has been openly negotiated and agreed, and as long as you stick to the operating schedule that the council has approved, nobody (in theory, at least) has a beef.

Another important provision of the 2003 Act was to split liquor licences in two, so you now have to have a personal licence as well as a premises licence. Ill-informed critics claimed this merely added another layer of bureaucracy to an already growing pile; in fact, it ended the cumbersome old system of licence transfers and made it much easier for licensees to migrate from pub to pub. In the old days, when there was just the one licence, a "protection order" had to be granted whenever a pub fell vacant in order to allow it to trade while a new licensee was appointed. This meant a day in court for someone – normally the brewery area manager – and another when a new tenant or manager was finally appointed. If the new appointee was found to be inadequate for some reason, the procedure had to be gone through again; if a pub was proving hard to let or was going through a rough patch, there could be several licence transfers in the space of a few months. Magistrates were often highly critical when they had to hear serial transfers for the same premises, which they occasionally even threatened to close down. Under the new system, the publican and the pub have their own licences, and the whole fiddle-faddle of transfers has been almost entirely eliminated.

Personal Licences

A personal licence is not all that hard to get, and it's worth your while holding one even if you have no immediate plans to acquire a licensed premises. Possession of a personal licence removes an otherwise almost insuperable bar to holding beer festivals or open days at the brewery; and it also allows you to retail liquor on other occasions – outdoor bars or beer-tents at the local village show, for instance. You can also hold the licence for a third party's beer festival – the local Lions Club or Rotary, or even the CAMRA branch of which you ought, by now, to be a member – where your beers will, of course, be prominently on sale. And if, as many local brewers do, you run outside bars at public functions on unlicensed premises – a ceilidh at the village hall, for instance – a personal licence is now indispensable. It is also usually necessary if you sell your beer at farmers' markets: very few of them are held in venues that have premises licences.

National Certificate for Personal Licence Holders

The personal licence is issued by your local council and costs £35. Before you apply, though, you'll need a National Certificate for Personal Licence Holders, and to get one of those you need to attend a one-day training course which has a 40-question multiple choice exam at the end. This will cost about £200. The British Institute of Innkeeping accredits some 500 training providers who run courses at over 5,000 locations all over the country – to find one, visit www.biiab.org. These training providers did not appear overnight when the Act came into force; for some time previously, magistrates had been stipulating that the requirement for applicants to demonstrate that they were "fit and proper persons" included not only a clean record but also some modicum of training. Many of those who set up as trainers were extremely experienced publicans or brewery staff, so the Act's training requirement really only codified what had been existing common practice. Many times the fear has been expressed that this compulsory training would prove a turn-off to the kind of individualistic mavericks who had often made top-class licensees in the past. But a quick look at the syllabus reveals it to be an extremely practical affair and one that even the most colourful of would-be publicans should be able to complete without having the individuality entirely squeezed out of them.

The sheer number of training providers means you can shop around for the cheapest, but before enrolling do make sure that they actually teach the one-day course, rather than just the five-day National

Certificate in Licensed Retailing course. Not all of them do. For general information on the subjects covered by the course click on "find a qualification" on the left-hand side of the BIIAB home page; then click on "search" under the drop-down box at the foot of the page; and finally click on "Level 2 National Certificate for Personal Licence Holders". To gain a personal licence you will also need a Criminal Records Bureau check, which costs £31. Your local authority's liquor licensing department should give you an application form before you apply for the licence itself.

The best explanation of the 2003 Licensing Act I have come across is at **www.opsi.gov.uk/ACTS/en2003/2003en17.htm**. It's long, but well worth studying in detail, because even if you have been too ambitious on your plans to acquire your own pub, there are other uses for a premises licence.

Premises Licences

Your personal licence gives you leeway to indulge in a limited amount of retail on your premises – you can apply to hold up to 12 brewery open days or beer festivals a year, for instance, by submitting a temporary events notice. Increasingly, though – and especially if you've found a location that is worth visiting in its own right – some more regular form of retail is part of the local brewer's business model. It might just be selling polypins of bright beer at the door; it might be running a brewery shop; it might be establishing a visitors' centre with its own bar; it might be running a thoroughgoing brewery tap open to all and sundry. For any of these activities, you will need a premises licence; for among its many revolutionary aspects, the Licensing Act 2003 swept away two previous distinctions that made a retail operation of limited scope an easy option.

Licence Definitions

The first change was to redefine the term "wholesale". Before the Act came in, this was an expression of quantity, which meant that Majestic Wines outlets didn't need to be licensed because they sold alcohol only in wholesale quantities – by the case. It also meant that breweries and cider farms could sell wholesale quantities – a 36-pint polypin or a five-gallon polycask – to members of the public without a licence. Not any more. The Act has redefined "wholesale" as sales by a producer or distributor to licensed retailers, not to the public – a definition which, you may feel, better reflects the real meaning of the word.

The second was to sweep away the distinctions between full on-licences (for pubs), restricted on-licences (for restaurants) and off-licences. Getting an off-licence used to be a virtual formality – an application supported by a site map would be followed by an inspection by a panel of magistrates, and then a five-minute appearance in court with the payment of a nominal fee. I myself held an off-licence for a while, and getting it was simplicity itself. But all the different categories have now been subsumed under the general heading "premises licence", and the procedure for getting one is the same whether you only intend to sell polypins at the back door or to turn part of your premises into a pub.

As mentioned earlier, the new-look licensing procedure is principally about consultation and ensuring that you understand your responsibilities. There's a fearsome-looking 16-page form to fill in, but if you download it from your local council's website and give it a thorough read you'll see that much of it isn't relevant to you. That's one of the (admittedly slight, in this case) drawbacks to the new one-stop licensing procedure: a single form has to cover every licensable activity including showing films, putting on plays, promoting boxing matches and so on, which makes it a pretty bulky document. But most of it isn't concerned with liquor licensing at all and need not concern you, and the trickier sections, such as those to do with carrying out the council's licensing objectives, will be covered in your training course. (By the way, and although the form doesn't say so, the section covering "performance of dance" specifically excludes folk dancers such as the local morris side and their more extreme variant, molly dancers.)

The form, then, is fairly simple – you shouldn't need a solicitor to help you fill it out, and it should only take you a day or at most two. But the form is only the last part of the application process. You will also need to advertise your application in the manner prescribed by the licensing authority (which is not a new requirement); you will need to submit an accurate and intelligible plan of the premises, although hiring a proper draughtsman to do it probably won't be necessary; and you will need to submit copies to the "responsible authorities" (fire, police and so on) as stipulated by the licensing authority. Just as important as correctly observing the formalities is to consult as widely as possible in advance, both with neighbours (and the term "interested parties" includes schools and businesses as well as residents) and with the "responsible authorities" and the council's licensing officers. As with your planning application, if you do your homework thoroughly the application you finally submit should be bomb-proof.

Council Licensing Officers

The key to dealing with council licensing officers, as with planning officers, is to be calm, rational and friendly, and not to regard them as jobsworths to be either bamboozled or placated or both. If treated with respect they can be – and indeed most of them want to be – extremely helpful, and can give you the advice that makes sure your application succeeds. They can suggest ways in which the four licensing policy objectives can be met, many of which you may not have thought of. They will give you many invaluable tips, such as how to present the general description of the premises (section 3 of the application form) and the more detailed operating schedule as flexibly as possible, so you don't have to keep varying your licence whenever you want to try something new. They will have issues of their own to raise, too: you might, for instance, be in or near an area where public drinking is prohibited, and they may ask what steps you propose to deter customers from buying takeaways to drink in the street. (The best answer to this is to stress that your shop, if you have one, won't be selling cheap cans of lager or cider but extremely expensive bottles of hand-crafted ale – although once you've made this commitment you will be bound by it.).They may very well also want to know what provisions you have in mind for parking, toilets, noise abatement, control of numbers and public disorder contingency planning should you be hoping to run on-site beer festivals or open days. Going into this kind of detail may seem a bit previous, but it's as well to have considered such matters as site layout, advance publicity and limiting numbers even at this early stage.

Brewery Tours

More and more, brewers large and small are treating brewery tourism – guided tours and visitors' centres – as profitable parts of their businesses.

Brewery tours have been with us for many years, but have traditionally been seen, by established mainstream brewers and newcomers, as a sort of bolt-on – something you do as an after-hours profile-raiser for your trade customers and local CAMRA branches.

Typically, the brewer or a member of the brewing staff would be detailed to conduct the tour, which would be followed by a few sandwiches and some free samples in any space that happened to be convenient and available. Larger breweries might take on a retired employee or simply a knowledgeable enthusiast to conduct the tours, but essentially the whole thing would be pretty ad-hoc. Some breweries

– especially those that had grown higgledy-piggledy over the years and were an obstacle course of steps, pipes, slippery bits and so on – were unhappy about giving tours at all and only did so as an occasional necessity. But demand grew, because breweries are interesting in a way that hosiery mills and plastic extrusions factories aren't (mainly because of the nature of the product!); slowly brewers came to see that here was a way of marketing themselves not just to key accounts and special interest groups but to the general public as well.

Planning the Tour

If you're going to run tours seriously you need to do it properly, which means devoting time, thought and investment to it. The very basics, in investment terms, are a comfortable and suitably decorated lounge where people can sit and enjoy their samples after the tour, and decent loos. And you need to devote some though to making the tour interesting, too: chances are, if you're a new brewery, that your site and plant aren't going to be too picturesque or have much history, so you need to work at making the tour as entertaining as possible. A confident, well-briefed and characterful tour guide who doesn't mumble but doesn't drone on and on either can make or break the whole thing (this is a branch of showbiz, after all). Samples of hops to smell and malt to chew are essential. Posters or wall-plaques explaining each step in the process create interest. Any "breweriana" you can rustle up adds atmosphere – difficult if you're a newcomer with no history of your own, but not impossible: CAMRA beer festivals usually host breweriana stalls and auctions.

The question with tours, though, is how far to take them. Are they still an after-hours bolt-on and even a bit of a distraction, or are you going to go the whole hog and organise and promote them properly? Often, demand for tours grows by word of mouth, and you may find yourself in the position of having a decision forced on you: are you going to turn down requests for guided tours you can't manage; or are you going to give in and provide the right facilities to meet the demand? And once you've done that, you have to sustain the level of demand, or whatever investment you've made could be wasted. The time to make up your mind about the potential of your brewery itself, then, is early on.

There's also the question of whether offering free samples as part of the charge for the tour constitutes a "licensable activity" under the 2003 Act. My local licensing authority thinks it isn't, because the main purpose of the activity is not the retail of alcohol. Others might think differently, and as there has yet to be a test case to decide who's right, it's wise to check what the local line of thinking is. And even those who think it isn't licensable will warn you against abusing their good will. If you are charging enough for a five-minute tour to cover seven or eight pints as well, it will be counted as "hidden retailing" – a tried and tested concept that goes way back to the days when shops were shut on Sunday except for the sale of perishables. (The test case was a furniture shop convicted of Sunday trading for selling bunches of carrots with free sofas thrown in.)

Brewery Shops

The next logical step is to add a small shop. Visitors like to have something to take away with them – a branded polo-shirt, perhaps, or a brewery beer glass. Or even some of its beer. A little souvenir and merchandise counter is easy enough to organise and adds value to the tour, from both the visitor's point of view and, of course, your own. But selling bottled beer or bright beer in polypins to take away is definitely a licensable activity and one for which you will require a premises licence as well as your own personal licence.

Many breweries now run a shop selling their own beers and branded merchandise and perhaps a little local produce – honey, cheeses, jam, chutneys and the like. It can be a valuable generator of cash-flow, but there are cost implications. The most obvious one is that you have to have someone to run it, even when it's not busy. And it has to be open at the stated times – nothing upsets shoppers, some of whom might have come a long way to buy your beer, like turning up when the shop says it's going to be open only to find it's shut. So someone has to be there all the time. If it's rarely busy, and you have someone at the brewery to run the office – to do the paperwork and take telephone orders – they might as well answer the bell too. The shop can also be the collection point for customers who have ordered polypins of bright beer for parties and other functions. The critical decision comes when the shop gets too busy to be run as an adjunct to your book-keeper's job description but not busy enough to justify a full-time employee. That's the point at which you have to decide which way to jump – whether to close down a profitable sideline, or invest in making it a separate profit-centre in its own right.

A handful of brewers have jumped in with both feet and expanded their on-site retail sideline into a thorough-going beer shop selling not only their own products but beers from other breweries, traditional ciders, country wines and imported speciality beers as well. It's a big investment, though: if you visit Hog's Back Brewery in Tongham, Surrey, or Tucker's Maltings in Newton Abbot, Devon (which is also home to the Teignworthy Brewery), you will find not hastily kitted-out sheds but beautifully designed and decorated boutiques with huge ranges of all sorts of wonders. They were not cheap to develop; they are not cheap to keep stocked; they are not cheap to staff; and they are not cheap to promote. Such places often acquire legendary status and generate significant custom simply by word of mouth, but you can't build that into your business plan!

Customer Car Parking

One key area for consideration when planning a shop is customer parking. There are two matters arising from this. One is that neighbours may object to your licence if they think it will mean customers will have to park in the surrounding streets. Another is that customers buy more when they can load their purchases straight into their car boots. Hackney Council's decision to ban parking in Pitfield Street was a contributory factor in the demise of the Beer Shop in Hoxton. But at Beers of Europe at Setchey near King's Lynn, which has enough parking for a fleet of trucks, customers browsing among the treasure-packed aisles aren't deterred from trying one of these and a couple of those by the knowledge that they've got to carry the whole lot 500 yards back to their cars. In addition, the knowledge that there's plenty of parking brings avid beer-lovers to Setchey from astonishing distances. So if you plan to make a shop a significant part of your business, you will need adequate customer car parking.

Brewery Visitors' Centres

The need to entertain tour parties saw sample-rooms morph first into shops and then into hospitality suites with bars and loos and displays of brewery memorabilia ferreted out from dark corners – when Greene King closed its maltings, for example, enough material came to light to fill an entire museum.

Breweries had often made a modest charge for tours, to cover the guide's pay and the cost of the free samples. But they had still seen tours more as a marketing tool than as a paying part of the business. Now they saw income from extra drinks for tour parties over and above the free samples, and sales of bottled beer and branded merchandise as profitable activities in their own right. In the last 15 years many established breweries have opened visitors' centres – there's a partial list in Chapter One – in a bid

CASE STUDY

A Brewery Licensed to the Hilt – High House Farm Brewery, Matfen, Northumberland

Steve Urwin and Sally never really planned to become brewers. They were quite happy running their 200-acre family farm on Hadrian's Wall, where they grew wheat and barley and ran small beef herd and a flock of 150 sheep. But then in 2001 foot and mouth disease struck, and the Urwins realised they needed another string to their bow if the farm was to survive.

At the time the Hadrian's Wall Walk had just been completed, and a public footpath ran to it through the farm, so the Urwins were quite used to casual visitors and serious ramblers hiking through their yard. And a lovely yard it is to hike through, too, with its early Victorian stone buildings, all Grade II listed, and its glorious setting in the rolling Northumbrian hills. It was those natural advantages that decided the Urwins not just to install a brewery and join the competition for the local free trade (although as things have turned out, High House has enjoyed enviable success in gaining local distribution). Instead, they wanted to share everything they had with the thousands of visitors who flock to the region.

The brewery opened in 2003, after an 18-month planning process in which the Urwins overcame the local council's deep reservations by lining up as much support as they could and getting their application honed to the last detail. Crucially, their supporters included English Heritage, which co-operated with their plans to reroof and repoint the sturdy farm buildings sympathetically to make room for a purpose-built 10-barrel brewery, and backed their application for listed building consent.

Creating a visitors' centre was always part of the business plan, says Steve, but while he originally thought it would take five years to get round to it, the brewery was so successful that in fact it only took three. And what a visitors' centre the Urwins have developed! There's scarcely a whistle or a bell that High House doesn't have except perhaps B&B and a campsite, which might have been pushing it a bit.

As well as tours at a nominal charge per head, there's an exhibition of the brewing process and the farm's history in the old malt loft (and incidentally, High House beers are all brewed from the farm's own barley, which is malted by Thomas Fawcett at Castleford in Yorkshire).

Then there's a well-patronised coffee-shop where light meals are also served; a shop that sells High House beers alongside beer mustard, beer chutney, beer pickles, beer chocolate and even beer ice-cream, all locally made; a bar which is open to the general public as well as brewery visitors; a farm walk; and a function room.

All this has cost the Urwins a pretty penny – £300,000 all told. But then they already owned the premises, and the buildings were in good condition to start with, which all kept the bills down. And it's paid off handsomely. Within minutes of opening in September 2006, High House had been named Rural Retailer of the Year in the Countryside Alliance's Diversification Awards and had won bronze in the North-East England Tourism Awards. And while the Urwins had reckoned on 3,500 visitors in their first year, they actually saw more than 9,000 in their first 10 months, despite the rotten weather.

And after a patchy relationship with the local planning authorities, Steve describes the licensing process as "one of the easier bits". "We got on very well with the licensing authority," he says. "They were happy to work with us, and we got a licence for off-sales, on-sales, entertainment and music, with opening hours from 10am to midnight every day."

Not that High House actually runs all the activities for which it has approval, but the Urwins took very good advice to apply for more than they needed. "Apply for more than you want because you never know how the operation will evolve, and it's easier to get it all in one than have to go back later on," says Steve. "Even though you might never use all the permissions you apply for, get them anyway. Having to go back is expensive and time-consuming, and you never know – you might get knocked back!"

www.highhousefarmbrewery.co.uk

to slot themselves into the wider tourism industry.

In time, then, your well-conducted tours and small shop counter might grow into a fully fledged visitors' centre, open throughout the day. This is a different level of operation – requiring paid staff, for one thing. Whether to take this step depends very much on your location and premises. If you're in a new unit on an industrial estate on the edge of a northern industrial town, the chances that you'll make much of a tourist attraction are pretty slim (although even northern industrial towns have their tourism initiatives these days, so don't dismiss the possibility out of hand). If you're in a historic barn on a picturesque farm in Devon, though, or in a converted 18th-century silk-mill in a quaint old Cotswold town, it's a very different proposition. Either way, before making plans, consult those who know. The local council, the local tourist board and Businesslink (never forget Businesslink!) will all be able to provide useful information on local visitor numbers, levels of investment in local tourism, and the possible availability of grants. The council and the tourist board will also be able to offer ongoing practical help in marketing and promoting your visitors' centre. You may even find that a sound and well thought out proposal to develop your brewery as a visitor attraction is a plus-point in your planning application – especially as promoting self-sustaining rural communities is one of the 2003 Licensing Act's stated objectives.

And finally, never forget the marketing value of welcoming people into your brewery and sending them away with a story to tell their drinking buddies and, hopefully, your logo emblazoned on their sweatshirts. You've entered the tourism industry now, and it's essential to send your visitors away satisfied. That means never doing anything half-heartedly. If visitors feel they've had a substandard experience at your brewery, the feeling of dissatisfaction will embrace your products as well – and they'll be telling their friends and neighbours bad things about you when they get home.

Your Pub

All this brings us back rather neatly to where we started – whether or not to invest in a pub of your own. Running a pub, as I have said, is a very different business from running a brewery:

- You will be face to face with the public, and if you're no good at it you'll soon come unstuck
- You will be hiring staff – maybe a manager for the pub, certainly bar staff and cleaners and possibly kitchen staff as

well – and employment law is a whole new area of knowledge in itself
- You will need to maintain investment in customer facilities – scuffed carpets and chipped urinals need to be replaced
- There is an unimaginable tangle of compliance issues – you wouldn't believe the amount of time the average publican spends on paperwork these days. It's not something you want to slide into

But just by running brewery tours, a shop and maybe a visitors' centre, you will be gradually acquiring the marketing, planning, budgeting, administrative and supervisory skills you'll need to run a pub – or pubs. One of the best ways to learn is to do, and gradually stepping up the retail side of your brewery will enable you to learn a step at a time, even leaving yourself leeway to fail without failing too disastrously. You are also, at each stage of development, increasing the sale value of your business without investing too much at a time.

Buying a pub may be your next logical step, but it's a big one. It's a very expensive and extremely complicated business and calls for detailed professional advice from one of the many specialist estate agents, correctly termed licensed trade property brokers, such as Christie & Co, Fleuret's, Guy Simmonds or Humberstone's. Like ordinary estate agents, they also offer financial advice, and in some cases run full licensed trade training courses as well. The best way of contacting them – and of browsing a huge number of pubs for sale – is to visit the sites of the licensed trade newspapers, the Morning Advertiser (www.morningadvertiser.co.uk) and The Publican (www.thepublican.com). It's a good idea anyway, as we saw in Chapter One, to subscribe to both of them.

Freehold or Leasehold

Buying outright has been made much more difficult by the fact that freehold prices have soared in recent years. This is largely thanks to the intervention of acquisitive pubcos, which snap up freeholds for silly prices. An alternative is to lease from a pubco instead; but although the ingoings are much, much lower, it's probably a bad idea. For a start, rents are sky-high, and you may soon find yourself struggling to pay them. And although few pubcos these days have a formal "upwards only" clause in the rent review section of their leases, in practice it is extremely difficult to get a reduction! Worse, you may well find that

CASE STUDY

When Owning a Pub Didn't Quite Work Out – Rebellion Brewery, Marlow, Buckinghamshire

It seemed a golden opportunity for Marlow's Rebellion Brewery when the lease of the Three Horseshoes just 500 yards away came up.

Rebellion had been running successfully for 10 years and had built up a big local following with its shop, tours, beer club and open days. It had already had to move once, but there was no scope to develop a brewery tap on the site, and the pub's sudden availability was a godsend. True, the Three Horseshoes was owned by Enterprise Inns, but property prices in the area made an outright freehold purchase prohibitive. And as the pub had been closed for three months when partners Tim Coombes and Mark Gloyens heard the lease was for sale, Enterprise was lenient with the premium.

It all looked pretty good. But four years later, Rebellion abandoned the experiment and sold the lease to pub manager Nigel Douglas, who still stocks the brewery's beer. What had gone wrong?

Well nothing, really. The relationship with Enterprise worked well. Nigel turned out to be a great manager. Locals were delighted that their neighbourhood brewery had its own tap at last and flocked there, and in 2005 the local CAMRA branch voted it Pub of the Year. Diners gave it their seal of approval, too: although the cooking had no pretensions to haute cuisine, it was all top-quality stuff at sensible prices. The pub quickly got busy.

Despite the pub's success, though, it was a constant worry.

"Nigel was a great manager, but it was always in the back of your mind that he might decide to move on, or fall under a bus," says Tim. "These things happen, and the thought of having to find a new manager as good as him was a constant source of stress.

"There are other worries with a pub, too: there's a massive amount of cash and some very high-value stock in any pub; no matter how good your control systems, if someone decides to start pilfering there's precious little you can do about it. We never had that problem. But it's all part of the risk, and running a pub is a big risk.

"Staff turnover is another headache. There's a very fast rate of staff turnover in the pub trade because the wages are so poor, but if you pay a decent wage your margin disappears. "It might seem a better idea to put in a tenant or lessee rather than a manager. That way, all the problems with staff and so on are his, not yours, but the trade-off is that you lose control over the style of operation. The lessee could turn it into a restaurant or a young people's pub and you'd still get your rent, but your ale sales would fall."

Running the pub was actually far more stressful than running the brewery, says Tim. It absorbed a huge amount of time and energy and, worse, it didn't make much money.

"The amount of profit it made for the brewery relative to its turnover was negligible," says Tim. "We paid better than average wages to get and keep good staff, and we invested heavily in beer quality. If you're setting out your stall as a showcase for your beers, the quality has to be spot-on all the time. We hired a cellar services contractor to clean and inspect all the lines every week, and that isn't cheap and we allowed Nigel unlimited ullage – So after four years Rebellion sold the lease to Nigel, with a tie, and got its original investment back without losing any sales – a good result all round. And Tim isn't entirely downbeat about running a pub of your own.

"Potentially there are huge advantages to having a flagship for your brewery," he says. "And lots of breweries have done really well with it. Some of them have even built up little tied estates of five or six really cracking pubs doing five barrels a week each, and if you can do that you're straight up to your 3,000-barrel duty relief limit with no further effort.

"But you need to go into it with your eyes open. Expect it to take up a third of your time at least. Expect to invest double the capital you'd originally planned, and then some. And budget for success – if you aim double the dining trade, say, the kitchen has to be able to cope. If you can get all that right, and get the right people in, there are potentially huge benefits – not least that it vastly increases the value of your brewery should you need or want to sell up.

"But it's a big investment in a specialist operation, and if you're not a specialist yourself you have entrust what is a huge part of your business to someone else, someone you probably don't know all that well. And in the end, we decided that it wasn't for us. It's all worked out okay in the end, but I don't think we'll ever get another pub."

www.rebellionbeer.co.uk

you can't even sell your own ales in your own pub. For these leases are invariably tied; and even if you negotiate an exemption, there's no guarantee it will last. One brewer I know did just that and built up a very successful and popular business – until the pubco sold the pub over his head to a regional brewer, which immediately ordered him to take his beers off the bar and revert to tied status. He tried to argue his corner, but legally he didn't have a leg to stand on and found himself saddled with an expensive lease that was no good to his brewery at all. His is not the only case of brewers getting into trouble by leasing pubs rather than buying them freehold. Having said that, there are a small number of free-of-tie leasehold pubs around, often belonging to a private owner. The National Trust, for instance, owns a scattering of historic pubs around the country, one of which – the 15th-century King's Head in Aylesbury – is run by the Chiltern Brewery. The opportunity to acquire a free-of-tie lease like this comes but rarely, although it's worth keeping an eye out in case one comes up.

A third course of action is to acquire an unlicensed premises and turn it into a pub. This is exactly how the JD Wetherspoon empire was founded: Tim Martin, a barrister by training, started out in the 1970s by acquiring old shoe-shops, garages and minimarts and turning them into characterful local pubs. His legal background stood him in good stead, because in those days it was extremely difficult to persuade magistrates to grant new licences. It's easier now, thanks in large part to him; indeed Wetherspoon's has made a speciality out of acquiring character properties such as disused banks and cinemas and turning them into splendid and often award-winning pubs: the Winter Gardens in Harrogate is a brilliant example. But it's almost certainly far too ambitious for anyone at entry level. Having said that, many smaller brewers have bought less grandiose historic properties, often in industrial heritage sites, and made wonderful pubs of them. Grainstore's brewery and pub next to the railway station in Oakham is an outstanding example, as is the old station at Codsall in the West Midlands, turned into a fantastic pub by local family brewer Holden's. In terms of cost, this may well be your best way into the pub trade. Certainly there are more than a few microbrewers who have successfully turned a part of their brewery into a successful bar. The drawback is that projects of this kind can easily become all-consuming in terms of your time and energy, and of capital, distracting you from the core business of running and growing your brewery.

Transfer of Premises Licence

Most of the steps described above assume that you're seeking a completely new premises licence to develop the retail activities based on your brewery site. If you're taking over an existing pub or shop, of course, you won't need a new licence; you can simply transfer the existing one.

I use the word "simply" here advisedly. The sections of the 2003 Act dealing with premises licence transfer look incredibly involved, largely because the Act is seeking to cover all possible eventualities. Actually the transfer is effected by the outgoing and incoming licensee signing a single form and one of them (you, probably) paying the council a small administration fee.

That's not the end of it, though. When you move in the existing premises licence will still be in force. That covers the opening hours, including seasonal variations as detailed in the operating schedule, and the previous licensee's choice of licensable activities – for instance, you may find you can have a disco but not a folk night, because the existing licence permits dancing but not live music. That means that if you want to change the way your new pub is run not just substantially but even slightly, you may very well need to change the terms of the premises licence too. You can get an application form to vary the licence from the council; when you do you'll find it's almost identical to an entirely new application.

Before you fill it in, it's wise to go through a thorough consultation exercise with the interested parties and responsible authorities. You might perhaps drop a letter through all the doors within reach asking residents if they had any reservations about the way the pub was run in the past and what improvements, if any, they would like to see. Check with the local police, too – there's normally an officer detailed to deal with licensing matters – to find whether the pub had a bad record and, if so, what steps they would recommend to clean it up.

Many might find this sort of consultation tedious or even an unwarranted intrusion. But remember: the best pubs are at the heart of their communities. Your neighbours are also going to be your regulars – unless, that is, you start off by picking a fight with them.

CASE STUDY

A Classic Brewpub – The Chequers, Binham, Norfolk

When Steve and Alex Chroscicki finally found the Chequers at Binham in the rural depths of North Norfolk, it was the end of a two-year hunt for a pub to site a brewery in.

Steve, an engineer by trade, had been made redundant by Parisa Café-Bars, an ambitious venture intended to be a chain of 250 boutique breweries across the world. Only one ever opened, in exotic Putney, and Steve ran it. By the time it had run its course, the brewing bug had bitten.

Steve and Alex took the lease of the Chequers – which is, rather unusually, owned by a local charity – in 2003 and originally planned to renovate its barn as the site of the planned brewery. After much havering, though, their bank wouldn't lend them enough on the security of an unassignable lease, so the Front Street Brewery (the pub is in Front Street) was installed in the old ladies' loo across the yard instead.

From the start, there were issues of size. A five-barrel brewery was squeezed into the building, but cask washing and bottling on a tiny hand-bottler have to be carried out in the pub's cellar. And as the local waterworks hadn't been expanded since it was built in the 1940s or 50s, the water company even tried to stop the brewery opening on the grounds that the drainage was inadequate. In the end Steve challenged the company to set him an effluent limit. It did – but such a small one that any spoiled beer has to be taken away to be disposed of elsewhere.

The local council, by contrast, was a delight. The planning officers even told the Chroscickis that they didn't need planning permission as brewing would be ancillary to pub use. A lot of councils probably wouldn't agree, but the only town hall official who came on site after that was the building inspector!

Size issues have continued to dog the brewery's progress. Binham is extremely rural, and there's no public transport, so the Chequers has never become the magnet for beer-hunters that more accessible brewpubs have. Trade is good in summer, but never quite good enough to justify hiring staff to cover the bar while Steve is brewing. And in winter sales are so low that Steve has to sell through wholesalers just to keep the whole thing ticking over. That means he has to keep a stock of nearly 100 casks at £50–£80 each, many of which spend months out in trade and some 10 per cent of which vanish every year.

As a result of all this, Steve is desperate to expand, off-site if necessary.

"It's been interesting and worthwhile but I think the lesson is not to start a brewery at a small rural pub," he says. "There is money to be made – your own beer is cheaper than beer you buy in, and you can use it to do swaps with other breweries too, but you need to be of a size where you can take on bar staff and concentrate on the brewing.

"Any good-sized wet-led town pub could do well out of its own brewery, but to make the best of the brewery here I definitely need to expand."

www.binhamchequers.co.uk

Appendix
Customs and Duty

I know of two microbrewers who have been jailed for failing to pay their duty and VAT bills. (I also know of one who was jailed for taking part in one of those fake export scams, but that's another story.) Two out of the 1,000 or so people who have set up microbreweries in the last 40 years may not seem all that many. But neither of them set out to commit fraud: cash-flow and other problems simply overwhelmed them. The harshness of their sentences shows, nevertheless, that Her Majesty's Revenue & Customs (HMRC) is not to be trifled with.

Fortunately, the world wide web makes HMRC much easier to understand and therefore satisfy. These days you no longer have to send away for copies of the various rules and regulations, or search them out in the public library and then spend hours and a small fortune photocopying them. All you have to do is visit **www.hmrc.gov.uk** and look up Notice 226 (April 2010) by clicking first on the "Excise & Other" link, then on "Information & Guides", then on "Alcohols", then on "A guide to alcohol duties & procedures", and finally on "Beer duty". This covers everything any brewer or group of breweries or packager or importer needs to know and even includes downloadable copies of all the forms you will need to fill in. It also gives access to all the relevant bits of legislation in their unintelligible full version. Unfortunately, Notice 226 runs to over 90 pages and is almost unreadable. What follows is therefore an attempt to condense and, to an extent, to translate (although not to interpret) its relevant sections. I have left out sections relating to importers, contract packagers and "connected" breweries; but I have presented my précis in the same order as the Notice (illogical though it often seems) to make it easier for you to check my version against the original.

Before we plunge in, though, to find out the actual rate of duty currently payable go back to "A guide to alcohol duties & procedures" and click on "Rates of duty on alcohol". This will tell you that the standard rate brought in by the 2011 budget is £18.57 per percentage point of alcohol by volume per hectolitre, or "Hl%". It changes every year, though, so keep up to date! Also from 1st October 2011 there will be introduced a new tax on high strength beers of over 7.5% abv (alcohol by volume) of 25% of the general beer duty.

The Basics

In the Alcoholic Liquor Duties Act 1979, beer is defined as including ale, porter, stout and any other description of beer, and "any liquor which is made or sold as a description of beer or as a substitute for beer" whose alcoholic strength exceeds 0.5 per cent alcohol by volume (although no duty is chargeable on beer that does not exceed 1.2 per cent). This includes mixtures such as shandy. Also classified as beer for duty purposes are certain mixtures of beer with alcoholic liquors or "other substances" of less than 5.5 per cent.

Brewers and their production premises must be registered with HMRC's National Registration Unit (address below).

Duty is charged on the quantity and alcoholic strength by volume (see Hl% above) and is normally calculated by reference to the quantity and alcoholic strength stated on the package label or invoice. It becomes payable when the beer is released from or consumed in registered premises or excise warehouses but may also be paid on the constructive removal of beer from "duty suspense" (that is, when it is held on or moved between registered premises or to excise warehouses on registered premises).

To calculate and pay the duty, you must:

- Keep records of all beer produced
- Keep records of all beer leaving registered premises, or otherwise passing the "duty point"
- Keep a beer duty account
- Complete a beer duty return and send it to the Beer Duty Accounting Centre; and
- Pay the duty by direct debit unless HMRC has agreed otherwise

If you qualify for small brewery relief, you must keep records in a beer production account as above, and records of all "small brewery beer" you handle in duty suspension.

Customs officers will make visits to ensure duty is being correctly assessed and accounted for, based on audits of your commercial, accounting and management control systems and on physical checks. Officers will also carry out physical checks on production, stock and movements of beer in duty suspension. They will normally make an appointment, and if they visit without one the attending officer must explain why. You have to grant officers access to any area of the registered premises provided the visit is "at a reasonable time". Your staff must be aware that officers showing the correct ID may visit without appointment.

If you fail to comply with the law or do not account for the correct amount of excise duty, HMRC's options include the issue of assessments and/or civil penalties. These are explained in Notice 208 "Excise assessments" and Notice 209 "Civil penalties: fixed, geared and daily". In many cases you will have the right to appeal. Full details of the appeals procedure are set out in Notice 990 "Excise and customs appeals".

Registration: General

The proprietor, partnership or company intending to brew beer must apply for registration at least 14 days before it starts brewing unless it is covered by one of the exemptions listed below. A separate registration is required for each premises in which beer is produced. Once any queries on the application are settled HMRC will send you a certificate of registration, which will confirm the date of registration and advise you of any conditions which are applicable. You should keep the certificate on the premises to which it refers as officers may ask to see it. Your registration will last until you stop production at the premises. You must notify HMRC in writing both if you intend to cease production and when you have actually ceased.

Except where registration is not required (if you are brewing for personal consumption or for research purposes), production of beer by an unregistered person or on unregistered premises is an offence for which there are severe penalties.

Conditions affecting registration may include:

- Limitation of the extent of registered premises
- Restriction of the classes of beer which may be received or held on registered premises
- Financial security for the duty on beer in duty suspension

If you need temporary additional storage space (for instance, due to seasonal demand) you can apply for temporary registration of any premises under your control within 5km of your registered premises.

To apply for registration, complete the form in Section 32 with a detailed plan of the proposed registered premises. If the whole site is to be registered premises, the plan need only show details of its perimeter. Send them together with a direct debit mandate to the:

National Registration Unit (Alcohol and Tobacco)
HM Revenue & Customs
Portcullis House, 21 India Street
Glasgow G2 4PZ
Tel: 0141 555 3369; **Fax:** 0141 555 3555

If you have an enquiry once your application form has been sent, contact the NRU. For enquiries before this stage, contact our National Advice Service.

To obtain a direct debit mandate form, contact the:

Beer Duty Accounting Centre
HM Revenue & Customs
Queens Dock
Liverpool L74 4AA
Tel: 0151 703 1375

You must make a reasonable estimate of the volume of beer that will be produced in the current calendar year in your brewery and enter details in question 9 of the application form. You must note in your business records how you have made the estimate.

If any changes affect your registration particulars you must write to the NRU giving details. The following changes would require a new application to be completed:

- Change of legal entity, for example, formation of a limited company or
- Change in the ownership or control of your business (in the case of a sole proprietor or partnership)

Other changes that must also be notified in writing

- Change of address of your registered and/or adjacent premises
- Change of name of the registered person, for example, change of company name
- Change of bank account details
- Change of your duty guarantor
- Cessation of production
- The production of other excisable goods on your registered premises;
- Financial difficulties/impending insolvency; or
- You become VAT registered or de-registered
- You must apply in writing to make any changes to the classes of beer you are registered to hold

Financial Security

The financial security that may be required as a condition of registration is a guarantee by an approved guarantor who undertakes to pay HMRC in the event of an irregularity. HMRC requires financial security to safeguard the duty on all duty-suspended intra-EU movements of beer and may also require guarantees:

- For beer held in duty suspension on registered premises
- For movement of beer within the UK
- To cover beer that has passed the duty point but upon which duty has not been paid. Further information can be obtained from our National Advice Service

To arranging financial security, contact the Financial Securities Centre (FSC) at the NRU. Guarantees are the only form of security acceptable to HMRC, which will issue the draft wording to you when you have agreed your level of security. Only HMRC-approved companies may act as guarantors: most banks and insurance companies have this approval, but if you want to check a particular company please ask the Financial Securities Centre.

Duty Liability

Beer becomes liable to duty at the point of production, normally when the earliest of any of the following events takes place: The beer

- Is put into any package
- Is removed from the brewery
- Is consumed
- Is lost (see Section 10)
- Reaches that state of maturity at which it is fit for consumption

However, duty only becomes payable when the beer passes the duty point: that is, when it leaves duty suspension. Duty ceases to be suspended when:

- The beer leaves registered premises, unless it is delivered to other registered premises, to an excise warehouse; for export or shipment as stores, or to entitled diplomats or members of visiting forces
- The beer is received by registered premises not entitled on their registered holder certificate to hold that class or description of beer
- The beer is constructively removed
- The beer is consumed
- The beer is lost

- The beer is irregularly diverted
- You are no longer registered
- The premises on which you are holding the beer cease to be registered or
- The beer is produced and you are not registered to hold beer in duty suspense

Normally, the duty should be paid by the 25th of the month following the calendar month in which the beer passed the duty point. But if it would help your business to account for duty on duty-suspended beer in advance of delivery from registered premises, you may do so. This is known as "constructive removal". The duty should be paid by the 25th of the month following the calendar month in which the beer was constructively removed.

To work out the duty at the end of the accounting period:

- Total the quantity of beer, at each particular strength, which has passed the duty point in the accounting period
- Convert the quantity to hectolitres (hl) to two decimal places.
- Multiply the result by the strength of the beer to arrive at the hectolitre percentage (hl%)
- Add together all the hl% totals for the beers of different strengths
- Multiply the total hl% by the applicable duty rate to find the duty figure

If in doubt, contact the National Advice Service.

Small Brewery Relief

Small brewery relief applies to beer, other than that brewed under licence, produced by eligible breweries whether in the UK or overseas. Beer produced under licence is always liable to the standard rate of duty.

A brewer with a single brewery not connected with other breweries must satisfy all of the following conditions for its beer to qualify as small brewery beer:

- No more than 60,000hl of beer may have been produced in the brewery in the previous calendar year
- Less than half of the beer produced in the last calendar year may have been produced under licence
- You must reasonably estimate that in the current calendar

year your brewery will produce no more than 60,000hl

These conditions are the same if your brewery is "connected to" other small breweries. The definition of "connected breweries" is determined by Section 839 of the Income and Corporation Taxes Act 1988. The definition of "connection" is very wide; but examples of the circumstances in which people are connected include individuals connected by marriage, family relationships and the like, and companies under the same effective control. This is to prevent, for example, a large business being artificially split into a number of small breweries, each of which would be below the 60,000hl limit.

Estimating the Current Calendar Year's Production

An acceptable estimate must take all relevant factors into account including existing contracts to supply beer and expansion plans. It is important that you make this estimate honestly. If you make an untrue estimate by stating the brewery will produce 50,000 hectolitres in the current calendar year when you know that a reasonable estimate would be 70,000 hectolitres, HMRC can recover all the underpaid duty. You only have to inform HMRC of your estimate when you apply for registration. You don't need to notify subsequent estimates but you must note in your business records each calendar year's estimate at the time you make it, together with details of how you arrived at the estimate. This information must be kept in the form of a beer production account, which you must keep under Schedule 3(a) of the Beer Regulations 1993.

Beer Production Account

Your beer production account must show your estimate of the current year's production with details of how you arrived at it; the previous calendar year's actual production, to be recorded each month; and the total quantity (in bulk hectolitres to two decimal places) of beer that you have produced in that calendar month, split if appropriate between your own beer and beer produced under licence. The quantity should not be reduced to take account of beer you subsequently determine to be spoilt or unfit for use or for any returned spoilt beer.

At the end of each calendar year, your annual production totals shown in the beer production account will determine your continued eligibility for small brewery relief scheme and the reduced rate for the following year. To be eligible for relief you must use the previous calendar year's production and your estimated production in the current year. Neither production total may exceed 60,000hl.

To qualify for small brewery relief, a new brewer must make a reasonable estimate of the amount of beer that will be produced in the current year which, where appropriate, must be grossed up to arrive at a full calendar year's figure using the formula E ÷ (365–N) x 365 where E = the amount of your estimate and N = the number of days in the calendar year before the brewery started production. For example, if you intend to start brewing on 1 August and you estimate, based on the size of the plant, business plans and workforce, that in the five months to 31 December the brewery will produce 6,400hl, the grossed-up estimate is 15,268hl.

Provided that you reasonably estimated that production would not exceed 60,000hl in the current year, then the first 60,000hl will be small brewery beer and subject to a reduced rate of duty. Any beer over that amount will not be small brewery beer and will be subject to the standard rate of duty.

If you share a brewery, the production figure for the purposes of small brewery relief will be the total production within the brewery in that year.

Calculating Duty Relief

If your actual or honestly estimated annual production figure is less than 5,000hl, the reduced rate is quite simply 50 per cent of the standard rate, rounded up to the nearest penny. (Remember, though, that any beer you brew under licence from a third party is charged at the full rate; if beer produced under licence exceeds half of your output you lose your eligibility for relief altogether.)

If your production figure or estimate (P) is between 5,000hl and 30,000hl, you can calculate the duty per hectolitre as follows: (P–2,500) over P times the standard rate. If your production figure is, say 7,500hl then the calculation is: 7,500 – 2,500 = 5,000; divided by 7,500 = 0.6 recurring; multiplied by £16.15 = £10.96 per hectolitre times the % ABV.

The calculation for production of 30,000-60,000hl is more complicated still. You will find it expressed as a proper equation on the HMRC website; translated into step-by-step sums, it goes like this.

First, subtract 30,000 from your production figure or estimate P. Let's say P = 45,000hl. Divide the result (15,000) by 100 and multiply by 8.33 to get 1,249.5. Subtract 1,249.5 from 2,500 to get 1,250.5. Subtract 1,250.5 from P (45,000hl, remember?) and get 43,749.5.

Divide 43,749.5 by P and you get 0.97221; which multiplied by the standard rate of £16.47 gives you a duty payable figure of £16 per hectolitre, a saving of 47p per hectolitre or, assuming the average strength of your output is about 4 per cent ABV, nearly £85,000 for the year. And even if you're useless at sums, £85,000 is well worth a few minutes' struggle with a calculator.

Irregularities, Losses and Deficiencies

As a registered brewer, you are responsible for the control of beer in your premises. You must have the necessary systems to control and safeguard your stocks and examine critically all losses and deficiencies.

Your records of production and processing should indicate how much beer you lose during routine operations, for example, the losses you normally incur during packaging. For accidental losses, regulation 14(3) of the Beer Regulations 1993 requires you to record the date and time the loss occurred; a description (product name) and the volume of beer lost, and the alcoholic strength if the loss occurred after production has been completed; and the reason why the accidental loss occurred. If beer cannot be accounted for after the start of production and there is no acceptable explanation, you are liable for duty on the missing beer. You cannot normally offset stock losses against production surpluses.

Normally there is no duty relief of beer that is lost after it has passed the duty point. But if duty suspended beer is spoilt, you do not have to pay the duty providing HMRC is satisfied that it was unintentionally spoilt, contaminated or otherwise rendered unfit for consumption in the registered premises and has not been consumed.

Measurement of Quantity

HMRC can require duty to be paid on the actual quantity of beer in each container as it passes the duty point. However most brewers do not measure the quantity in each container, but use the "average system" common throughout Europe.

Under this arrangement, the average contents of packages must not be less than the declared contents marked on the can, bottle or label or, in the case of draught beer, on the invoice or delivery note. Within limits the actual contents of each container may be more or less than the declared contents. Packagers using the system must conform to weights and measures codes of practice issued by the British Beer & Pub Association and agreed with Trading Standards. SIBA has a similar code of practice on the contents of kegs and casks. Packagers are obliged to monitor and record the actual quantity of beer in a proportion of packages to ensure they fulfil the code's requirement.

Small Packages

Under the average system of quantity control, duty will normally be charged on the quantity declared on the container for beer delivered in small packages such as cans and bottles. Evidence of compliance with weights and measures legislation will be sufficient to accept the labelled contents as the duty base, unless there are grounds for believing that deliberate duty avoidance is involved. Packagers should exercise "due diligence" to ensure not only that the can or bottle contains on average at least the amount of beer stated on its label, but also that volume in excess of the declared contents is minimised.

To demonstrate that "due diligence" has been observed, you must monitor the filling process to ensure that the quantity put into the package does not regularly or excessively exceed the amount declared on the container. You should record these checks and provide an adequate audit trail to satisfy our officer that due diligence is being exercised. Where there is evidence of consistent excessive overfilling, additional duty will be due.

Large Pack

To ascertain the quantity in large pack – containers from 10–400 litre capacity such as kegs and casks – packagers customarily fill the containers in accordance with the average contents rules. The quantity of beer in a container will be treated as the average of the samples taken for the purpose of complying with these rules, which permit some latitude in the sampling regime adopted. However, in determining volume for duty purposes, packagers will be expected to maintain a minimum sampling rate of one container per filling head per operating day or 0.1 per cent of a production run of 4,000 containers or more; the samples to be representative of the mix of containers filled. A copy of the sampling protocol you intend to use to assure compliance with average contents rules and for duty purposes should be sent to HM Revenue & Customs.

Some smaller packagers do not take samples but use their containers as capacity measures. To meet HMRC's tolerance requirements, they should ensure that the average capacity of their containers is such that they are operating within the set limits.

Sediment in Cask-conditioned Beer

Duty need not be charged on undrinkable sediment in cask-conditioned beer provided the customer (for example, the publican) is made fully aware in writing, at or before the time of receipt, of the quantity of beer on which duty has been charged. If, for example, a barrel (163.7 litres) is likely to contain 2.3 litres of sediment, the customer must be made aware by a statement on the label, delivery note or price list etc. that duty has been charged on 161.4 litres (a copy of the notification to customers must be retained). Any sediment on which duty has not been charged cannot be included in any subsequent claims for relief on spoilt beer.

There is no prescribed method for calculating the proportion of undrinkable sediment, but you must be able to satisfy HMRC that the method you use gives equitable results. Sediment levels for each quality of beer and container size must be regularly monitored by brewers and reviewed/amended (as necessary) at least annually and agreed with HMRC. Any changes to recipes/ingredients etc. during the year, that would significantly affect sediment levels must be notified to HMRC and the allowance adjusted accordingly.

There is no undrinkable sediment allowance for polypins and mini-pins sold by brewers to the public or on bottle-conditioned beer.

Alcoholic Strength

For duty purposes alcoholic strength is the percentage of alcohol by volume (ABV) in the beer expressed to one decimal place, for example, 4.19 per cent ABV becomes 4.1 per cent ABV. Ignore figures after the first decimal place.

You may use any method you wish to measure the strength of beer as long as it produces results that agree with those that would be achieved using one of HMRC's preferred reference methods (described below). Additionally, if using these or any other methods not based on laboratory analysis, an independent analyst must test the ABV of each of your products at least annually to ensure consistency with calculated results. The results of the independent analyses must be held in your business records.

You must continuously monitor and record your ABV results, which should normally fall randomly on either side of the target strength. The average of your results should equate closely with the declared strength.

It is recognised that ABV may occasionally vary, but provided appropriate action is taken quickly to return the strength of the beer to within its normal specification, due diligence will have been demonstrated. You must keep records of action taken to maintain product strength within control limits.

You must establish the strength of each batch for each of your products. Where beer from one batch is packaged into different container types, for example, cans and bottles, you may combine the results. If you can demonstrate to HMRC that based on available information and experience, due care was taken when deciding target ABVs for new and/or infrequently brewed products (and that all decisions, actions etc were properly recorded), HMRC will accept the label/invoice/delivery note strength for duty purposes. Customs officers will examine your results and your record of actions taken. Where the results have consistently fallen above your target, they will wish to confirm that action was taken as soon as the problem was identified to bring the process back into control or to change the declared ABV. If you have failed to take such action, an assessment will be raised for the additional duty due. If there is a dispute over the strength, officers may take samples of beer to be analysed to establish the actual dutiable strength of the beer.

Cask and bottle conditioned beers will continue to ferment after removal from registered premises, resulting in an increase in strength. You must account for duty on the strength at which you expect the beer to be when it is consumed. This is also the strength that must be shown on the label/invoice/delivery note. And in addition to procedures to establish the ABV at packaging, you must regularly monitor and record the actual strength of each quality of cask and bottled conditioned beer at the expected time of consumption, to establish its alcoholic strength. The precise method and frequency of checking is a matter for you, but you must be able to satisfy HMRC of the accuracy of your results.

Beer Duty Account

A beer duty account is a summary of the beer duty due in each accounting period and must contain the following information:

- The amount of duty due on all beer which leaves duty suspension
- The amount of duty reclaimed on spoilt beer which has been reprocessed or destroyed
- The amount of duty reclaimed on drawback.
- The amount of any underdeclarations and overdeclarations from previous periods

- The net amount of duty due for the period and the date and method of payment

You should keep the beer duty account in a specific book or ledger opening or on computer, providing you can print a satisfactory legible copy of the account when required.

Records

As a revenue trader you need to keep and preserve certain records and accounts as evidence of your business activities (Revenue Traders (Accounts and Records.) Regulations 1992). You also have to allow HMRC access to all your business records.

You have to keep a record of:

- Production
- Stock
- Handling
- Buying
- Selling
- Importation
- Exportation of beer

You must normally keep your records for six years. If, however, this causes problems, ask our National Advice Service if you can keep some records for a shorter period. You must get HMRC's agreement before destroying any business records that are less than six years old. You can keep your records on any form of storage technology, provided that copies can be easily produced and there are adequate facilities for allowing a Customs officer to view them when required. Obtain agreement from the National Advice Service before transferring records. You may be required to operate the old and new systems side by side for a limited period, and HMRC may refuse or withdraw approval if its requirements are not met.

You can keep your records on a computer provided they can be readily converted into a satisfactory form and made available to HMRC when required. If you do keep your records on a computer HMRC requires access to it.

Beer Duty Returns

The beer duty return (EX46) is the form on which you declare your liability for beer duty in each accounting period. You must make a return of your beer duty liability each accounting period, which will normally be a calendar month. If your duty liability for the month is nil you must still send a return. If you fail to submit a return on time you will be liable to penalties. HMRC may estimate the duty which would have been due and to pursue the debt through the civil courts.

If you foresee any problems, you should immediately contact the Beer Duty Accounting Centre.

EX46s are routinely sent out to all registered brewers and packers. If you do not receive a return, you should contact the Beer Duty Accounting Centre.

Each copy of the form has full instructions on how to complete it. You must complete all boxes, writing "none" where necessary. Returns must be completed in ink and any changes must be initialled and dated by the person who signs the declaration – the proprietor, a partner, the company secretary, a director of the company, or any other authorised person.

You must ensure that beer duty returns are completed accurately. Failure to exercise due care in completing returns may result in civil penalties. If you deliberately make a false beer duty return you may face prosecution for the offence and incur heavy penalties. You must submit your return so that it arrives not later than the 15th of the month following the accounting period. When the 15th falls at a weekend or on a public holiday the return must be received by the previous working day.

Mixing Beers from Different Brews and of Different Strengths

- Beer that is not in duty suspension may not be mixed with duty-suspended beer
- Beers of different strengths that have passed the duty point may not be mixed until they are sold to the consumer
- You may mix duty-suspended beers only on registered premises or in an excise warehouse
- There are no restrictions on the mixing of beers of the same strength that have passed the duty point
- You may mix small brewery beer with beer of a different rate of duty in duty suspension
- The mixture will not be small brewery beer and therefore will be liable to the standard rate of duty

Additions to Beer

Beer can be primed with sugar to boost its alcohol content on registered premises. A registered brewer planning to prime beer on unregistered premises must seek HMRC approval and keep a record of any such premises at which priming takes place. If you are not a registered brewer or registered holder you may not prime beer unless HMRC has approved the arrangements. For approval write to the National Advice Service stating the address of the premises at which you wish to undertake priming; the addresses of registered premises from which the beer will be received; estimated annual quantities of the beer to be primed; and the destination of the primed beer. Your records must include the date and time of the priming; the type of beer (product name) that is primed; the quantity and strength of beer that is primed; the type and quantity of primings added; and the quantity and expected strength of the resulting product. These records should be kept on the premises where priming is carried out and made available to Customs officers when requested.

Beer in duty suspension on registered premises can be diluted with water. If you wish to add water to beer elsewhere, before it is sold or supplied to the consumer, you must obtain approval by writing to the National Advice Service. However, if you add water to small brewery beer after it has left the registered premises where it was produced, it will no longer be small brewery beer.

You can add finings (provided they do not contain alcohol) to beer on any premises.

If you wish to mix beer with any substance, including non-alcoholic beverages, after the duty point, you must seek prior approval from the National Advice Service. However, if you only intend to produce drinks whose alcoholic strength is not more than 1.2 per cent ABV, you do not need approval. Shandy of more than 1.2 per cent and its variants are classified as beer. These include shandy made with lemonade, lemon cordial, lemon flavourings, lemon juice,or lemon squash; lager and lime made with lime cordial, lime flavouring, lime juice, lime squash or limeade; ginger beer shandy, shandygaff or a mixture of beer and ginger, ginger cordial, ginger, ginger flavouring, ginger squash or unfermented ginger beer; beer and fruit cordial, fruit flavourings, fruit-flavoured carbonated water, fruit juice or fruit squash; or beer and any alcoholic liquor or substance other than spirits.

Mixtures containing alcohol other than spirits – for instance, alcopops based on beer – where the end strength is 5.5 per cent or more are liable at the appropriate made-wine rate of duty. Beer mixed with spirits is liable to the spirits rate of duty. This applies even if the majority of the product is beer rather than spirit. There are only two instances where an element of spirit may be used without incurring the spirits rate. These are where there are no spirits in the final product – for example, the spirits are used as a carrier for a flavouring added at an early stage of production – or where the spirits are added in such a small quantity they do not increase in the alcoholic strength of the beer.

If you are unsure of the duty category that will apply to any blend or mixture you intend to produce, you should consult the National Advice Service beforehand. It will also advise you of any additional approvals you may need (for example, a made-wine producer's licence.)

You can claim duty relief on any beer that has been charged with duty and has subsequently been spoilt or is otherwise unfit for use, provided it is presented for duty reclaim in the same container in which it left duty suspension (unless delivered in bulk). However, where this is not possible for health and safety or other practical reasons, and subject to prior approval from HMRC, the beer may be transferred to other containers.

If a registered brewer buys duty-paid beer that becomes spoilt and he wishes to destroy it and claim spoilt beer relief, he must apply to HMRC for approval to do so. In the case of beer imported on payment of duty, only the person who accounted for the duty on importation will normally be entitled to claim relief, but the National Advice Service will be able to provide further information in these circumstances. NB: spoilt beer does not have to be destroyed if duty relief is to be claimed. It can also be reprocessed – mixed with other beer on registered premises or filtered and/or repasteurised.

You can only claim for the actual quantity of beer destroyed or reprocessed on which duty has been charged. This will depend on whether you use a spoilt beer vessel; destroy directly from cask, keg or other package; or a combination of both. If you wish to decant beer to determine the volume for relief, you must use a gauged and tabulated spoilt beer vessel or establish the volume by using a properly calibrated meter. If you use a spoilt beer vessel to claim relief, you must take a sample for laboratory testing to establish its alcoholic strength. This must be recorded in support of your claim. If you destroy or decant for reprocessing directly from cask, keg or other containers, you can only claim for the actual quantity of beer returned on which duty has been charged. In the case of unbroached containers, you should use the declared contents. Broached containers and depot leakers, which may be unbroached, should have their actual contents measured for duty reclaim purposes. Remember to exclude any undrinkable sediment allowance from the quantity claimed.

To make a claim, total the entries in the spoilt beer record (including those for destructions and reprocessings of spoilt beer) and at the end of the accounting period transfer the total to your beer duty account. To claim relief, enter this total (that is, the spoilt beer relief claimed) in box 12 of your beer duty return.

The following are ineligible for relief:

- Unconsumable beer (for example, sediment in cask conditioned beer) on which duty was not charged
- Diluted beer (unless HMRC approved the addition of water)
- Adulterated beer (beer containing additions which HMRC did not approve)
- Where no satisfactory audit trail is available

Destroying Spoilt Beer

You must destroy the beer in a way acceptable to HMRC and which makes it unsaleable as a beverage. A Customs officer will advise you if you need to give notice; otherwise you can destroy spoilt beer whenever you wish provided that the following conditions are met:

- A full audit trail is maintained
- The requirements of other regulatory authorities (e.g. the Environment Agency) are observed; and
- Proper control practices are maintained, including appropriate action at management and supervisory levels

If you are destroying the beer away from your registered premises, the operation must be supervised by a representative of the brewery.

Repayment of the duty will normally require evidence of a full credit of the duty paid value, or replacement of the goods to your customer (or owner of the goods at the time they became spoilt). In cases where an abatement is made to the amount of credit given, you must be able to demonstrate that the abatement is not due to accidental loss, adulteration, dilution after the duty point or similar circumstances.

The minimum amount of relief that you can claim in any accounting period is £50. If you destroy beer on which the amount of relief is less than £50 during an accounting period, wait until the total amount of relief reaches at least £50 before making a claim. This means that one claim may cover beer destroyed during several accounting periods. If the accounting periods cover a change in duty rates, separate entries must be made in the spoilt beer record for beer charged with duty at the old and new rates.

HMRC does not object to bulking of beer at two rates in one destruction, so long as a system to apportion the total has been agreed in advance.

When beer is either returned to the brewery for destruction or destroyed remotely, Regulation 33(1) of the Beer Regulations 1993 requires you to enter the following details in the destruction section of your spoilt beer record:

- The total volume of spoilt beer destroyed
- The strength of the spoilt beer destroyed
- The date and time of the destruction
- The volume and strength of the beer in each container from which the spoilt beer was directly destroyed
- Evidence of duty charged or paid
- The amount of remission or repayment claimed
- The description of the beer returned by each purchaser for which a claim is made
- The name and address of each purchaser
- The numbers and sizes of each container in which the beer was returned by each purchaser returning the beer

When beer is returned for reprocessing, enter the following particulars in a reprocessing section of your spoilt beer record:

- The date and time the beer was returned for reprocessing
- The volume and strength of the beer, in each container, in which it was returned for processing
- The total volume and strength of the beer reprocessed
- Evidence of the duty charged or paid
- The amount of remission or repayment claimed
- The description of the beer returned by each purchaser for which a claim is made
- The name and address of each purchaser
- The numbers and sizes of each container in which the beer was returned by each purchaser returning the beer

Unless HMRC allows a longer period, the spoilt beer record must be completed within one hour of reprocessing taking place.

Returning Duty-paid Beer to Duty Suspenison

Duty-paid beer cannot be returned to duty suspension unless it is spoilt and comes back for reprocessing. If duty-paid beer is returned to registered premises because of a failed delivery, e.g. because a customer is unwilling or unable to accept a delivery or an erroneous delivery has been made, the beer must be separately identified from duty-suspended stock in your records, for example in a failed delivery account.

Beer on which duty has been paid can, subject to certain conditions, be stored with duty suspended beer in your registered premises. No physical segregation of duty suspended and duty-paid beer is necessary. The change of status of any beer from duty suspended to duty paid must be recorded and the product(s) clearly identified in your records.

Exports to EU Member States

You may remove either duty-suspended beer or duty-paid beer to other member states, but you must follow the detailed procedures set out in Notice 197 "Excise goods: holding and movement". These include the following requirements:

- You must have a minimum level of financial security (guarantee) of £20,000 in place
- You must ensure that you are sending the beer to a warehouse which the fiscal authorities have approved to receive that type of beer
- You must use specific documentation
- You must receive a valid certificate of receipt confirming that the consignee received the beer

If you despatch duty-paid beer to the EU and want to reclaim the duty, you must observe the conditions of the drawback system contained in Notice 207, "Excise duty: drawback". (Drawback is a relief for the repayment of duty paid on goods that have not and will not be consumed in the UK.) You should offset the amount of drawback claimed during the accounting period by completing Box 13 on duty return (EX46). The claim should be for completed movements only and supported by the necessary documentary evidence as required by Notice 207.

You may send duty-suspended beer to a non-registered trader, but before you send the beer you must ensure that the consignee has paid or secured the destination member state's duty on the goods. You must hold original evidence before you despatch the beer and you must ensure that the evidence of payment travels with the consignment.

Exports of Beer to Non-EU Countries

To export duty-suspended beer to non-EU countries, you must follow the procedures set out in Notice 502, "A brief guide to export procedures". You must use the Single Administrative Document (SAD/Form C88). Information and guidance on completion of the SAD is contained in the Tariff, Volume 3, parts 1 and 2.

HMRC expects you to produce commercial evidence of exportation for any beer you have removed from your registered premises for export in duty suspension to non-EU countries. Detailed guidance on the type of export evidence it accepts is in Notice 703, "Exports and removals of goods from the UK". Guidance on the documents to be completed is in Notice 197 "Excise goods holding and movement".

You may export duty-paid beer to non-EU countries, but to reclaim the duty you must observe the conditions of the drawback system contained in Notice 207 "Excise duty: drawback".

Unless you can produce evidence that the beer left the EU, you will be liable for duty and VAT at the appropriate rates.

Determining the Strength of Beer

Method One

(a) A representative sample is to be taken and, after first being cleared of sediment and gas by filtration in an approved manner, a definite quantity thereof by measure at the temperature of 20° C shall be distilled.

(b) The distillate shall be made up at the temperature of 20° C with distilled water to the original measure of the quantity before distillation.

(c) The strength of the distillate made up in accordance with paragraph (b) shall be ascertained by determining its density in air at the temperature of 20° C by means of an approved pycnometer used in an approved manner; and

(d) The strength of beer shall be taken to be the percentage of alcohol by volume in the table entitled "Laboratory Alcohol Table" which corresponds to the density determined in accordance with paragraph (c), except that where the density so determined is between two consecutive numbers in the table aforesaid the strength shall be determined by linear interpolation.

NB: Where the result ascertained by the method specified above is rendered inaccurate by the presence of substances other than alcohol, that method shall be adjusted in such manner as may be approved for the purpose of producing an accurate result.

Method Two

If you have no or minimal laboratory facilities, you may calculate the alcoholic strength of your beer by multiplying the number of degrees by which the beer has attenuated by a factor. In order to ensure that your calculations are accurate, it is essential that the original gravity be established as soon as possible after collection and before fermentation commences, which will normally be within one hour of filling the fermenting vessel.

NB: If you add priming sugar to promote secondary fermentation you will need to calculate the alcoholic strength of the finished product as outlined below.

Step action:

1. Measure the original gravity (OG) within one hour of collection
2. Wait until fermentation is completed – for cask-conditioned beer this will be after secondary fermentation in the casks; then
3. Measure the present gravity (PG - also known as the specific or final gravity)
4. When you have taken your readings, calculate the alcoholic strength using the formula:

(OG – PG) x f = a% ABV where:

- OG is the original gravity of the beer
- PG is the present gravity of the beer
- a is the beer's alcoholic strength; and
- f is the factor connecting the change in gravity to alcoholic strength

The value of f is not constant because the yield of alcohol is not constant for all fermentations. In lower-strength beers, more of the sugars available for fermentation are consumed in yeast reproduction than in producing alcohol.

Complaints and Suggestions

If you have a complaint you cannot resolve it on the spot with the Customs officer, or have a suggestion about how HMRC can improve its service, contact one of its Regional Complaints Units. You will find the telephone number under Revenue & Customs or Customs & Excise in your local telephone book. Ask for a copy of the code of practice "Complaints and putting things right" (Notice 1000). You will find further information on the HMRC website, **www.hmrc.gov.uk**. If HMRC is unable to resolve your complaint to your satisfaction you can ask the Adjudicator to look into it. The Adjudicator, whose services are free, is a fair and unbiased referee whose recommendations are independent of HMRC.

You can contact the Adjudicator at:

The Adjudicator's Office
Haymarket House
28 Haymarket
LONDON
SW1Y 4SP
Tel: (020) 7930 2292; **Fax:** (020) 7930 2298
E-mail: adjudicators@gtnet.gov.uk
Internet: www.adjudicatorsoffices.gov.uk

Further Information

The British Beer & Pub Association has a series of Guidelines offering advice to brewers and packagers. Copies from:

Brewing Publications Ltd,
Market Towers,
1 Nine Elms Lane,
London SW8 5NQ
Tel: 0207 627 9191; **Fax:** 0207 627 9123,
or visit **www.beerandpub.com.**

Further Help and Advice

If you need general advice or more copies of HM Revenue & Customs notices, ring the National Advice Service on 0845 010 9000, 8am-8pm Mon-Fri.

If you have hearing difficulties, please ring the Textphone service on 0845 000 0200.

To speak to someone in Welsh, Tel: 0845 010 0300, 8am-6pm, Mon-Fri.

Directory of Services and Supplies

Please note: the following list is as complete as we could make it but not exhaustive, and we apologise to any providers of supplies and services that have been missed out. Many of the firms listed offer services and supplies in more than one category, but space prevents us from listing each company more than once. It is therefore always worth a thorough check of their websites to see just how wide and varied their activities are. If you would like to be listed in the next edition please email **info@paragraphpublishing.com**.

ASSOCIATIONS, CLUBS & SOCIETIES

Brewing, Food & Beverages Suppliers Association (BFBi)
3 Brewery Road, Wolverhampton,
West Midlands WV1 4JT
Web: www.bfbi.org.uk
Phone: 01902 422 303

British Beer and Pub Association
Market Towers, 1 Nine Elms Lane,
London, Greater London SW8 5NQ
Web: www.beerandpub.com
Phone: 020 7627 9191

Campaign for Real Ale
230 Hatfield Road, St Albans,
Hertfordshire AL1 4LW
Web: www.camra.org.uk
Phone: 01727 867 201

Campden BRI (Brewing Research International)
Centenary Hall, Nutfield,
Surrey RH1 4HY
Web: www.bri-advantage.com
Phone: 01737 822 272

Cask Marque
Seedbed Centre, Severalls Park,
Colchester, Essex CO4 9HT
Web: www.cask-marque.co.uk
Phone: 01206 752 212

Institute of Brewing & Distilling
33 Clarges Street, London,
Greater London W1J 7EE
Web: www.ibd.org.uk
Phone: 020 7499 8144

International Centre for Brewing and Distilling
Heriot-Watt University, Riccarton,
Edinburgh EH14 4AS
Web: www.icbd.hw.ac.uk
Phone: 0131 451 3183

National Brewing Library
Journals Office, Oxford Brookes
University Library, Gipsy Lane Campus,
Oxford, Oxfordshire OX3 0BP
Web: www.brookes.ac.uk/library/speccoll/brewing
Phone: 01283 561 626

SIBA
PO Box 136, Rippon,
North Yorkshire HG4 5WW
Web: www.siba.co.uk
Phone: 0845 337 9158

The Association of Licensed Multiple Retailers (ALMR)
9b Walpole Court, Ealing Studios,
London, Greater London W5 5ED
Web: www.almr.org.uk
Phone: 020 8579 2080

BOTTLING MACHINES & SUNDRIES

Ardagh Glass Ltd
Headlands Lane, Knottingley,
West Yorkshire WF11 0HP
Web: www.ardaghgroup.com
Phone: 01977 674 111

Carlson Filtration
Butts Mill, Barnoldswick,
Lancashire BB18 5HP
Web: www.carlson.co.uk
Phone: 01282 811 000

Enterprise Tondelli
Unit 7 College Farm Buildings,
Barton Road, Pulloxhill,
Bedfordshire MK45 5HP
Web: www.enterprisetondelli.com
Phone: 01525 718 288

Krones UK Ltd.
Westregen House, Great Bank Road,
Wingates Industrial Park, Bolton,
Lancashire BL5 3XB
Web: www.krones.co.uk
Phone: 01942 845 000

LINPAC Packaging
A1 Business Park, Knottingley,
West Yorkshire WF11 0BL
Web: www.linpac.com
Phone: 01977 671 111

The OPM Group
Speedprint House, Halifax Road,
Cross Roads, Keighley, West
Yorkshire BD22 9DH
Web: www.opmlabels.com
Phone: 01535 642 528

Vetreria Etrusca Ltd.
16 Beckside, Plumpton, Penrith,
Cumbria CA11 9PD
Web: www.vetreriaetrusca.it
Phone: 01768 894 044

William Croxson and Son Ltd.
Alpha Place, Garth Road, Morden,
Surrey SM4 4LX
Web: www.croxsons.com
Phone: 020 8337 2945

BREWING CONSULTANCY & INSTALLATION

Advanced Bottling UK Limited (ABUK)
Brealey Works, Station Street, Misterton,
Nottinghamshire DN10 4DD
Web: www.abuk.co.uk
Phone: 01427 890 099

Brewlab Ltd
Unit 1, West Quay Court, Sunderland
Enterprise Park, Sunderland,
Tyne & Wear SR5 2TE
Web: www.brewlab.co.uk
Phone: 0191 549 9450

D A Smith Brewing Services & Consultancy
6 Church Street, Copmanthorpe,
North Yorkshire YO23 3SE
Web: www.brewingservices.co.uk
Phone: 01904 706 778

FlavorActiV Ltd
Sanderum House, Oakley Road, Chinnor,
Oxfordshire OX39 4TW
Web: www.flavoractiv.com
Phone: 01844 354 154

Iceni Brewery
3 Fouldon Road, Ickburgh,
Norfolk IP26 5BJ
Web: www.icenibrewery.co.uk
Phone: 01842 878 922

Johnson Brewing Designs
32 Kirklees Street, Tottington, Bury,
Lancashire BL8 3NJ
Web: www.johnsonbrewing.co.uk
Phone: 01204 887 754

MossBrew Brewing Systems
Ministry of Ale, 9 Trafalgar Street,
Burnley, Lancashire BB11 1TQ
Web: www.mossbrew.co.uk
Phone: 0114 249 2309

PBC (Brewery Installations) Ltd
Unit 4B, Britannia Mill, Cobden Street,
Bury, Lancashire BL9 6AW
Web: www.pbcbreweryinstallations.com
Phone: 01706 214 021

BREWING PLANT & EQUIPMENT

Abbott & Co (Newark) Ltd
Newark Boiler Works, Newark,
Nottinghamshire NG24 2EJ
Web: www.air-receivers.co.uk
Phone: 01636 704 208

Alan Ruddock Engineering
Unit 6 Shepherds Grove Ind.
Estate, Stanton, Bury St Edmunds,
Suffolk IP31 2AR
Web: www.brewing-equipment.co.uk
Phone: 01359 250 989

Axflow
Orion Park, Northfield Ave, Ealing,
London, Greater London W13 9SJ
Web: www.axflow.com
Phone: 020 8579 2111

Brupaks
Unit 12, Honley Business Centre,
New Mill Road, Honley, Holmfirth,
West Yorkshire HD9 6QB
Web: www.brupaks.com
Phone: 0131 629 9788

Buhler Ltd.
20 Atlantis Avenue, London,
Greater London E16 2BF
Web: www.buhlergroup.com
Phone: 020 7055 6650

CBI Fleetwood Ltd.
Plumtree Farm Industrial Estate,
Bircotes, Doncaster,
South Yorkshire DN11 8EW
Web: www.centralbottling.com
Phone: 01302 711 056

Dixon Group Europe Ltd.
350 Walton Summit Centre , Preston,
Lancashire PR5 8AS
Web: www.dixoneurope.co.uk
Phone: 01772 323 529

Don Valley Engineering Company Ltd
Sandall Stones Road, Doncaster,
South Yorkshire DN3 1QR
Web: www.donvalleyeng.com
Phone: 01302 881 188

Eastfield Process Equipment
Eastfield Farm, Tickhill, Doncaster,
South Yorkshire DN11 9JD
Web: www.eastfieldprocessequipment.co.uk
Phone: 01302 751 444

Fulton Boiler Works (Great Britain) Ltd.
Broomhill Road, Bristol,
Bristol BS4 4TU
Web: www.fulton.com
Phone: 0117 972 3322

Hi-Line Services
56 Britannia Way, Lichfield,
Staffordshire WS14 9UY
Web: www.hilineservices.co.uk
Phone: 01543 258 741

J & E Hall International
Questor House, 191 Hawley Road,
Dartford, Kent DA1 1PU
Web: www.jehall.com
Phone: 01322 394 420

Johnson Controls Inc.
72 Buckingham Avenue, Slough,
Berkshire SL1 4PN
Web: www.johnsoncontrols.co.uk
Phone: 01753 693 919

Moeschle U.K.
The Quadrant, 99 Parkway Avenue,
Parkway Business Park, Sheffield,
South Yorkshire S9 4WG
Web: www.moeschle.com
Phone: 01142 270 536

Samson Controls Ltd
Perrywood Business Park, Honeycrock
Lane, Redhill, Surrey RH1 5JQ
Web: www.samsoncontrols.co.uk
Phone: 01737 766 391

Seepex UK Ltd.
3 Armtech Row, Houndstone Business
Park, Yeovil, Somerset BA22 8RW
Web: www.seepex.com
Phone: 01935 472 376

Verder (UK) Ltd
3 California Drive, Castleford,
West Yorkshire WF10 5QH
Web: www.verder.co.uk
Phone: 01924 221 001

Windsor Fabrications Ltd
Park Works, Park Road, Dukinfield, Cheshire SK16 5LX
Web: www.windsor-fabrications.co.uk
Phone: 0161 330 5471

CASKS

CypherCo Limited
Unit 19 Rural Enterprise Centre, Vincent Carey Road, Rotherwas Industrial Estate, Hereford, Herefordshire HR2 6FE
Web: www.cypherco.com
Phone: 01432 343 340

Eurobung
Roe Mead Mill, Far Common Road, Mirfield, West Yorkshire WF14 0DG
Web: www.eurobung.co.uk
Phone: 01924 496 671

Kammac plc
Suite 39 Anglesey House, Anglesey Road, Burton-upon-Trent, Staffordshire DE14 3NT
Web: www.kammac.com
Phone: 01283 743 734

Rotech (Swindon) Ltd
10 Blackworth Industrial Park, Highworth, Swindon, Wiltshire SN6 7NA
Web: www.rotechkeg.co.uk
Phone: 01793 764 700

COMPUTER SERVICES

Chatsworth Computers
Chatsworth House, Millennium Way, Chesterfield, Derbyshire S41 8ND
Web: www.chatsworth.co.uk
Phone: 01246 457 150

Premier Systems Limited
Whiteside Farm, Fritham, Lyndhurst, Hampshire SO43 7HH
Web: www.premiersystems.ltd.uk
Phone: 02380 811 100

The Durham Brewery
Unit 6a, Bowburn North Ind Est., Bowburn, Durham DH6 5PF
Web: www.durhambrewery.co.uk
Phone: 0191 377 1991

Vintner Computer Systems
16 Station Road, Chesham, Buckinghamshire HP5 1DH
Web: www.vintner.co.uk
Phone: 01494 792 539

DISPENSE SYSTEMS

Alumasc Dispense
Unit 3 Acan Business Park, Garrard Way, Telford Way Industrial Estate, Kettering, Northamptonshire NN16 8TD
Web: www.alumascdispense.com
Phone: 01536 311 860

England Worthside Ltd
Hope Mills, Hope Place, Keighley, West Yorkshire BD21 5LJ
Web: www.worthside.co.uk
Phone: 01535 682 222

Hallamshire Brewery Services
Liverpool Street, Sheffield, South Yorkshire S9 2PU
Web: www.hallamshire.co.uk
Phone: 0114 243 1721

IMI Cornelius UK Ltd
Russell Way, Brighouse, Leicestershire HD6 4LX
Web: www.cornelius.com
Phone: 01484 714 584

John Guest
Horton Road, West Drayton, Middlesex UB7 8JL
Web: www.johnguest.com
Phone: 01895 449 233

ENGINEERING SERVICES

ABM Ltd
Pitt Street, Widnes, Cheshire WA8 0TG
Web: www.abm.ltd.uk
Phone: 0151 420 2829

Alectia
Garrett House, 24 Windmill Road, Brentford, Middlesex TW8 0QA
Web: www.penborn.com
Phone: 020 8569 7979

Beverage Process Ltd
PO BOX 25, Ledbury,
Herefordshire HR8 1JF
Web: www.beverageprocess.com
Phone: 01531 631 948

Briggs
Richard Sizer, Briggs House, Derby Street, Burton-upon-Trent,
Staffordshire DE14 2LH
Web: www.briggsplc.co.uk
Phone: 01283 566 661

Endress+Hauser Ltd
Floats Road, Manchester,
Greater Manchester M23 9NF
Web: www.uk.endress.com
Phone: 0161 286 5000

FMA Process Engineering
Three Spires House, Station Road,
Lichfield, Staffordshire WS13 6HX
Web: www.fma.uk.com
Phone: 01543 255 152

GEA Process Engineering Limited
Leacroft House, Leacroft Road,
Birchwood, Warrington,
Cheshire WA3 6JF
Web: www.geaprocess.co.uk
Phone: 01925 812 650

Gilbert Gilkes & Gordon Ltd.
Canal Head North, Kendal,
Cumbria LA9 7BZ
Web: www.gilkes.com
Phone: 01539 720 028

Hanovia
780 Buckingham Avenue, Slough,
Berkshire SL1 4LA
Web: www.hanovia.com
Phone: 01753 515 300

LH Stainless Ltd.
Towiemore, Drummuir, Keith,
Banffshire AB55 5JA
Web: www.l-h-s.co.uk
Phone: 01466 792 222

Logistex
2700 Kettering Parkway, Kettering,
Northamptonshire NN15 6XR
Web: www.logistex.com
Phone: 01536 480 600

Lorien Engineering Solutions Limited
Millennium Court, First Avenue, Centrum 100, Burton-upon-Trent,
Staffordshire DE14 2WH
Web: www.lorienengineering.com
Phone: 01543 444 244

Microdat
Lowfields Road, Leeds,
West Yorkshire LS12 6BS
Web: www.microdat.co.uk
Phone: 0113 244 5225

Moody plc
West Carr Road, Retford,
Nottinghamshire DN22 7SN
Web: www.moodyplc.com
Phone: 01777 701 141

Orapi Applied Limited
Spring Road, Smethwick,
West Midlands B66 1PT
Web: www.orapiapplied.com
Phone: 0121 525 4000

Pall Food and Beverage
Europa House, Havant Street,
Portsmouth, Hampshire PO1 3PD
Web: www.pall.com
Phone: 023 9230 3303

Parker Hannifin Ltd
Warwick, Warwickshire CV34 6TU
Web: www.parker.com
Phone: 01926 317 828

Scott Process Technology Ltd.
41 Main Street, Upper Largo, Leven,
Fife KY8 6EN
Web: www.scottprotec.com
Phone: 01333 360 225

Standfast Precision Engineering
Victoria Street, Craigellachie,
Banffshire AB38 9SR
Web: www.standfast-engineering.com
Phone: 01340 881 371

Sterling Fluid Systems (UK) Ltd
Atlantic Street, Altrincham,
Cheshire WA14 5DH
Web: www.sterlingfluid.com
Phone: 0161 928 6371

Tomlinson Hall & Co Ltd.
Lagonda Road, Billingham,
North Yorkshire TS23 4JA
Web: www.tomlinson-hall.co.uk
Phone: 01642 379 500

Veolia Water Solutions & Technologies
Aqua House, 2620 Kings Court ,
Birmingham Business Park,
Birmingham, West Midlands B37 7YE
Web: www.veoliawaterst.co.uk
Phone: 01628 896 900

EQUIPMENT

Alfa Laval Ltd
7 Doman Road, Camberley,
Surrey GU15 3DN
Web: www.alfalaval.com
Phone: 01276 633 83

Cockayne Ltd.
Scottish Enterprise Technology Park,
East Kilbride, Glasgow G75 0QU
Web: www.cockayne.co.uk
Phone: 01355 272 305

Ferrier Pumps
Unit 5, Portlethen Industrial Estate,
Aberdeen, Aberdeenshire AB12 4PF
Web: www.ferrierpumps.co.uk
Phone: 01224 782 022

MDM Pumps Ltd.
Spring Lane, Malvern,
Worcestershire WR14 1BP
Web: www.mdmpumps.co.uk
Phone: 01684 892 678

Mettler-Toledo Ltd.
64 Boston Road, Beaumont Leys,
Leicester, Leicestershire LE4 1AW
Web: www.mt.com
Phone: 0116 235 7070

Stainless Tanks Ltd
Faraday Close, Snape Lane Ind Est,
Harworth, Doncaster,
South Yorkshire DN11 8RU
Web: www.stainlesstanksltd.co.uk
Phone: 01302 752 010

Vigo Limited
Dunkeswell, Honiton,
Devon EX14 4LF
Web: www.vigoltd.com
Phone: 01404 890 262

HOP MERCHANTS

Botanix
Hop Pocket Lane, Paddock Wood,
Kent TN12 6DQ
Web: www.botanix.co.uk
Phone: 01892 833 415

Charles Faram & Co. Ltd.
The Hopstore, Monksfield Lane,
Newland, Malvern,
Worcestershire WR13 5BB
Web: www.charlesfaram.co.uk
Phone: 01905 830 734

Lupofresh Ltd.
Benover Road, Yalding, Maidstone,
Kent ME18 6ET
Phone: 01622 815 720

Steiner Hops Limited
319A High Street, Epping,
Essex CM16 4DA
Web: www.hopsteiner.com
Phone: 01992 572 331

LABELLING

Darley Limited
Wellington Road, Burton-upon-Trent,
Staffordshire DE14 2AD
Web: www.darleylimited.co.uk
Phone: 01283 564 936

Gavin Watson & Reid Printers
79-109 Glasgow Road, Glasgow,
Glasgow G72 0LY
Web: www.gavinwatson.co.uk
Phone: 01698 826 000

Logopak International Ltd
Grovewood House, Kettlestring Lane, Clifton Moor, York, North Yorkshire YO30 4XF
Web: www.logopak.com
Phone: 01904 692 333

M-Real UK Ltd
Zanders, New Thames Mill, Kemsley, Sittingbourne, Kent ME10 2SG
Web: www.zanders.com
Phone: 01628 411 833

NSD International
Dalkeith, Midlothian EH22 4AF
Web: www.nsdinternational.co.uk
Phone: 0131 654 2800

Weber Marking System UK
Macmerry Industrial Estate, Macmerry, East Lothian EH33 1HD
Web: www.webermarking.com
Phone: 01875 611 111

LABORATORY, ANALYSIS & TECHNICAL SUPPORT

Anton Paar
13 Harforde Court, John Tate Road, Hertford, Hertfordshire SG13 7NW
Web: www.anton-paar.com
Phone: 01992 514 730

Bellingham + Stanley Ltd.
Longfield Road, North Farm Industrial Estate, Tunbridge Wells, Kent TN2 3EY
Web: www.bellinghamandstanley.com
Phone: 01892 500 400

Cara Technology
Randalls Road, Leatherhead, Surrey KT22 7RY
Web: www.cara-online.com
Phone: 01372 822 218

Gas Measurement Instruments Ltd
Inchinnan Business Park, Renfrew, Renfrewshire PA4 9RG
Web: www.gmiuk.com
Phone: 0141 812 3211

GSPK Design Limited
Knaresborough Technology Park, Manse Lane, Knaresborough, North Yorkshire HG5 8LF
Web: www.gspkdesign.ltd.uk
Phone: 01423 865 641

Hach Ultra Analytics
Unit 4 Holmewood Business Park, Chesterfield Road, Holmewood, Chesterfield, Derbyshire S42 5US
Web: www.hachultra.com
Phone: 01246 599 760

Integrated Scientific Ltd
Unit 3, Aspen Court, Aspen Way, Centurion Business Park, Templeborough, Rotherham, South Yorkshire S60 1FB
Web: www.integsci.com
Phone: 01709 830 493

Skalar
Breda House, Millfield Industrial Estate, Wheldrake, York, North Yorkshire YO19 6NA
Web: www.skalar.com
Phone: 01904 444 800

Stevenson Reeves Ltd
40 Oxgangs Bank, Edinburgh, Edinburgh EH13 9LH
Web: www.stevenson-reeves.co.uk
Phone: 0131 445 7151

Strathkelvin Instruments Limited
Rowantree Avenue, Motherwell, North Lanarkshire ML1 5RX
Web: www.strathkelvin.com
Phone: 01698 730 400

The Tintometer Limited
Lovibond House, Solar Way, Solstice Park, Amesbury, Wiltshire SP4 7SZ
Web: www.tintometer.com
Phone: 01980 664 800

MALTS, SUGARS & ADJUNCTS

Bairds Malt Ltd.
Elliot Industrial Estate, Arbroath, Angus DD11 2NJ
Web: www.bairds-malt.co.uk
Phone: 01241 870 431

Crisp Malting Group Limited
Great Ryburgh, Fakenham,
Norfolk NR21 7AS
Web: www.crispmalt.com
Phone: 01328 829 391

D.D. Williamson (UK) Ltd.
Trafford Park, Manchester,
Greater Manchester M17 1PA
Web: www.caramel.com
Phone: 0161 886 3345

French & Jupp's
The Maltings, Stanstead Abbotts, Ware,
Hertfordshire SG12 8HG
Web: www.frenchandjupps.com
Phone: 01920 870 015

Greencore Malt Ltd
24 Eastern Way, Bury St Edmunds,
Suffolk IP32 7AD
Web: www.greencore.ie
Phone: 01284 772 000

Micronized Food Products
Standard Way, Northallerton,
North Yorkshire DL6 2XA
Web: www.micronizedfoodproducts.co.uk
Phone: 01609 751 000

Muntons plc
Cedars Maltings, Stowmarket,
Suffolk IP14 2AG
Web: www.muntons.com
Phone: 01449 618 300

Ragus Sugars (Manufacturing) Ltd.
193 Bedford Avenue, Slough,
Berkshire SL1 4RT
Web: www.ragus.co.uk
Phone: 08000 279 769

Simpsons Malt Limited
Tweed Valley Maltings, Berwick-upon-Tweed, Northumberland TD15 2UZ
Web: www.simpsonsmalt.co.uk
Phone: 01289 330 033

Thomas Fawcett & Sons Ltd.
EastField Lane, Castleford,
West Yorkshire WF10 4LE
Web: www.fawcett-maltsters.co.uk
Phone: 01977 552 460

Tuckers Maltings
Teign Road, Newton Abbot,
Devon TQ12 4AA
Web: www.tuckersmaltings.com
Phone: 01626 334 002

Warminster Maltings Ltd
39 Pound Street, Warminster,
Wiltshire BA12 8NN
Web: www.warminster-malt.co.uk
Phone: 01985 212 014

PACKAGING SUPPLIERS

Amcor
83 Tower Road North, Warmley,
Bristol BS30 8XP
Web: www.amcor.com
Phone: 0117 975 3200

Garthwest Ltd.
Rotterdam Road, Sutton Fields,
Kingston Upon Hull, East Riding of
Yorkshire HU7 0XA
Web: www.garthwest.com
Phone: 01482 825 121

Graphic Packaging
Filwood Road, Bristol,
Bristol BS16 3SB
Web: www.graphicpkg.com
Phone: 01179 656 222

PROCESS AIDS

I S P (International Speciality Products)
Waterfield, Pitwood Park, Tadworth,
Surrey KT20 5JL
Web: www.ispcorp.com
Phone: 01737 377 000

Murphy & Son Limited
Alpine Street, Old Basford, Nottingham,
Nottinghamshire NG6 0HQ
Web: www.murphyandson.co.uk
Phone: 0115 978 5494

PQ Silicas UK Limited
Bank Quay, PO Box 26, Warrington,
Cheshire WA5 1AB
Web: www.caramel.com
Phone: 01925 416 100

PROMOTIONAL ITEMS

Festival Glass Ltd
Unit 6 Calderdale Business Park, Club Lane, Ovenden, Halifax, West Yorkshire HX2 8DB
Web: www.festivalglass.co.uk
Phone: 01422 382 696

Invicta Plastics Limited
86-90 Scudamore Road, Leicester, Leicestershire LE3 1UQ
Web: www.invictagroup.co.uk
Phone: 01162 817 159

Mosaic Board Printers Ltd
1-2, Pytchley Lodge Road, Pytchley Lodge Road Industrial Estate, Kettering, Northamptonshire NN15 6JQ
Web: www.mosaic-boardprint.com
Phone: 01536 312 800

Riedel Crystal
22, Imex Technology Park, Trentham Lakes South, Stoke-on-Trent, Staffordshire ST4 8LJ
Web: www.riedel.com
Phone: 01782 646 105

The Beer Essentials
30a East Street, Horsham, Sussex RH12 1HL
Web: www.thebeeressentials.co.uk
Phone: 01403 218 890

The Beer Mat Company
Cariocca Business Park, 2 Sawley Road, Manchester, Greater Manchester M40 8BB
Web: www.beermatcompany.co.uk
Phone: 0161 401 0151

The Katz Group (Katz Quarmby)
Creative House, Units 22 – 23 Howley Park Business Village, Howley Park Road, Morley, West Yorkshire LS27 0BZ
Web: www.thekatzgroup.com
Phone: 0113 393 6390

SUPPORT SERVICES

CGA Strategy Ltd
Waterloo Place, Watson Square, Stockport, Cheshire SK1 3AZ
Web: www.cgastrategy.co.uk
Phone: 0161 476 8330

Klenzan Ltd
2 Cameron Court, Winwick Quay, Warrington, Cheshire WA2 8RE
Web: www.klenzan.co.uk
Phone: 01925 234 696

Simon Storage Ltd
Seal Sands Terminal, Seal Sands, Middlesborough, Cleveland TS2 1UB
Web: www.simonstorage.com
Phone: 01642 546 775

THE MICROBREWERS' DIRECTORY

We have done our best to provide the most comprehensive list of microbreweries in the UK, at the time of going to press (June 2013). The Society of Independent Brewers (SIBA) defines a microbrewer as one that produces up to 5,000 HL per annum. It's not an exact science, but we have tried to be as accurate as we possibly can in compiling this list. Inevitably some breweries may have slipped through the net, and some may be reproduced here that fall outside of SIBA's definitions. We hope you will let us know.

ENGLAND

BEDFORDSHIRE

B & T Brewery Ltd
The Brewery, Shefford,
SG17 5DZ
01462 815 080
www.banksandtaylor.com

White Park Brewery
Perry Hill Farm, Bourne End Road, Cranfield, MK43 0BA
01223 911 357
www.whiteparkbrewery.co.uk

BERKSHIRE

Binghams Brewery Ltd
Unit 10 Tavistock Industrial Estate, Ruscombe Lane, Ruscombe, Reading, RG10 9NJ
01189 344 376
www.binghams.co.uk

Butts Brewery Ltd
Unit 6a Northfield Farm Industrial Estate, Wantage Road, Great Shefford, Hungerford, RG17 7BY
01488 648 133
www.buttsbrewery.com

Siren Craft Brew Ltd
Unit 1 Weller Drive, Finchampstead, Wokingham, RG40 4QZ
0118 973 0929
www.sirencraftbrew.com

Two Cocks Brewery
Christmas Farm, Enborne, Newbury, RG20 0HB
07876 594 501
www.twococksbrewery.com

West Berkshire Brewery Co
The Old Bakery, Yattendon, Thatcham, RG18 0UE
01635 202 968
www.wbbrew.co.uk

Windsor & Eton Brewing Company Limited
Unit 1 Vansittart Estate, Windsor, SL4 1SE
01753 854 075
www.webrew.co.uk

Zerodegrees Reading
9 Bridge Street, Reading, RG1 2LR
01189 597 959
www.zerodegrees.co.uk

BRISTOL

Arbor Ales Ltd
The Old Tavern, Blackberry Hill, Bristol, BS16 1DB
07970 920 627
www.arborales.co.uk

Bath Ales
Units 3-7 Caxton Industrial Estate, Tower Road North, Bristol, BS30 8XN
01179 474 797
www.bathales.com

Bristol Beer Factory
The Beer Factory, Unit A The Old Brewery, Durnford Street, Bristol, BS3 2AW
01179 026 317
www.bristolbeerfactory.co.uk

Butcombe Brewery Ltd
Butcombe, Bristol, BS40 5PA
01275 472 240
www.butcombe.com

Towles' Fine Ales
Unit 11 Circuit 32, Easton Road, Bristol, BS5 0DB
0117 321 3188
www.towlesfineales.co.uk

Twisted Oak Brewery Ltd
The Brewery, Yeowood Farm, Iwood Lane, Wrington, BS40 5NU
01275 874 865
www.twistedoakbrewery.co.uk

Zerodegrees Bristol
53 Colston Street, Bristol, BS1 5BA
01179 252 706
www.zerodegrees.co.uk

BUCKINGHAMSHIRE

Buckingham Brewery Ltd
Unit 3 Hillcrest Way, Buckingham Industrial Estate, Buckingham, MK18 1HJ
01280 422 830
www.buckingham-brewery.co.uk

Chiltern Brewery
Nash Lee Road, Terrick, Aylesbury, HP17 0TQ
01296 613 647
www.chilternbrewery.co.uk

Concrete Cow Brewery
59 Alston Drive, Bradwell Abbey, Milton Keynes, MK13 9HB
01908 316 794
www.concretecowbrewery.co.uk

Hopping Mad Brewery
42 Yardley Road, Olney, MK46 5ED
01234 919 200
www.hoppingmad.com

Malt The Brewery Ltd
Collings Hanger Farm, 100 Wycombe Road, Prestwood, Great Missenden, HP16 0HP
01494 865 063
www.maltthebrewery.co.uk

Old Luxters Farm Brewery
Chiltern Valley Vineyard, Hambleden, Henley-on-Thames, RG9 6JW
01491 638 330
www.chilternvalley.co.uk/brewery.html

Oxfordshire Ales
12 Pear Tree Farm Industrial Units, Bicester Road, Marsh Gibbon, Bicester, OX27 0GB
01869 278 765
www.oxfordshireales.co.uk

Rebellion Beer Company
Bencombe Farm, Marlow Bottom, Marlow, SL7 3LT
01628 476 594
www.rebellionbeer.co.uk

Vale Brewery Co
Tramway Business Park, Brill, HP18 9TY
01844 290 008
www.valebrewery.co.uk

XT Brewing Company
Unit 27 Notley Farm, Chearsley Road, Long Crendon, Aylesbury, HP18 9ER
01844 208 310
www.xtbrewing.com

CAMBRIDGESHIRE

Bexar County Brewery
8 Belgic Square, Peterborough, PE1 5XF
07934 722 584
www.bexarcountybrewery.com

BlackBar Brewery
B3 Button End Industrial Estate, Harston, Cambridge, CB22 7GX
01223 872 131
www.blackbar.co.uk

Cambridge Moonshine Brewery
28 Radegund Road, Cambridge, Cambridge, CB1 3RS
07906 066 794

City of Cambridge Brewery
Ely Road, Chittering, Cambridge, CB5 9PH
01223 864 864
www.cambridge-brewery.co.uk

Devil's Dyke Brewery
8 Fair Green, Reach, Cambridge, CB25 0JD
01638 743 816

Hereward Brewery
50 Fleetwood, Ely, CB6 1BH
01353 666 441

Lord Conrad's Brewery
Unit 21 Dry Drayton Industries, Scotland Road, Dry Drayton, Cambridge, CB23 8AT
07736 739 700
www.lordconradsbrewery.co.uk

Mile Tree Brewery Ltd
Secret Garden Park, Miletree Lane, Wisbech, PE13 4TR
07858 930 363

Milton Brewery
Unit 111 Norman Industrial Estate, Cambridge Road, Milton, Cambridge, CB4 6AT
01223 226 198
www.miltonbrewery.co.uk

Oakham Ales
2 Maxwell Road, Woodston, Peterborough, PE2 7JB
01733 370 500
www.oakhamales.com

Son of Sid Brewery
The Chequers, 71 Main Street, Little Gransden, Sandy, SG19 3DW
01767 677 348
www.sonofsid.co.uk

The Red Brewery Company Limited
The Orchard, Garden Farm, Great Staughton, PE19 5BE
07827 294 229
www.redbrewery.co.uk

Tydd Steam Brewery
Kirkgate, Tydd St. Giles, Wisbech, PE13 5NE
01945 871 020
www.tyddsteam.co.uk

CHESHIRE

4T's Brewery Limited
Unit 15 EBI Centre, Picow Farm Road, Runcorn, WA7 4UA
07917 730 184
www.4tsbrewery.co.uk

Beartown Brewery
Bromley House, Spindle Street, Congleton, CW12 1QN
01260 299 964
www.beartownbrewery.co.uk

Blacker & Son Brewery
3 Chester Street, Saltney, Chester, CH4 8BL
01244 676 765
www.blacker-brewery.co.uk

Blakemere Brewery
Blakemere Brewery, Blakemere Craft Centre, Chester Road, Sandiway, CW8 2EB
01606 301 000
www.norbrew.co.uk

Blue Ball Brewery
Units 11-12 EBL Centre, Picow Farm Road, Runcorn, WA7 4UA
01928 238 330
www.blueballbrewery.com

Bollington Brewing Company Ltd
c/o Vale Inn, Adlington Road, Bollington, SK10 5JT
01625 575 147
www.valeinn.co.uk

Brimstage Brewing Co Ltd
Home Farm, Brimstage Lane, Brimstage, CH63 6LY
0151 342 1181
www.brimstagebrewery.com

Dunham Massey Brewing Company
100 Oldfield Lane, Dunham Massey, Altrincham, WA14 4TY
0161 929 0663
www.dunhammasseybrewing.co.uk

Frodsham Brewery
Lady Heyes Craft Centre, Kingsley Road, Frodsham, WA6 6SU
01928 787 917
www.frodshambrewery.co.uk

Front Row Brewing
Unit 1 Hopkins Close, Greenfield Farm Industrial Estate, Congleton, CW12 4TR
07861 718 673
www.frontrowbrewing.co.uk

Happy Valley Brewery
8 Hazelhurst Drive, Bollington, Macclesfield, SK10 5QT
07758 512 080
www.happyvalleybrewery.co.uk

Merlin Brewing Company Ltd
Unit 3 Spring Bank Farm, Congleton Road, Arclid, Sandbach, CW11 2UD
01477 500 893
www.merlinbrewing.co.uk

Offbeat Brewery Ltd
Unit 6 Thomas Street, Crewe, CW1 2BD
07502 096 438
www.offbeatbrewery.com

Quantum Brewing Company
Unit 4 Victoria Works, Hempshaw Lane, Stockport, SK1 4LG
07976 032 465
www.quantumbrewingcompany.co.uk

Redwillow Brewery Ltd
Sutton Mill, Gunco Lane, Macclesfield, SK11 7JL
01625 502 315
www.redwillowbrewery.com

Spitting Feathers Brewery
Common Farm, Common Lane, Waverton, Chester, CH3 7QT
01244 332 052
www.spittingfeathers.org

Storm Brewing Company Ltd
2 Waterside, Macclesfield, SK11 7HJ
01625 431 234

Tatton Brewery Ltd
Unit 7, Longridge Trading Estate, Knutsford, WA16 8PR
07738 150 898
www.tattonbrewery.co.uk

The Cheshire Brewhouse
Unit 5 Daneside Business Park, Riverdane Road, Congleton, CW12 1UN
07830 304 929
www.cheshirebrewhouse.co.uk

The Coach House Brewing Company Ltd
Wharf Street, Warrington, WA1 2DQ
01925 232 800
www.coach-house-brewing.co.uk

WC Brewery
3 Micklegate, Mickle Trafford, Chester, CH2 4TF
www.wcbrewery.com

Weetwood Ales Ltd
Weetwood Grange, Weetwood, Torporley, CW6 0NQ
01829 752 377
www.weetwoodales.co.uk

Woodlands Brewing Company
Unit 3 Meadow Lane Farm, London Road, Stapeley, Nantwich, CW5 7JU
01270 841 511
www.woodlandsbrewery.co.uk

CORNWALL

Atlantic Brewery
Treissaac Farm, Treissaac, Newquay, TR8 4DX
0870 042 1714
www.atlanticbrewery.com

Blue Anchor
50 Coinagehall Street, Helston, TR13 8EL
01326 562 821
www.spingoales.com

Coastal Brewery
Unit 9b Cardrew Industrial Estate, Redruth, TR15 1SS
07875 405 407
www.coastalbrewery.co.uk

Cornish Crown Brewery
Badgers Cross Farm Units, Badgers Cross, Gulval, Penzance, TR20 8XE
01736 351 070

Driftwood Spars
Trevaunance Cover, Quay Road, St Agnes, TR5 0RT
01872 552 428
www.driftwoodspars.com

Frys Brewery Ltd
Trerice, Boyton, Launceston, PL15 8NU
01566 785 840
www.frysbrewery.co.uk

Hogswood Brewing Company
Higher Goshen, Mithian, St Agnes, TR5 0QE
01872 554 224
www.hogswood.co.uk

Keltek Brewery
Candela House, Cardrew Industrial Estate, Redruth, TR15 1SS
01209 313 620
www.keltekbrewery.co.uk

Lizard Ales Ltd
Old Bunker, Treleaver, Coveack, Helston, TR12 6SE
01326 281 135
www.lizardales.co.uk

Longhill Brewery
Longhill Cottage, Whitstone, Holsworthy, EX22 6UG
01288 341 466

Padstow Brewing Company llp
The Brewery, Unit 4a Trecerus Industrial Estate, Padstow, PL28 8RW
01841 532 169
www.padstowbrewing.co.uk

Penpont Brewery
Inner Trenarrett, Altarnun, Launceston, PL15 7SY
01566 86 069
www.penpontbrewery.co.uk

Penzance Brewing Company Ltd
Star Inn, Crowlas, Penzance, TR20 8DX
01736 740 375

Skinner's Brewing Co. Ltd
Riverside, Newham Road, Truro, TR1 2DP
01872 271 885
www.skinnersbrewery.com

The Harbour Brewing Company Ltd
Trekillick Farm, Bodmin, PL30 5BB
01208 832 131
www.harbourbrewing.com

The Rebel Brewing Company
West End Industrial Estate, Penryn, TR10 8RT
01326 378 517

Wooden Hand Brewery
Unit 3 Grampound Road Industrial Estate, Grampound Road, Truro, TR2 4TB
01726 884 596
www.woodenhand.co.uk

CUMBRIA

Barngates Brewery Ltd
Barngates, Ambleside, LA22 0NG
01539 436 575
www.barngatesbrewery.co.uk

Beckstones Brewery
Upper Beckstones Mill, The Green, Millom, LA18 5HL
01229 775 294
www.beckstonesbrewery.co.uk

Blackbeck Brewery
Blackbeck Hotel, Blackbeck, Egremont, CA22 2NY
01946 841 661
www.blackbeckbrewery.co.uk

Bowness Bay Brewery Ltd
Silver Badge Works, Crag Brow, Bowness-on-Windermere, Windermere, LA23 3BX
www.bownessbaybrewery.co.uk

Castle Brewery
Cockermouth, CA13 9NE
01900 823 214
www.jenningsbrewery.co.uk

Cumberland Breweries Ltd
The Forge, Great Corby, Carlisle, CA4 8LR

Cumbrian Legendary Ales
Old Hall Brewery, Hawkshead, Ambleside, LA22 0QF
01539 436 436
www.cumbrianlegendaryales.com

Dent Brewery
Hollins, Cowgill, Sedburgh, LA10 5TQ
01539 625 326
www.dentbrewery.co.uk

Derwent Brewery
Units 2a-2b Station Road Industrial Estate, Silloth, Wigton, CA7 4AG
01697 331 522

Eden Brewery Ltd
Brougham Hall, Brougham, Penrith, CA10 2DE
01768 210 565
www.edenbrewery.com

Fell Brewery Ltd
Unit 27 Airfield Approach Business Park, Grange-over-Sands, Flookburgh, LA11 7NG
01539 558 980
www.fellbrewery.co.uk

Foxfield Brewery
Prince of Wales, Foxfield, Broughton in Furness, LA20 6BX
01229 716 238
www.princeofwalesfoxfield.co.uk

Geltsdale Brewery Ltd
Unit 6 Old Brewery Yard, Craw Hall, Brampton, CA8 1TR
01697 741 541
www.geltsdalebrewery.com

Hardknott Brewery
The Woolpack Inn, Boot, Holmrook, CA19 1TH
01946 723 230
www.woolpack.co.uk

Hawkshead Brewery Ltd
Mill Yard, Staveley, LA8 9LR
01539 822 644
www.hawksheadbrewery.co.uk

Hesket Newmarket Brewery
Old Crown Barn, Hesket Newmarket, CA7 8JG
01697 478 066
www.hesketbrewery.co.uk

Keswick Brewing Co
The Old Brewery, Brewery Lane, Keswick, CA12 5BY
01768 780 700
www.keswickbrewery.co,uk

Kirkby Lonsdale Brewery
Old Station Yard, Kirkby Lonsdale, LA6 2HP
01524 272 221
www.kirkbylonsdalebrewery.com

Loweswater Brewery
Kirkstile Inn, Loweswater, Cockermouth, CA13 0RU
01900 85 219
www.kirkstile.com

Strands Brewery
Strands Hotel, Wasdale, Seascale, CA20 1ET
01946 726 237
www.strandshotel.com

The Bitter End Pub & Brewery
15 Kirkgate, Cockermouth, CA13 9PJ
01900 828 993
www.bitterend.co.uk

The Coniston Brewing Company Co.Ltd
Coppermines Road, Coniston, LA21 8HL
01539 441 133
www.conistonbrewery.com

The Ennerdale Brewery Ltd
Croasdale Farm Barn, Ennerdale, Cleator, CA23 3AT
01946 861 755

The Great Gable Brewing Co Ltd
Wasdale Head Inn, Wasdale Head, Seascale, CA20 1EX
01946 726 229
www.greatgablebrewing.com

Ulverston Brewing Company
Diamond Buildings, Pennington Lane, Lindal-in-Furness, LA12 0LE
01229 584 280
www.ulverstonbrewing.co.uk

Unsworth's Yard Brewery Ltd
Unit 4 Unsworth's Yard, Cartmel, LA11 6PN
07966 681 706
www.unsworthsyardbrewery.co.uk

Watermill Brewing Company
Watermill Inn, Ings, Kendal, LA8 9PY
01539 821 309
www.watermillinn.co.uk

Yates Brewery
Ghyll Farm, Westnewton, Wigton, CA7 3NX
01697 321 081
www.yatesbrewery.co.uk

DERBYSHIRE

Amber Ales Ltd
PO Box 7277, Ripley, DE5 4AP
01773 512 804
www.amberales.co.uk

Ashover Brewery
1 Butts Road, Ashover, Chesterfield, S45 0EW
01246 590 888
www.ashoverbrewery.com

Barlow Brewery
Unit 6 Church Farm, Hackney Lane, Barlow, Dronfield, S18 7TR
01142 891 767
www.barlowbrewery.co.uk

Blue Monkey Brewing Ltd
Unit 1 Enterprise Court, Manners Ave, Manners Industrial Estate, Ilkeston, DE7 8EW
07500 555 595

Bottle Brook Brewery
10 Church Street, Kilburn, Belper, DE56 0LU
07971 189 915

Brampton Brewery Ltd
Unit 5 Chatsworth Business Park, Chatsworth Road, Chesterfield, S40 2AR
01246 221 680
www.bramptonbrewery.co.uk

Bumpmill Brewery
The Cottage, Hallfieldgate, Shirland, Alfreton, DE55 6AG
01773 830 431
www.bumpmillbrewery.co.uk

Buxton Brewery Company Ltd
Unit 7 Staden Business Park, Staden Lane, Buxton, SK17 9RZ
01298 244 200
www.buxtonrealale.co.uk

Coppice Side Brewery Ltd
Unit 3 Small Businees Centre, Adams Close, Heanor Gate Industrial Estate, Heanor, DE75 7SW
07790 305 682
www.coppicesidebrewery.com

Dancing Duck Beer Ltd
Unit 1 John Cooper Buildings, Payne Street, Derby, DE22 3AZ
01332 205 582
www.dancingduckbrewery.com

Derby Brewing Co Ltd
Masons Place, Nottingham Road, Derby, DE21 6AQ
01332 242 888
www.derbybrewing.co.uk

Derventio Brewery
Trusley Brook Farm, Sutton on the Hill, Ashbourne, DE6 5JP
07816 878 129
www.derventiobrewery.co.uk

Falstaff Brewery
24 Society Place, Derby, DE23 6UH
01332 342 902
www.falstaffbrewery.co.uk

Funfair Brewing Company
34 Spinney Road, Ilkeston, DE7 4LH
07971 540 186
www.funfairbrewingcompany.com

Haywood Brewery
Callow Top Holiday Park, Buxton Road, Sandybrook, Ashbourne, DE6 2AQ
01335 344 020
www.callowtop.co.uk

Howard Town Brewery Ltd
Unit 10 Howard Town Mill, Mill Street, Glossop, SK13 8PT
01457 869 800
www.howardtownbrewery.co.uk

John Thompson Inn and Brewery
Ingleby Lane, Ingleby, Melbourne, DE73 1HW
01332 862 469
www.johnthompsoninn.com

Leadmill Brewery Co
Park Hall, Park Hall Road, Ripley, DE5 8PX
01332 883 577
www.leadmillbrewery.co.uk

Leatherbritches Brewery
Bentley Brook, Fenny Bentley, Ashbourne, DE6 1LF
01335 350 278
www.bentleybrookinn.co.uk

North Star Brewing Company Ltd
Unit 6 Gallows Industrial Estate, Furness Road, Ilkeston, DE7 5EP
07521 961 881
www.northstarbeers.co.uk

Nutbrook Brewery Ltd
6 Hallam Way, West Hallam, DE7 6LA
0800 458 2460
www.nutbrookbrewery.com

Peak Ales
The Barn Brewery, Chatsworth, Bakewell, DE45 1EX
01246 583 737
www.peakales.co.uk

Raw Brewing Company Limited
Unit 3 & 4 Silver House, Adelphi Way, Staveley, Chesterfield, S43 3LS
01246 475 445
www.rawbrew.co.uk

Spire Brewery
Unit 2-3 Gisbourn Close, Ireland Business Park, Staveley, Chesterfield, S43 3JT
01246 476 005
www.spirebrewery.co.uk

The Brunswick Brewery Ltd
1 Railway Terrace, Derby, DE1 2RU
01332 290 677
www.brunswickinn.co.uk

The Hope Valley Brewing Company Ltd
Castleton Youth Hostel, Hope Valley, S33 8WG
0845 371 9628

The Marlpool Brewing Company Limited
5 Breach Road, Heanor, DE75 7NJ
01773 711 285
www.marlpoolbrewing.co.uk

The Shottle Farm Brewery
School House Farm, Lodge Lane, Shottle, Belper, DE56 2DS
01773 550 056
www.shottlefarmbrewery.co.uk

Thornbridge Brewery
Thornbridge Hall, Ashford-in-the-Water, Bakewell, DE45 1NZ
01629 640 617
www.thornbridgehall.co.uk

Tollgate Brewery
Unit 8 Viking Business Centre, High Street, Woodville, Swadlincote, DE11 7EH
01283 229 194

Townes Brewery
Speedwell Inn, Lowgates, Staveley, Chesterfield, S43 3TT
01246 472 252

Wentwell Brewery
15 Wingfield Drive,
Chaddesden, Derby, DE21 4PW
07900 475 755
www.wentwellbrewery.com

Whim Ales Ltd
Whim Farm, Hartington,
Buxton, SK17 0AX
01298 849 91
www.whimales.co.uk

Wirksworth Brewery
25 St John Street,
Wirksworth, DE4 4DR
01629 824 011
www.wirksworthbrewery.co.uk

DEVON

Barum Brewery Ltd
c/o The Reform Inn, Reform
Street, Barnstaple, EX31 1PD
01271 329 994
www.barumbrewery.co.uk

Bays Brewery Ltd
Yalberton Industrial Estate,
Aspen Way, Paignton, TQ4 7QR
01803 555 004
www.baysbrewery.co.uk

Branscombe Vale Brewery
Great Seaside Farm,
Branscombe, EX12 3DP
01297 680 511

Clearwater Brewery
2 Devon Units, Hatchmoor
Industrial Estate, Torrington,
EX38 7HP
01805 625 242

Country Life Brewery
The Big Sheep, Abbotsham,
Bideford, EX39 5AP
01237 420 808
www.countrylifebrewery.com

Dartmouth Brewery
Dartmouth Inn, 63 East Street,
Newton Abbot, TQ12 2JP
01626 353 451

Devon Earth Brewery
7 Fernham Terrace, Torquay
Road, Paignton, TQ3 2AQ
07927 397 871
www.devonearthbrewery.co.uk

Exe Valley Brewery
Land Farm, Silverton,
Exeter, EX5 4HF
01392 860 406

Exmoor Ales
Golden Hill Brewery,
Wiveliscombe, TA4 2NY
01984 623 798
www.exmoorales.co.uk

Forge Brewery
Ford Hill Forge, Hartland,
Bideford, EX39 6EE
01237 440 015
www.forgebrewery.co.uk

Gargoyles Brewery
Court Farm, Holcombe Village,
Dawlish, EX7 0JT
07773 444 501

Holsworthy Ales
Unit 5 Circuit Business Park,
Clawton, Holsworthy, EX22 6RR
01566 783 678
www.holsworthyales.co.uk

Hunter's Brewery Ltd
Glebe Acres, Orley Road,
Ipplepen, Newton Abbot,
TQ12 5SA
01803 814 399

Jollyboat Brewery (Bideford) Ltd
The Coach House, Buttgarden
Street, Bideford, EX39 2AU
01237 424 343

Noss Beer Works
Unit 6 Ash Court, East Way,
Lee Mill Industrial Estate,
Ivybridge, PL21 9GE
07977 479 634
www.nossbeerworks.co.uk

O'Hanlon's Brewing
Great Barton Farm,
Whimple, EX5 2NY
01404 822 412
www.ohanlons.co.uk

Otter Brewery
Mathayes, Luppit,
Honiton, EX14 4SA
01404 891 285
www.otterbrewery.com

Plymouth Beer Company
City College Plymouth,
Kings Road, Devonport,
Plymouth, PL1 5QG
07899 730 036
www.plymouthbeercompany.com

Princetown Breweries Ltd
The Brewery, Station Road,
Princetown, PL20 6QX
01822 890 789
www.princetownbreweries.co.uk

Red Rock Brewery Ltd
Higher Humber Farm, Humber,
Teignmouth, TQ14 9TD
01626 879 738
www.redrockbrewery.co.uk

Ringmore Craft Brewery
Ringmore Barn, Higher
Ringmore Road, Shaldon,
TQ14 0HG
01626 873 114
www.ringmorecraftbrewery.co.uk

South Hams Brewery Co Ltd
Stokeley Barton Barns,
Stokenham, Kingsbridge,
TQ7 2SE
01548 581 151
www.southhamsbrewery.co.uk

Summerskills Brewery
Unit 15, Pomphlett Farm
Industrial Estate, Billacombe,
Plymouth, PL9 7BG
01752 481 283
www.summerskills.co.uk

Tavy Ales
c/o Jasmine Cottage, Peter
Tavy, Tavistock, PL19 9NN
07966 522 266
www.tavyales.co.uk

Teignworthy Brewery
The Maltings, Teign Road,
Newton Abbott, TQ12 4AA
01626 332 066
www.teignworthybrewery.com

The Beer Engine
Newton St Cyres, Exeter,
EX5 5AX
01392 851 282
www.thebeerengine.co.uk

The Bridgetown Brewery
The Albert Inn, Bridgetown
Close, Totnes, TQ9 5AD
01803 863 214

The Devon Brewing Company
Lower Yelland Farm, Yelland,
Barnstaple, EX31 3EN
07976 724 243
www.devonbeer.co.uk

The Exeter Brewery Ltd
Unit 5 Lionsrest Estate,
Station Road, Exminster,
Exeter, EX6 8DZ
01392 823 013
www.theexeterbrewery.co.uk

Two Beach Brewing Co Ltd
Cafe-ODE, Ness Drive, Shaldon,
Teignmouth, TQ14 0HP

Union Brewery
Fore Street, Holbeton,
Plymouth, PL8 1NE
01752 830 460
www.dartmoorunion.co.uk

Wizard Ales
Unit 4 Lundy View, Mullacott
Cross Industrial Estate,
Ilfracombe, EX34 8PY
01271 865 350
www.wizardales.co.uk

DORSET

Art Brew
North End Farm, Venn Lane,
Chideock, Bridport, DT6 6JY
07881 783 626
www.artbrew.co.uk

Dorset Brewing Company
Brewers Quay, Hope Square,
Weymouth, DT4 8TR
01305 777 515
www.dbcales.com

Isle of Purbeck Brewery
Bankes Arms Hotel, Manor
Road, Studland, Swanage,
BH19 3AU
01929 450 227

Mighty Hop Brewery Ltd
Silverdale, Woodmead Road,
Lyme Regis, DT7 3AD
01297 445 358
www.mightyhopbrewery.co.uk

Sherborne Brewery Ltd
257 Wesbury, Sherbourne,
DT9 3EH
01935 817 307
www.sherbornebrewery.co.uk

Small Paul's Brewery
27 Briar Close, Gillingham,
SP8 4SS
01747 823 574

Sunny Republic Brewing Company
The Old Grain Barns, North West Farm, West Street, Winterborne Kingston, DT11 9AT
01929 471 600
www.sunnyrepublic.com

Town Mill Brewery
Mill Lane, Lyme Regis, DT7 3PU
01297 444 354
www.townmillbrewery.com

DURHAM

Consett Ale Works
The Grey Horse, 115 Sherburn Terrace, Consett, DH8 6NE
01207 591 540
www.thegreyhorse.co.uk

Double Maxim Beer Company Ltd
1 Gadwall Road, Houghton Le Spring, DH4 5NL
0191 584 8844
www.maximbrewery.co.uk

Hill Island Brewery
Unit 7 Fowlers Yard, Back Silver Street, Durham, DH1 3RA
07740 932 584

Jarrow Brewing Company
Robin Hood Inn, Primrose Hill, Jarrow, NE32 5UB
0191 483 6792
www.jarrowbrewery.co.uk

Just a Minute Brewery
Coulson Street, Spennymoor, DL16 7RS
07586 896 091
www.justaminutebrewery.co.uk

Temptation Brewing Company
Unit 18D Cherry Way, Dubmire Industrial Estate, Houghton Le Spring, DH4 5RJ
07932 774 745
www.temptationbrewingcompany.co.uk

The Four Alls Brewery
Four Alls Hotel, Ovington, Richmond, DL11 7BP
01833 627 302
www.thefouralls-teesdale.co.uk

The Gambling Man Brewing Company
21 Buchanan Street, Hebburn, NE31 1NB
07545 464 968
www.gamblingmanbrewco.com

The Yard of Ale Brewing Company Ltd
Surtees Arms, Chilton Lane, DL17 0DH
01740 655 724
www.thesurteesarms.co.uk

Wear Valley Brewery Ltd
The Grand Hotel, Holdforth Crest, Bishop Auckland, DL14 6DU
01388 601 956
www.the-grand-hotel.co.uk

EAST RIDING OF YORKSHIRE

Big River Brewery Limited
48 Grange Park, Brough, HU15 1AA
07737 820 922
www.bigriverbrewery.co.uk

Bird Brain Brewery Ltd
30 Hailgate, Howden, Goole, DN14 7SL
01430 432 166

Brass Castle Brewery
Market Street, Pocklington, York, YO42 2AB
www.brasscastlebrewery.co.uk

Jolly Sailor Brewery
Riccall Business Park, Selby Road, Riccall, York, YO19 6QR

Old Mill Brewery
Mill Street, Snaith, Goole, DN14 9HU
01405 861 813
www.oldmillbrewery.demon.co.uk

The Great Newsome Brewery Ltd
Great Newsome Farm, Winestead, Hull, HU12 0NR
01964 612 201
www.greatnewsomebrewery.co.uk

Whalebone Brewery
The Whalebone, 163 Wincolmlee, Hull, HU2 0PA
01482 226 648

Yorkshire Brewing Company Limited
Brewery Wharf, The Old Fruit Market, Humber Street, Kingston-upon-Hull, HU1 1TU
01482 618 000
www.yorkshirebrewing.co.uk

EAST SUSSEX

1648 Brewing Co Ltd
Mill Lane, East Hoathley, Lewes, BN8 6DR
01825 840 830
www.1648brewing.co.uk

Beachy Head Brewing Co Ltd
c/o The Dipperays, Upper Street, East Dean, Eastbourne, BN20 0BS
01323 423 906
www.beachyhead.org.uk

Black Cat Brewery
Top Lodge, Eridge Road, Groombridge, Tunbridge Wells, TN3 9NJ
07948 387 718
www.blackcat-brewery.com

Fallen Angel
PO Box 95, Battle, TN33 0XF
01424 777 996
www.fallenangelbrewery.com

Full Moon Brewery
Sharpes Farm, Henley Down, Catsfield, Battle, TN33 9BN
07832 220 745

Hastings Brewery
12 Moorhurst Road, St Leonards-on-Sea, TN38 9NB
07708 259 342
www.hastingsbrewery.co.uk

Isfield Brewing Company
New Place Farm, Pump Lane, Framfield, Uckfield, TN22 5RH
01825 750 633
www.isfieldbrewing.co.uk

Kempton Brewery
Hand in Hand, 33 Upper St. James's Street, Brighton, BN2 1JN
01273 693 070

Pin-up Beers Ltd
Unit 2 Rocks Farm Business Park, Stone Cross, Crowborough, TN6 3SJ
01892 611 411
www.pinupbeers.com

Rother Valley Brewing Company
Gate Court Farm, Station Road, Northiam, Rye, TN31 6QT
01797 252 922

The FILO Brewing Co Ltd
14-15 High Street, Hastings, TN34 3EY
01424 425 079
www.thefilo.co.uk

The Kitchen Garden Brewery
The Walled Garden, Sheffield Park, Uckfield, TN22 3QX
01825 790 775
www.kitchengardenbrewery.co.uk

The Long Man Brewery
Church Farm, Litlington, Polegate, BN26 5RA
07976 777 992
www.longmanbrewery.com

White Brewing Company
The 1066 Country Brewery, Pebsham Farm Industrial Estate, Pebsham Lane, Bexhill, TN40 2RZ
01424 731 066
www.white-brewing.co.uk

ESSEX

Brentwood Brewing Company
Frieze Hall Farm, Coxtie Green Road, Brentwood, CM14 5RE
01277 375 577
www.brentwoodbrewing.co.uk

Colchester Brewery Ltd
Viaduct Brewhouse, Unit 16 Wakes Hall Business Centre, Colchester Road, Colchester, CO6 2DY
01787 829 422
www.colchesterbrewery.com

Crouch Vale
23 Haltwhistle Road, Rochford, CM3 5ZA
01245 322 744
www.crouch-vale.co.uk

Deverell's Brewery
Globe Industrial Estate, Rectory Road, Grays, RM17 6ST
07843 627 791

Farmer's Ales
The Stable Brewery, The Blue Boar Stable Yard, Silver Street, Maldon, CM9 4QE
01621 851 000
www.maldonbrewing.co.uk

Felstar
Felsted Vineyard, Crix Green, Felsted, CM6 3JT
01245 361 504
www.felstarbrewery.co.uk

George's Brewery
Mill Barn, Common Road, Great Wakering, Southend-on-Sea, SS3 0AG
07771 871 255
www.hopmonster.co.uk

Harwich Town Brewing Company
Station Approach, Harwich, CO12 3NA
01255 551 155
www.harwichtown.co.uk

Indian Summer Brewing Co Ltd
Unit 3 Commercial Centre, Ashdon Road, Saffron Walden, CB10 2NQ
07986 637 826
www.bombayblonde.co.uk

Mersea Island Brewery
Mersea Vineyard, Rewsalls Lane, East Mersea, CO5 8SX
01206 385 900
www.merseawine.com

Nethergate
Growler Brewery, The Street, Pentlow, CO10 7JJ
01787 283 220
www.nethergate.co.uk

Red Fox Brewery Ltd
The Chicken Sheds, Upphall Farm, Salmons Lane, Coggeshall, CO6 1RY
01376 563 123
www.redfoxbrewery.co.uk

Round Tower Brewery
Unit 11a Robjohns House, Navigation Road, Chelmsford, CM2 6ND
01245 807 343
www.roundtowerbrewery.co.uk

Sticklegs Brewery
The Cross Inn, Ardleigh Road, Great Bromley, Colchester, CO7 7TL
01206 230 282
www.bromleycrossinn.co.uk

The Famous Railway Tavern
58 Station Road, Brightlingsea, CO7 0DT
01206 302 581
www.geocities.com/famousrailway

The Hart of Stebbing Brewery Ltd
White Hart, High Street, Stebbing, Dunmow, CM6 3SQ
01371 856 383
www.hartofstebbingbrewery.co.uk

The Mighty Oak Brewing Co Ltd
14b West Station Yard, Maldon, CM9 6TW
01621 843 713

The Shalford Brewery
c/o 3 Broome Close Villas, Shalford, Braintree, CM7 5EY
01371 850 925
www.shalfordbrewery.co.uk

Vens Brewing Co Ltd
Unit 3 Clovelly Works, Chelmsford Road, Rawreth, Wickford, SS11 8SY
07715 410 489
www.vensbrewing.co.uk

Wibblers Brewery Ltd
Joyces Farm Buildings, Southminster Road, Mayland, CM3 6EB
01621 789 003
www.wibblers.com

GLOUCESTERSHIRE

Ashley Down Brewery
15 Wathen Road, Bristol, BS6 5BY
01179 836 567

Battledown Brewery
Keynsham Works, Keynsham Street, Cheltenham, GL52 6EJ
07734 834 104
www.battledownbrewery.com

Corinium Ales Ltd
22 Bowling Green Avenue, Cirencester, GL7 2HB
07716 826 467
www.coriniumales.co.uk

Cotswold Lion Brewery Ltd
Grain Store 5, Downmans Farm, Coberley, Cheltenham, GL53 9QY
01242 870 164
www.cotswoldlionbrewery.co.uk

Cotswold Spring Brewery Ltd
Dodington Spring, Dodington Ash, Chipping Sodbury, BS37 6RX
01454 323 088
www.cotswoldbrewery.com

Festival Brewery
Unit 17 Malmesbury Road, Kingsditch Trading Estate, Cheltenham, GL51 9PL
01242 521 444
www.festivalbrewery.co.uk

Gloucester Brewery
Llanthony Warehouse, The Docks, Gloucester, GL1 2EH
01452 690 420
www.gloucesterbrewery.co.uk

Goff's Brewery Ltd
9 Isbourne Way, Winchcombe, Cheltenham, GL54 5NS
01242 603 383
www.goffsbrewery.co.uk

May Hill Brewery
Holly Bush Farm, Ross Road, Longhope, GL17 0NG
01452 830 222
www.mayhillbrewery.com

Nailsworth Brewery Ltd
The Village Inn, Bath Road, Nailsworth, Stroud, GL6 0HH
07878 448 377
www.nailsworth-brewery.co.uk

Severn Vale Brewing Company
Lower Knapp Farm, Woodend Lane, Cam, Dursley, GL11 5HS
01453 547 550
www.severnvalebrewing.co.uk

Stanway Brewery
Stanway, Cheltenham, GL54 5PQ
01386 584 320
www.stanwaybrewery.co.uk

Stroud Brewery Ltd
c/o 141 Thrupp Lane, Thrupp, Stroud, GL5 2DQ
07891 995 878
www.stroudbrewery.co.uk

Terrace Brewery
Sandford Terrace, Aylburton, Lydney, GL15 6DW
01594 840 100
www.terracebrewery.co.uk

The Bespoke Brewing Company Limited
Unit 5 The Mews, Mitcheldean, GL17 0SL
01594 546 557
www.bespokebrewery.co.uk

The Halfpenny Brewery
The Crown Inn, High Street, Lechlade, GL7 3AE
01367 252 198
www.halfpennybrewery.co.uk

Uley Brewery Ltd
The Old Brewery, 31 The Street, Uley, Dursley, GL11 5TB
01453 860 120
www.uleybrewery.com

Whittingtons Brewery
Three Choirs Vineyard, Welsh House Lane, Castle Tump, Newent, GL18 1LS
01531 890 223
www.whittingtonbrewery.co.uk

Wickwar Brewing Company
The Old Brewery, Station Road, Wickwar, GL12 8NB
01454 294 168
www.wickwarbrewing.co.uk

GREATER LONDON

Belleville Brewing Company
Unit 36 Jaggard Way, London, SW12 8SG
07712 298 273
www.bellevillebrewing.co.uk

Brew By Numbers
66 Southwark Bridge Road, London, SE1 0AS
07528 684 105
www.brewbynumbers.com

Brew Wharf
Brew Wharf Yard, Stoney Street, London, SE1 9AD
020 7940 8333
www.brewwharf.com

Brockley Brewing Company Limited
31 Harcourt Road, London, SE4 2AJ
brockleybrewingco.tumblr.com

Brüpond Brewery
Unit 004 15 Argall Avenue, London, E10 7QE
020 8556 7438
www.brupond.com

By The Horns Brewing Company
25 Summerstown, London, SW17 0BQ
020 3417 7338
www.bythehorns.co.uk

Clarence & Fredericks Brewing Co
35a Neville Road, Croydon, CR0 2DS
07725 437 396
www.cfbrewing.co.uk

East London Brewing Company Limited
Unit 45 Fairways Business Centre, Lammas Road, London, E10 7QB
020 8539 0805
www.eastlondonbrewing.com

Ellenberg's Brewery Ltd
Unit 5 Boston Business Park, Trumpers Way, London, W7 2QA
020 3261 1089
www.ellenbergsbrewery.co.uk

Hackney Brewery
Arch 358 Railway Arches, Laburnum Street, London, E2 8BB
020 3489 9595
www.hackneybrewery.co.uk

Hoppy Collie Brewery
106 Fulham Palace Road, London, W6 9PL
www.hoppycollie.com

Horseshoe Brewery
The Horseshoe, 28 Heath Street, London, NW3 6TE
020 7431 7206

Late Knights Brewery Ltd
Southey Street, London, SE20 7JD
www.lateknightsbrewery.co.uk

Little Brew
Unit 21 43 Carol Street, London, NW1 0HT
07817 001 376
www.littlebrew.co.uk

London Fields Brewery
Arch 374, Helmsley Place, Hackney, E8 3SB
07917 322 550
www.londonfieldsbrewery.co.uk

Moncada Brewery Ltd
Unit 1 Buspace Studios, Conlan Street, London, W10 5AP
07795 511 505
www.moncadabrewery.co.uk

Partizan Brewing Limited
Railway Arch 8 Almond Road, London, SE16 3LR
020 8127 5053
www.partizanbrewing.co.uk

Redemption Brewing Company Ltd
Unit 2/33 West Road, London, N17 0XJ
020 8885 5227
www.redemptionbrewing.co.uk

Rocky Head Brewery Ltd
Unit 16 Glenville Mews, London, SW18 4NJ
020 8875 9917
sites.googlecom/site/rockyheadbrewery

Sambrook's Brewery Limited
Units 1 & 2 Yelverton Road, London, SW11 3QG
020 7228 0598
www.sambrooksbrewery.co.uk

The Cronx Brewery
Unit 6 Vulcan Business Centre, Vulcan Way, New Addington, Croydon, CR0 9UG
01689 809 093
www.thecronx.com

The Five Points Brewing Company Ltd
3 Institute Place, London, E8 1JE
020 8533 7746

The Florence Brew House
133 Dulwich Road, London, SE24 0NG
020 7326 4987
www.florencehernehill.com

The Redchurch Brewery Ltd
273 Railway Arches , Poyser Street, London, E2 9RF
07968 173 097
www.theredchurchbrewery.com

Twickenham Fine Ales
Ryecroft Works, Edwin Road, Twickenham, TW2 6SP
020 8241 1825
www.twickenham-fine-ales.co.uk

Zerodegrees Blackheath
29-31 Montpelier Vale, Blackheath, London, SE3 0TJ
020 8852 5619
www.zerodegrees.co.uk

GREATER MANCHESTER

Blackedge Brewing Company Ltd
Engine House No 2, Hampson Street, Horwich, Bolton, BL6 7JH
07719 438 587
www.blackedgebrewery.co.uk

Blackjack Beers Ltd
36 Gould Street, Manchester, M4 4RN
0161 819 2767

Brightside Brewing Company Ltd
1 New George Street, Bury, BL8 1NW
0161 705 2625
www.brightsidebrewing.co.uk

Deeply Vale Brewery
Unit 24 Peel Industrial Estate, Chamberhall Street, Bury, BL9 0LU
0161 761 7334
www.deeplyvalebrewery.com

Hornbeam Brewery
Unit 1-1c Grey Street, Denton, Manchester, M34 3RU
0161 320 5627
www.hornbeambrewery.com

Marble Brewery
73 Rochdale Road, Manchester, M4 4HY
0161 819 2694
www.marblebeers.co.uk

Phoenix Brewery Ltd
Green Lane, Heywood, Manchester, OL10 2EP
01706 627 009
www.phoenixbrewery.co.uk

Privateer Beers Ltd
80 Temperance Street, Manchester, M12 6HU
07969 771 102

Ramsbottom Craft Brewery
1 Heapworth Avenue, Ramsbottom, Bury, BL0 9EH
07976 263 344
www.ramsbottombrewery.com

Ringway Brewery Ltd
16 Station Road Industrial Estate, Reddish, Stockport, SK5 6ND
0161 443 1818
www.ringwaybrewery.co.uk

The Hay Rake Brewery Ltd
The Rake Mediterranean Tapas Restaurant, Blackstone Edge Old Road, Littleborough, OL15 0JX
01706 379 689
www.raketapas.co.uk

The Ticketybrew Company
Unit 25 Waterloo Court, Waterloo Road, Stalybridge, SK15 2AU
07970 093 665
www.ticketybrew.co.uk

Wilson Potter Brewery llp
Unit E2 Hanson Park, Hanson Close, Middleton, Manchester, M24 2QZ
0161 654 6446
www.wilsonpotterbrewery.co.uk

HAMPSHIRE

Alfred's Brewery Ltd
5b Scylla Industrial Estate, Winnall Valley Road, Winchester, SO23 0LD
01962 859 999
www.alfredsbrewery.co.uk

Andwell Brewing Company
Lodge Farm, Hook Road, North Warnborough, Hook, RG29 1HA
01256 704 412
www.andwells.com

Ballard's Brewery
The Old Sawmill, Nyewood, Petersfield, GU31 5HA
01730 821 362
www.ballardsbrewery.org.uk

Botley Brewery
Botley Mills, Botley, Southampton, SO30 2GB
01489 784 867
www.botleybrewery.co.uk

Bowman Ales Ltd
Wallop Woods, Sheardsley Lane, Droxford, Southhampton, SO32 3QY
01489 878 110
www.bowman-ales.com

Flack Manor Brewery Ltd
8 Romsey Industrial Estate, Greatbridge Road, Romsey, SO51 0HR
01794 518 520
www.flackmanor.co.uk

Irving & Co Brewers Ltd
Unit G1 Railway Triangle, Walton Road, Portsmouth, PO6 1TH
023 9238 9988
www.irvingbrewers.co.uk

Itchen Valley Brewery Ltd
Unit D Prospect Commercial Park, Prospect Road, Alresford, SO24 9QF
01962 735 111
www.itchenvalley.com

Longdog Brewery
Unit 1a Moniton Trading Estate, West Ham Lane, Basingstoke, RG22 6NQ
01256 324 286
www.longdogbrewery.co.uk

Oakleaf Brewing Company Ltd
Unit 7 Mumby Road, Clarence Wharf Industrial Estate, Gosport, PO12 1AJ
023 9251 3222
www.oakleafbrewing.co.uk

Red Shoot Brewery
Linwood, Ringwood, BH24 3QT
01425 475 792

Sherfield Village Brewery
Goddards Farm, Goddards Lane, Sherfield-On-Loddon, Hook, RG27 0EL
07906 060 429
www.sherfieldvillagebrewery.co.uk

The Emsworth Brewery
16-18 West Street, Emsworth, PO10 7DY
01243 430 020

The Flowerpots Brewery
Flowerpots Inn, Cheriton, Alresford, SO24 0QQ
01962 771 534
www.flowerpots.f2s.com

Triple FFF Brewing Co.
Unit 3, Old Magpie Works, Alton, GU34 5AN
01420 561 422
www.triplefff.com

Wild Weather Ales Ltd
Unit 19 Easter Park, Benyon Road, Silchester, Reading, RG7 2PQ
0118 970 1837
www.wildweatherales.com

HEREFORDSHIRE

Arrow Brewery
The Wine Vaults, 37 High Street, Kington, HR5 3BJ
01544 230 685

Ledbury Real Ales
Gazerdine House, Hereford Road, Ledbury, HR8 2PZ
01531 671 184
www.ledburyrealales.co.uk

Oast House Brewery
The Farm, Brockhampton Estate, Whitbourne, Worcester, WR6 5SH
07974 371 294

Saxon City Ales
Glebe Farm Industrial Estate, Stoke Edith, Hereford, HR1 4HG
01432 890 602
www.saxoncityales.co.uk

Shoes Brewery
Three Horseshoes Inn, Norton Canon, Hereford, HR4 7BH
01544 318 375

The Mayfields Brewery
Unit 8 Croft Business Park, Southern Avenue, Leominster, HR6 0QF
01568 611 197
www.mayfieldsbrewery.co.uk

The Mulberry Duck Brewing Company Ltd
c/o Elan Portway, Burghill, Hereford, HR4 8NF
07740 468 675
www.mulberryduck.co.uk

The Wobbly Brewing Company
Unit 22a Beech Business Park, Tillington Road, Hereford, HR4 9QJ

Wye Valley Brewery
Stoke Lacy, HR7 4HG
01885 490 505
www.wyevalleybrewery.com

Buntingford Brewery Company Ltd
Greys Brewhouse, Therfield Road, Royston, SG8 9NW
01763 250 749
www.buntingford-brewery.co.uk

Red Squirrel Brewery
14b Minram Road, Hertford, SG14 1NN
01992 501 100
www.redsquirrelbrewery.co.uk

Saffron Brewery
Unit 2 Pledgdon Hall Farm, Henham, Bishop's Stortford, CM22 6BJ
01279 850 923
www.saffronbrewery.co.uk

Sawbridgeworth Brewery
Gate Inn, 81 London Road, Sawbridgeworth, CM21 9JJ
01279 722 313
www.the-gate-pub.co.uk

The 3 Brewers Ltd
The Potato Shed, Symonds Hyde Farm, Symonds Hyde Lane, Hatfield, AL10 9BB
01707 271 636
www.3brewers.co.uk

The Hertford Brewery
26 Old Cross, Hertford, SG14 1RD
01992 584 911
www.mcmullens.co.uk

The Old Cross Tavern Brewery
Old Cross Tavern, 8 St Andrew Street, Hertford, SG14 1JA
01992 583 133

Tring Brewery Co Ltd
81-82 Akeman Street, Tring, HP23 6AF
01442 890 721
www.tringbrewery.co.uk

ISLE OF MAN

Bushys Brewery
The Mount Murray Brewing Co Ltd, The Hop Garden, Mount Murray, Douglas, IM4 1JE
01624 661 244
www.bushys.com

ISLE OF WIGHT

Goddard's Brewery
Barnsley Farm, Bullen Road, Ryde, PO33 1QF
01983 611 011
www.goddards-brewery.co.uk

Island Brewery
Dinglers Farm, Yarmouth Road, Shalfleet, Newport, PO30 4LZ
01983 821 731
www.islandbrewery.co.uk

ISLES OF SCILLY

Ales of Scilly Brewery
2b Porthmellon Industrial Estate, St Mary's, TR21 0JY
01720 423 233

JERSEY

Pocket Brewery
La Croix, La Rue de la Croix, St Ouen, JE3 2HA
07797 771 931

KENT

Goody Ales Ltd
Bleangate House, Braggs Lane, Herne, CT6 7NP
01227 361 555
www.goodyales.co.uk

Hop Fuzz Brewery Ltd
Unit 8 Riverside Industrial Estate, West Hythe Road, West Hythe, Hythe, CT21 4NB
01303 230 304
www.hopfuzz.co.uk

Hopdaemon Brewery Ltd
Unit 1, Parsonage Farm, Seed Road, Newnham, ME9 0NA
01795 892 078
www.hopdaemon.com

Kent Brewery
Birling Place Farm, Stangate Road, Birling, West Malling, ME19 5JN
01634 780 037
www.kentbrewery.com

Larkins Brewery
Chiddingstone, Edenbridge, TN8 7BB
01892 870 328

Mad Cat Brewery Ltd
Brogdale Farm, Brogdale Road, Faversham, ME13 8XU
07960 263 615
www.madcatbrewery.co.uk

Nelson Brewery
Unit 2, Building 64, Historic Dockyard, Chatham, ME4 4TE
01634 832 828
www.nelsonbrewingcompany.co.uk

P & DJ Goacher
Unit 8, Tovil Green Buiness Park, Maidstone, ME15 6TA
01622 682 112
www.goachers.com

Ramsgate Brewery Ltd
Unit 1 Hornet Close, Pysons Road Industrial Estate, Broadstairs, CT10 2YD
01843 594 758
www.ramsgatebrewery.co.uk

Ripple Steam Brewery Limited
Parsonage Farm, Vale Road, Sutton, Dover, CT15 5DH
07917 037 611
www.ripplesteambrewery.co.uk

Royal Tunbridge Wells Brewing Company Limited
Spa Brewery, 18h Chapman Way, Tunbridge Wells, TN2 3EF
01892 618 140
www.royaltunbridgewellsbrewing.co.uk

Spencers Beers
Unit 5 Ashford Works, Brunswick Road, Cobbs Wood Industrial Estate, Ashford, TN23 1EH

The Canterbury Ales
Unit 7 Stour Valley Business Park, Ashford Road, Chartham, Canterbury, CT4 7HF
01227 732 541
www.canterbury-ales.co.uk

The Millis Brewing Company Ltd
St Margrets Farm, St Margrets Road, South Darenth, Dartford, DA4 9LB
01322 866 233

The Old Dairy Brewing Co Ltd
The Old Parlour, Benenden Road, Rolvenden, Cranbrook, TN17 4JD
01580 243 185
www.olddairybrewery.com

The Swan Microbrewery
Swan on the Green, The Green, West Peckham, Maidstone, ME18 5JW
01622 812 271
www.swan-on-the-green.co.uk

Tonbridge Brewery
White Oaks, Tudeley Road, Tonbridge, TN11 0NW
01732 366 770
www.tattonbrewery.co.uk

Westerham Brewery Company Ltd
Grange Farm, Pootings Road, Crockham Hill, Edenbridge, TN8 6SA
01732 864 427
www.westerhambrewery.co.uk

Whitstable Brewery
Little Telpits Farm, Woodcock Lane, Grafty Green, Maidstone, ME17 2AY
01622 851 007
www.whitstablebrewery.info

LANCASHIRE

AllGates Brewery
The Old Brewery, Brewery Yard, Wallgate, Wigan, WN1 1JU
01942 234 976
www.allgatesbrewery.com

Arkwright's Brewery
Real Ale Shop, 47 Lovat Road, Preston, PR1 6DQ
01772 201 591

Bank Top Brewery
The Pavilion, Ashworth Lane, Bank Top, Bolton, BL1 8RA
01204 595 800
www.banktopbrewery.com

Barlick Brewery
The Greyhound, Manchester Road, Barnoldswick, Colne, BB18 5PW
07722 388 155
www.barlick-brewery.co.uk

Bluestone Brewery
Whitworth,
www.bluestonebrewery.co.uk

Boggart Hole Clough Brewery
Unit 13 Brookside Works, Clough Road, Moston, M9 4SP
0161 277 9666
www.boggart-brewery.co.uk

Chorley Brewhouse Limited
F12/13 Coppull Enterprise Centre, Mill Lane, Coppull, Chorley, PR7 5BW
07875 965 164
www.chorleybrewhouse.com

Cross Bay Brewery
Unit 1 Newgate, White Lund Industrial Estate, Morcambe, LA3 3PT
01524 39 481
www.crossbaybrewery.co.uk

Derby Brewery
Empire Street, Cheetham, M3 1JD
0161 834 3285
www.joseph-holt.com

Fallons Exquisite Ales
Unit 15 Darwen Enterprise Centre, Railway Road, Darwen, BB3 3EH
07905 246 810
www.fallonsales.com

George Wright Brewing Company
11 Diamond Business Park, Sandwash Close, Rainford Industrial Estate, St Helens, WA11 8LU
01744 886 686
www.georgewrightbrewing.co.uk

Green Mill Brewery
Queensway Snooker Club, Green Mill, Well I' Th'Lane, Rochdale, OL11 2LS
07896 702 328
www.greenmillbrewery.co.uk

Greenfield Real Ale Brewery Ltd
Unit 8 Waterside Mills, Chew Valley Road, Greenfield, Oldham, OL3 7PF
01457 879 789
www.greenfieldrealale.co.uk

Hart Brewery
Cartford Hotel, Cartford Lane, Little Eccleston, PR3 0YP
01995 671 686
www.hartbreweryltd.co.uk

Higson's Brewery Ltd
Unit 17 Maritime Enterprise Park, Atlas Road, Liverpool, L20 4DY
0151 922 5640
www.higsonbrewery.co.uk

Hopstar Brewery
11 Pole Road, Darwen, BB3 3LD
01254 703 389

Irwell Works Brewery Ltd
Irwell Works, Square Street, Ramsbottom, Bury, BL0 9YQ
01706 825 019
www.irwellworksbrewery.co.uk

Lancaster Brewery Company Ltd
Unit 19 Lansil Walk, Lansil Industrial Estate, Lancaster, LA1 3PQ
01524 844 610
www.lancasterbrewery.co.uk

Lytham Brewery Ltd
Unit 11 Lidun Park Industrial Estate, Boundary Road, Lytham St Annes, FY8 5HU
01253 737 707
www.lythambrewery.co.uk

Millstone Brewery
Unit 4 Vale Mill, Micklehurst Road, Mossley, Ashton-under-Lyne, OL5 9JL
01457 835 835
www.millstonebrewery.co.uk

Moonstone Brewery
Ministry of Ale, 9 Trafalgar Street, Burnley, BB11 1TQ
01282 830 909
www.moonstonebrewery.co.uk

Moorhouse Brewery
The Brewery, Moorhouse Street, Bucksburn, BB11 5EN
01282 422 864
www.moorhouses.co.uk

Old School Brewery
Holy Bank Barn, Crag Road, Warton, Carnforth, LA5 9SD
01524 735 005
www.oldschoolbrewery.co.uk

Outstanding Brewing Co Ltd
Unit 4b Britannia Mill, Cobden Street, Bury, BL9 6AW
0161 764 7723
www.outstandingbeers.com

Pennine Ale
Rosendale Brewery, The Griffin Inn, Haslingden, BB4 5AF
01706 214 021
www.pbcbreweryinstallations.com

Pictish Brewing Co Ltd
Unit 9 Canalside Industrial Estate, Woodbine Street East, Rochdale, OL16 5LB
01706 522 227
www.pictish-brewing.co.uk

Prospect Brewery Ltd
120 Wigan Road, Standish, Wigan, WN6 0AY
01257 421 329
www.prospectbrewery.com

Reedley Hallows Brewing Company
Unit 12 Farrington Court, Burnley, BB11 5SS
07749 414 513
www.reedley-hallows-brewery.co.uk

Saddleworth Brewery
Church Inn, Running Hill Gate, Uppermill, Oldham, OL3 6LW
01457 820 902

Southport Brewery
Unit 3 Southport Enterprise Centre, Russell Road, Southport, PR9 7RF
07748 387 652
www.southportbrewery.co.uk

Star Brewery
Syke Street, Blackburn, BB1 5BU
01254 686 868
www.thwaites.co.uk

Stringers Beer
Unit 3 Low Mill Business Park, Ulverston, LA12 9EE
01229 581 387
www.stringersbeer.co.uk

The Bowland Beer Company
The Bowland Brewery, Bashall Barn, Twitter Lane, Clitheroe, BB7 3LQ
01200 428 825
www.bowlandbrewery.com

Three Bs Brewery
Unit 5 Laneside Works, Stockclough Lane, Feniscowles, Blackburn, BB2 5JR
01254 207 686
www.threebsbrewery.co.uk

Wellcross Brewing Company
Wellcross Farm, 123 Tower Hill Road, Upholland, Skelmersdale, WN8 0DS

LEICESTERSHIRE

Barrowden Brewing Company
c/o Exeter Arms, Barrowden, Rutland, LE1 5EQ
01572 747 247
www.exeterarms.com

Bees Brewery
c/o Branstons Petroleum Services, 1487 Melton Road, Queniborough, Leicester, LE7 3FP
01162 607 715

Belvoir Brewery
Crown Park, Station Road, Old Dalby, LE14 3NQ
01664 823 455
www.belvoirbrewery.co.uk

Dow Bridge Brewery
2-3 Rugby Road, Catthorpe, Lutterworth, LE17 6DA
01788 869 121

Golden Duck Brewery
Unit 2 Redhill Farm, Top Street, Appleby Magna, Swadlincote, DE12 7AH
www.goldenduckbrewery.com

Langton Brewery
Grange Farm, Thorpe Langton, Market Harborough, LE16 7TU
01858 545 483

Parish Brewery
6 Main Street, Burrough On The Hill, Somerby, LE14 2JQ
01664 454 782

Shardlow Brewing Co Ltd
Old Brewery Stables, British Waterways Yard, Cavendish Bridge, DE72 2HL
01332 799 188
www.shardlowbrewery.co.uk

The Grainstore Brewery
Station Approach, Oakham, Rutland, LE15 6RE
01572 770 065
www.grainstorebrewery.com

Wicked Hathern Brewery Ltd
33 Green Hill, Hathern, Loughborough, LE12 5LD
01509 842 585
www.wicked-hathern.co.uk

LINCOLNSHIRE

8 Sail Brewery
Heckington Windmill, Hale Road, Heckington, Sleaford, NG34 9JW
01529 469 308

Austendyke Ales
The Beeches, Austendyke Road, Weston Hills, Spalding, PE12 6BZ
07866 045 778

Axholme Brewing Company
2 Garthorpe Road, Luddington, Scunthorpe, DN17 4QT
07551 910 040
www.axholmebrewing.co.uk

Bacchus Brewing Co
The Bacchus Hotel, 17 High Street, Sutton-on-Sea, Mablethorpe, LN12 2EY
01507 441 204
www.bacchushotel.co.uk

Blue Bell Brewery Ltd
Blue Bell Inn, Cranesgate South, Whaplode St Catherine, Holbeach, PE12 6SN
01406 540 300
www.bluebellbrewery.co.uk

Brewsters Brewing Company
Unit 5, Burnside Industrial Estate, Turnpike Close, Grantham, LE14 4HR
01476 566 000
www.brewsters.co.uk

Cathedral Ales
Green Dragon Hotel, Broadgate, Lincoln, LN2 5DH
01522 567 155
www.greendragonpub.co.uk

DarkTribe Brewery
Dog & Gun, High Street, East Butterwick, Scunthorpe, DN17 3AJ
01724 782 324
www.darktribe.co.uk

Fulstow Brewery
13 Thames Street, Louth, LN11 7AD
01507 608 202
www.fulstowbrewery.co.uk

Fuzzy Duck Brewery
c/o 6 Northway, Fulstow, Louth, FY6 0ES
07904 343 729
www.fuzzyduckbrewery.co.uk

Grafters Brewery
Half Moon, 23 High Street, Willingham By Stow, DN21 5JZ
01427 788 340
www.graftersbrewery.com

Highwood Brewery Ltd
Unit 3, Birchams Way, Grimsby, DN31 2SG
01472 255 500
www.tom-wood.com

Leila Cottage Brewery
c/o The Countryman, Chapel Road, Ingoldmells, Skegness, PE25 1ND
01754 872 268

Newby Wyke Brewery
Willoughby Arms Cottages, Little Arms Cottages, Little Bytham, Grantham, NG33 4RA
01780 411 119
www.newbywyke.co.uk

Oldershaw Brewery
12 Harrowby Lane, Harrowby, Grantham, NG31 9HB
01476 572 135
www.oldershawbrewery.com

Poachers Brewery
439 Newark Road, North Hykeham, Lincoln, LN6 9SP
01522 807 404
www.poachersbrewery.co.uk

Sleaford Brewery
Hop Me Up Ltd, 21 Pride Court, Enterprise Park, Sleaford, NG34 8GL
07854 829 718
www.sleafordbrewery.co.uk

Willys Brewery Ltd
17 High Cliff Road, Cleethorpes, DN35 8RQ
01472 602 145

MERSEYSIDE

Liverpool Craft Beers Ltd
10 Love Lane, Liverpool,
L3 7DD
0151 236 9400
www.liverpoolcraftbeer.com

Liverpool One Brewery Limited
82-84 Vauxhall Road,
Liverpool, L3 6DL
07948 918 740
www.liverpoolonebrewery.co.uk

Mad Hatter Brewing Company
Unit D22 Ayrton House,
Commerce Way, Liverpool,
L8 7BA
0151 739 1702
www.madhatbrew.co.uk

Stamps Brewery
17 Boundary Street,
Liverpool, L5 9UB
07779 000 094
www.stampsbrewery.co.uk

The Melwood Beer Company
Home Farm, Knowsley Park,
Prescot, L34 4AQ
0151 214 3340
www.themelwoodbeercompany.co.uk

NORFOLK

Beeston Brewery Ltd
Fransham Road Farm, Beeston,
King's Lynn, PE32 2LZ
01328 700 844
www.beestonbrewery.co.uk

Buffy's Brewery Ltd
Rectory Road, Tivetshall St
Mary, NR15 2DD
01379 676 523
www.buffys.co.uk

Chalk Hill Brewery
Rosary Road, Norwich,
NR1 4DA
01603 477 078

Elmtree Beers
Unit 10 Oakwood Industrial
Estate, Harling Road,
Snetterton, NR16 2JU
01953 887 065
www.elmtreebeers.co.uk

Fat Cat Inns Ltd
49 West End Street, Norwich,
NR2 4NA
01603 624 362
www.fatcatpub.co.uk

Fox Brewery
Fox and Hounds, Station Road,
Heacham, King's Lynn,
PE31 7EX
01485 570 345
www.foxbrewery.com

Front Street Brewery
The Chequers Inn, Front Street,
Binham, NR21 0AL
01328 830 297
www.frontstreetbrewery.co.uk

Golden Triangle Brewery
Unit 9, Watton Road,
Norwich, NR9 4BG
01603 757 763
goldentrianglebrewery.co.uk

Grain Brewery
South Farm, Alburgh,
Harleston, IP20 0BS
01986 788 884
www.grainbrewery.co.uk

Humpty Dumpty
Church Road, Reedham,
NR13 3TZ
01493 701 818
www.humptydumptybrewery.co.uk

Iceni Brewery
3 Fouldon Road, Ickburgh,
IP26 5BJ
01842 878 922
www.icenibrewery.co.uk

JoC's Ales
The Old Store, Walsingham
Road, West Barsham,
Fakenham, NR21 9NP
01328 863 854
www.jocsnorfolkale.co.uk

Norfolk Brewhouse
Moon Gazer Barn, Harvest
Lane, Hindringham, Fakenham,
NR21 0PW
01328 878 495
www.norfolkbrewhouse.co.uk

Norfolk Cottage Brewing
22 The Green, North
Bulingham, NR13 3DJ
01603 270 520

Norfolk Square Brewery
Unit 7 Estcourt House,
Estcourt Road, Great Yarmouth,
NR30 4JQ
01493 854 484
www.norfolksquarebrewery.co.uk

Old Chimney's Brewery
Hopton End Farm, Church Road,
Market Weston, Diss, IP22 2NX
01359 221 013
www.oldchimneysbrewery.com

Ole Slew Foot Brewing Company Ltd
3 Pollard Road, Hainford,
Norwich, NR10 3BE
0845 1162 204
www.oleslewfootbrewery.co.uk

Panther Brewery
Unit 1, Collers Way, Reepham,
NR10 4SW
07766 558 215
www.pantherbrewery.co.uk

Redwell Brewing
7 The Arches, Bracondale,
Norwich, NR1 2EF
01603 624 072
www.redwellbrewing.co.uk

The Why Not Brewery
17 Cavalier Close, Thorpe St
Andrew, Norwich, NR7 0TE
01603 300 786
www.thewhynotbrewery.co.uk

Tipples Brewery
Unit 6, Damgate Lane Industrial
Estate, Acle, NR13 3DJ
01493 741 007
www.tipplesbrewery.com

Uncle Stuart's Brewery
Antoma, Pack Lane, Lingwood,
NR13 4PD
01603 211 833
www.littlebeershop.co.uk

Wagtail Brewery
New Barn Farm, Wilby Warrens,
Old Buckenham, NR17 1PF
01953 887 133
www.wagtailbrewery.com

Waveney Brewing Co.
Queen's Head, Station Road,
Earsham, Bungay, NR35 2TS
01986 892 623

Winter's Brewery
8 Keelan Close, Norwich,
NR6 6QZ
01603 787 820
www.wintersbrewery.com

Wissey Valley Brewery
The Grey House, Lynn Road,
Stoke Ferry, PE33 9FW
01366 500 767
www.wisseyvalleybrewery.com

Yetman's Brewery
Bayfield Farm Barns, Bayfield
Brecks, Holt, NR25 7DZ
07774 809 016
www.yetmans.net

NORTH YORKSHIRE

Barkston Brewery
c/o Orchard House, Saw Wells
Court, Barkston Ash, Tadcaster,
LS24 9UJ
0845 224 6324
www.barkstonbrewery.com

Brown Cow Brewery
Brown Cow Road, Barlow, Selby,
YO8 8EH
01757 618 947
www.browncowbrewery.co.uk

Captain Cook Brewery Ltd
White Swan, 1 West End,
Stokesley, Middlesbrough,
TS9 5BL
01642 710 263
www.thecaptaincookbrewery.co.uk

Conquest Brewery
Unit 2B Larpool Lane Industrial
Estate, Larpool Lane,
Whitby, YO22 4LX
07790 502 035

Copper Dragon Brewery Ltd
Snaygill Industrial Estate,
Keighley Road, Skipton,
BD23 2QR
01756 702 130
www.copperdragon.uk.com

Cropton Brewery
Cropton Pickering, YO18 8HH
01751 417 330
www.croptonbrewery.co.uk

East Coast Brewing Company
3 Clay House Yard, Mitford
Street, Filey, YO14 9DX
01723 514 865

Great Heck Brewing Company Ltd
Harwinn House, Main Street, Heck, Goole, DN14 0BQ
01977 661 430
www.greatheckbrewery.co.uk

Hambleton Ales
Unit 2a Barker Business Park, Melmerby Green Lane, Melmerby, Ripon, HG4 5NB
01765 640 108
www.hambletonales.co.uk

Litton Ale Brewery
Queen's Arms, Litton, Skipton, BD23 5QJ
01756 770 208
www.yorkshirenet.co.uk/stayat/queensarms

Marston Moor Brewery Co
Crown Hotel, Church Street, Kirk Hammerton, York, YO26 8DD
01423 330 341
www.marstonmoorbrewery.co.uk

Mithril Ales
Mithril, East End, Aldbrough St John, Richmond, DL11 7TL
01325 374 817
www.mithrilales.co.uk

North Yorkshire Brewing Co
Pinchinthorpe Hall, Pinchinthorpe, Guisborough, TS14 8HG
01287 632 000
www.nybrewery.co.uk

Redscar Brewery Ltd
The Cleveland Hotel, 9-11 High Street West, Redcar, TS10 1SQ
01642 513 727
www.theclevelandhotel.co.uk

Richmond Brewing Company Limited
The Station Brewery, Station Yard, Richmond, DL10 4LD
07912 347 946
www.richmondbrewing.co.uk

Scarborough Brewery Limited
1b Barrys Lane, Falsgrave, Scarborough, YO12 4HA
01723 367 506
www.scarboroughbrewery.co.uk

Settle Brewing Company Ltd
Unit 8a The Sidings Industrial Estate, Settle, BD24 9RP

Three Peaks Brewery
7 Craven Terrace, Settle, BD24 9DB
01729 822 939

Treboom Brewery
Millstone Yard, Main Street, Shipton-by-Beningbrough, YO30 1AA
01904 471 569
www.treboom.co.uk

Truefitt Brewing Company Ltd
3 Carcut Road, Middlesbrough, TS3 6QL
07883 072 389
www.truefittbrewing.co.uk

Wainstones Brewery
Unit 9 Terry Dicken Industrial Estate, Station Road, Stokesley, Middlesbrough, TS9 7AE
07885 240 226
www.stokesleybrewing.co.uk

Wold Top Brewery
Hunmanby Grange, Wold Newton, Driffield, YO25 3HS
01723 892 222
www.woldtopbrewery.co.uk

York Brewery Co Ltd
12 Toft Green, Micklegate, YO1 6JT
01904 621 162
www.yorkbrew.co.uk

Yorkshire Dales Brewing Company Limited
Seata Barn Elm Hill, Main Street, Askrigg, Leyburn, DL8 3HG
01969 622 027
www.yorkshiredalesbrewery.com

Yorkshire Heart Brewery
The Vineyard, Pool Lane, Nun Monkton, York, YO26 8EL
01423 330 716
www.yorkshireheart.com

Daleside Brewery
Camwal Road, Starbeck, Harrogate, HG1 4PT
01423 880 041
www.dalesidebrewery.co.uk

Roosters Brewing Co
Unit 3/4 Grimbald Park, Wetherby Road, Knaresborough, HG5 8LJ
01423 865 959
www.roosters.co.uk

NORTHAMPTONSHIRE

Alcazar Brewing Company
33 Church Street, Old Basford, Nottingham, NG6 0GA
01159 785 155
www.alcazarbrewery.co.uk

Brigstock Brewhouse
7 Park Walk, Brigstock, NN14 3HH
01536 373 428
www.brigstockbrewhouse.co.uk

Caythorpe Brewery
Walnut Cottage, Boat Lane, Hoveringham, Nottingham, NG14 7JP
01159 664 933

Digfield Ales
North Lodge Farm, Thurning Road, Barnwell, Peterborough, PE8 5RJ
01832 293 295

Frog Island Brewery
The Maltings, Westbridge, St James Road, Northampton, NN5 5HS
01604 587 772
www.frogislandbrewery.co.uk

Great Oakley Brewery
Bridge Farm, 11 Brooke Road, Great Oakley, Corby, NN18 8HG
01536 744 888
www.greatoakleybrewery.com

Gun Dog Ales Ltd
5b Great Centre Way, Woodford Halse, Daventry, NN11 3PZ
01327 830 803
www.gundogales.co.uk

Hart Family Brewers Ltd
The 1833 Brewery, Unit 21 Nene Court, 27 The Embankment, Wellingborough, NN8 1LD
01933 228 324
www.hartfamilybrewers.com

Hoggleys Brewery
Unit 12 Litchborough Trading Estate, Northampton Road, Litchborough, NN12 8JB
01604 831 762
www.hoggleys.co.uk

Julian Church Brewing Company
c/o 38 Nunnery Avenue, Rothwell, NN14 6JJ
07794 289 559

Latimer Ales Ltd
13 South Folds Road, Corby, NN18 9EU
07812 450 988

Milestone Brewing Co
Great North Road, Cromwell, Newark, NG23 6JE
01636 822 255
www.milestonebrewery.co.uk

Nene Valley Brewery Limited
Oundle Wharf, Station Road, Oundle, PE8 4DE
01832 272 776
www.nenevalleybrewery.com

Nobby's Brewery
c/o The Ward Arms, High Street, Guisborough, NN6 8PY
01604 740 785
www.nobbysbrewery.co.uk

Potbelly Brewery Ltd
c/o Corium Leather Co Ltd, 25-31 Durban Road, Kettering, NN16 0JA
01536 410 818
www.potbelly-brewery.co.uk

Rockingham Ales
c/o 25 Wansford Road, Elton, Peterborough, PE8 6RZ
01832 280 722
www.rockinghamales.co.uk

The Copper Kettle Craft Brewery
Bencroft Grange, Bedford Road, Rushden, NN10 0SE
07930 105 301
ckcb.webs.com

The Silverstone Brewing Company Ltd
Kingshill Farm, Kingshill, Syresham, Brackley, NN13 5TH
01280 850 629
www.silverstonebrewingcompany.com

Tom Smith Ales Ltd
15 Lindsey Street, Kettering, NN16 8RG
01536 399 859
www.pigsandbeer.co.uk

Whistling Kite Brewery Ltd
35a Buccleuh Street, Kettering, NN16 9ED

Whittlebury Brewery Ltd
Home Farm Yard, Church Way, Whittlebury, Towcester, NN12 8XS
07812 366 369
www.whittleburybrewery.com

NORTHUMBERLAND

Allendale Brew Co. Ltd
Allen Mill, Allendale, Hexham, NE47 9EQ
01434 618 686
www.allendalebrewco.co.uk

Anarchy Brew Co
Unit 5 Whitehouse Farm Centre, Stannington, Morpeth, NE61 6AW
01670 789 755
www.anarchybrewco.com

Bear Claw Brewery
Unit 3 Meantime Workshops, North Greenich Road, Spittal, Berwick-upon-Tweed, TD15 1RG

Coquetdale Brewery Ltd
Unit 4c Station Road, Rothbury, Morpeth, NE65 7QJ
01669 621 411

Cullercoats Brewery Ltd Unit 19
Unit 19, Maurice Road Industrial Estate, Wallsend, NE28 6BY
07895 692 881
www.cullercoatsbrewery.co.uk

Gundog Brewery Ltd
2 Castle Island Way, North Seaton, Ashington, NE63 0XL
07707 703 182

Hadrian & Border Brewery
Unit 5 Preserving Works, Newburn Industrial Estate, Shelley Road, Newcastle-Upon-Tyne, NE15 9RT
0191 264 9000
www.hadrian-border-brewery.co.uk

Hexhamshire Brewery
Leafields, Hexham, NE46 1SX
01434 606 577
www.hexhamshire.co.uk

High House Farm Brewery
High House Farm, Matfen, NE20 0RG
01661 886 069
www.highhousefarmbrewery.co.uk

Northern FC Brewery
McCracken Park, Great North Road, Gosforth, Newcastle-Upon-Tyne, NE3 2DT
0191 236 3369
www.northernfootballclub.co.uk

Northumberland Breweries Ltd
Accessory House, Barington Road, Bedlington, NE22 7AP
01670 822 112
www.northumberlandbrewery.co.uk

Ouseburn Valley Brewery
The Brandling Villa, Haddricks Mill Road, Newcastle-Upon-Tyne, NE3 1QL
07932 677 899
www.ouseburnvalleybrewery.co.uk

Out There Brewing Co
Unit 4 Foundry Lane Industrial Estate, Newcastle-Upon-Tyne, NE6 1LH
07946 579 534
www.outtherebrewing.com

Redburn Brewery
Roselea, Redburn, Hexham, NE47 7EA
01434 344 656

The Ship Inn Brewery
The Ship Inn, The Square, Newton-by-the-Sea, Alnwick, NE66 3EW
01665 576 262
www.shipinnnewton.co.uk

V.I.P. Brewery
Village Inn Pub Company, Unit E Hawkshill Business Park, Lesbury, Alnwick, NE66 3PG
07545 885 352
www.thevillageinnpub.co.uk

Wylam Brewery Ltd
South Houghton Farm, Heddon on the Wall, Newcastle, NE15 0EZ
01661 853 377
www.wylambrew.co.uk

NOTTINGHAMSHIRE

Castle Rock Brewery
Queensbridge Road, Nottingham, NG2 1NB
01159 851 615
www.castlerockbrewery.co.uk

Dukeries Brewery Ltd
Unit 6 Peppers Warehouse, Blythe Road, Worksop, S81 0TP
07584 305 027

Flipside Brewery
Unit 3 East Link Trading Estate, Private Road 2, Colwick Industrial Estate, Nottingham, NG4 2JR
0115 987 7500
www.flipsidebrewery.co.uk

Full Mash Brewery
17 Lower Park Street, Stapleford, NG9 8EW
01159 499 262

Grafton Brewing Company
c/o 8 Oak Close, Worksop, S80 1BH
01909 476 121

Justice Brewery
15 Heron Way, Mansfield, NG19 6EQ
07954 989 594

Kings Clipstone Brewery
The Keepers Bothy, Main Road, Kings Clipstone, Mansfield, NG21 9BT
01623 823 589
www.kingsclipstonebrewery.co.uk

Lincoln Green Brewing Company Ltd
Unit 5 Enterprise Park, Wigwam Lane, Hucknall, Nottingham, NG15 7SZ
0115 963 4233
www.lincolngreenbrewing.co.uk

Magpie Brewery
4 Ashling Court, Ashling Street, Nottingham, NG2 3JA
01159 611 556
www.magpiebrewery.com

Maypole Brewery
North Laithes Farm, Kneesall, Newark, NG22 0AN
07971 277 598
www.maypolebrewery.co.uk

Medieval Brewery
Home Farm, New Road, Colston Bassett, Nottingham, NG12 3FQ
07552 798 027
www.medievalbeers.co.uk

Newark Brewery Ltd
77 William Street, Newark, NG24 1QU
07879 885 000

Nomad Brewery llp
Unit 1 Beggarlee Park, Engine Lane, Moorgreen Industrial Park, Newthorpe, NG16 3RN
www.nomadbrewery.co.uk

Nottingham Brewery
The Plough Inn, 17 St Peters Street, Radford, Nottingham, NG7 3EN
01159 422 649
www.nottinghambrewery.com

Pheasantry Brewery
High Brecks Farm, Lincoln Road, East Markham, Newark, NG22 0SN
01777 872 728
www.pheasantrybrewery.co.uk

Prior's Well Brewery
The Old Kennels, Hardwick Village, Clumber Park Estate, Worksop, S80 3PB
07971 277 592
www.priorswell.co.uk

Springhead Brewery
Old Great North Road, Sutton on Trent, Newark, NG23 6QS
01636 821 000
www.springhead.co.uk

The Idle Brewery
The White Hart Inn, Main Street, West Stockwith, DN10 4EY
01427 890 176

Welbeck Abbey Brewery
Lower Motor Yard, Welbeck, Worksop, S80 3LR
01909 512 539

White Dog Brewery
Unit 9 Sycamore House, Moorgreen Industrial Park, Engine Lane, Newthorpe, NG16 3QU
07984 935 700
www.whitedogbrewery.co.uk

OXFORDSHIRE

Adkin Brewery
Wantage
07709 086 149
www.adkinbrewery.co.uk

Appleford Brewery Company Ltd
c/o Iron Bridge House, St Peter's Court, Appleford, OX14 4YA
01235 848 055
www.applefordbrewery.co.uk

Bellinger's Brewery
Station Road, Grove, Wantage, OX12 0DH
01235 772 255
www.bellingersbrewery.co.uk

Complete Pig Brewery
Red Lion Farm, Britwell Salome, Watlington, OX49 5LG
07742 861 882
www.thecompletepig.co.uk

Hen House Brewery
The Old Dairy, Walliscote Farm, Whitchurch-on-Thames, Reading, RG8 7EP
www.henhousebrewery.co.uk

Loddon Brewery Ltd
Dunsden Green Farm, Church Lane, Dunsden, Reading, RG4 9QD
01189 481 111
www.loddonbrewery.com

Lovibonds Brewery Ltd
19-21 Market Place, Henley-on-Thames, RG9 2AA
01491 576 596
www.lovibonds.com

The Cotswold Brewing Company
Foxholes Lane, Foscot, OX7 6RL
01608 659 631
www.cotswoldbrewingcompany.co.uk

The Loose Cannon Brewing Company Ltd
6 Suffolk Way, Abingdon, LL20 8DD
01235 531 141
www.lcbeers.co.uk

The Old Bog Brewery
Masons Arms, 2 Quarry School Place, Headington, Oxford, OX3 8LH
01865 764 579
www.masonsquarry.co.uk

White Horse Brewery Co Ltd
3 Ware Road, White Horse Business Park, Stanford-in-the Vale, SN7 8NY
01367 718 700
www.whitehorsebrewery.com

SHROPSHIRE

Dickensian Brewery
Roden Lane, Roden, Telford, TF6 6BP
07976 724 243
www.dickensianbrewery.co.uk

Ironbridge Brewery Ltd
Unit 7 Merrythought, The Wharfage, Ironbridge, Telford, TF8 7NJ
01952 433 910
www.ironbridgebrewery.co.uk

Joule's Brewery Ltd
The Brewery, Great Hales Street, Market Drayton, TF9 1JP
01630 654 400
www.joulesbrewery.co.uk

Lion's Tale Brewery
Red Lion Hotel, High Street, Cheswardine, Market Drayton, TF9 2RS
01630 661 234

Offa's Dyke Brewery
Chapel Lane, Trefonen, SY10 9DX
01691 656 889
www.offasdykebrewery.com

Salopian Brewing Co Ltd
67 Mytton Oak Road, Copthorne, Shrewsbury, SY3 8UQ
01743 248 414
www.salopianbrewery.co.uk

Six Bells Brewery
Church Street, Bishop's Castle, SY9 5AA
01588 638 930
www.bishops-castle.co.uk/SixBells/brewery

Stonehouse Brewery
Weston, Oswestry, SY10 9ES
01691 676 457
www.stonehousebrewery.co.uk

The Ludlow Brewing Co Ltd
Kingsley Garage Yard, 105 Corve Street, Ludlow, SY8 1DJ
01584 873 291
www.theludlowbrewingcompany.co.uk

Three Tuns Brewery
Salop Street, Bishop's Castle, SY9 5BW
01588 638 392
www.threetunsbrewery.co.uk

Wood Brewery Ltd
Wistanstow, Craven Arms, SY7 8DG
01588 672 523
www.woodbrewery.co.uk

SOMERSET

Berrow Brewery
Coast Road, Berrow, Burnham-on-sea, TA8 2QU
01278 751 345

Blindmans Brewery Ltd
Talbot Farm, Leighton, Frome, BA11 4PN
01749 880 038
www.blindmansbrewery.co.uk

Cheddar Ales
Unit 5 Winchester Farm, Draycott Road, Cheddar, BS27 3RP
01934 744 193
www.cheddarales.co.uk

Cotleigh Brewery
Ford Road, Wiveliscombe, TA4 2RE
01984 624 086
www.cotleighbrewery.co.uk

Glastonbury Ales
Grovers Brewhouse, Unit 10 Wessex Park, Bancombe Road, Somerton, TA11 6SB
01458 272 244
www.glastonburyales.com

Kubla Brewery Ltd
The Source Building, Tower Farm, Lydeard St Lawrence, Taunton, TA4 3QN
07855 342 208
www.kubla.co.uk

Matthews Brewing Co Ltd
Unit 7 Timsbury Workshop Estate, Hayeswood Road, Timsbury, Bath, BA2 0HQ
01761 472 242
www.matthewsbrewing.co.uk

Milk Street Brewery
The Griffin, Milk Street, Frome, BA11 3DB
01373 467 766
www.milkstreetbrewery.co.uk

Moor Beer Co
Whitley Farm, Stagman Lane, Ashcott, Bridgewater, TA7 9QW
01458 210 050
www.moorbeer.co.uk

North Curry Brewery
The Old Coach House Gwyon House, Church Road, North Curry, Taunton, TA3 6LH
07928 815 053
www.thenorthcurrybrewerycouk.com

Odcombe Brewery
The Masons Arms, 41 Lower Odcombe, Yeovil, BA22 8TX
01935 862 591
www.masonsarmsodcombe.co.uk

Ordnance City Brewery
The Old Brewery, Whitley Lane, Ashcott, Bridgwater, TA7 9QW
01458 210 280

Quantock Brewery
Unit E Monument View, Summerfield Avenue, Chelston Business Park, Wellington, TA21 9ND
07824 860 282
www.quantockbrewery.co.uk

RCH Brewery
West Hewish, Weston-super-Mare, BS24 6RR
01934 834 447
www.rchbrewery.com

Six Trees Brewing Co
Triscombe House, Triscombe, Bishops Lydeard, Taunton, TA4 3HG
01984 617 000
www.6trees.co.uk

Stocklinch Ales Limited
Manor Farm, Owl Street, Stocklinch, Ilminster, TA19 9JG
07711 479 917

Stowey Brewery
The Old Cider House, 25 Castle Street, Nether Stowey, TA5 1LN
01278 732 228
www.stoweybrewery.co.uk

The Abbey Brewery
Camden Row, Bath, BA1 5LB
01225 444 437
www.abbeyales.co.uk

The Cottage Brewing Company Ltd
The Old Cheese Dairy, Blue Farm, Lovington, Castle Cary, BA7 7PP
01963 240 551

The Wild Beer Company Ltd
Lower Westcombe Farm, Westcombe, Shepton Mallet, BA4 6ER
01749 838 742
www.wildbeerco.com

Yeovil Ales Brewery
Bofors Park, Artilllery Road, Lufton, Yeovil, BA22 8YH
01935 414 888
www.yeovilales.com

SOUTH YORKSHIRE

Abbeydale Brewery
Unit 8, Aizlewood Road, Sheffield, S8 0YX
01142 818 712
www.abbeydalebrewery.co.uk

Acorn Brewery
Unit 3 Aldham Industrial Estate, Mitchell Road, Wombwell, Barnsley, S73 8HA
01226 270 734
www.acorn-brewery.co.uk

Blue Bee Brewery Ltd
Unit 30 Hoyland Road, Sheffield, S3 8AB
07791 662 484

Bradfield Brewery
Watt House Farm, High Bradfield, Sheffield, S6 6LG
01142 851 118
www.bradfieldbrewery.com

Chantry Brewery Ltd
Unit 1 Callum Court, Gateway Industrial Estate, Parkgate, Rotherham, S62 6NR
01709 711 866
www.chantrybrewery.co.uk

Crown Brewery
Hillsborough Hotel, 54-58 Langsett Road, Sheffield, S6 2UB
01142 322 100
www.hillsborough-hotel.co.uk

Doncaster Brewery Ltd
Unit 4a Coopers Mill Business Park, Clay Lane West, Doncaster, DN2 4QR
07770 958 394
www.doncasterbrewery.co.uk

Geeves Brewery Ltd
Unit 12 Grange Lane Industrial Estate, Carrwood Road, Barnsley, S71 5AS
07859 039 259
www.geevesbrewery.co.uk

Glentworth Brewery
Glentworth House, Crossfield Lane, Skellow, Doncaster, DN6 8PL
01302 725 555

Harthill Village Brewery
Beehive Farm, Union Street, Sheffield, S26 7YH
www.harthillbrewery.co.uk

Kelham Island Brewery
23 Alma Street, Sheffield, S3 8SA
01142 494 804
www.kelhambrewery.co.uk

Little Ale Cart Brewing Company
The Wellington, 1 Henry Street, Sheffield, S3 7EQ
01142 492 295

Oakwell Brewery
Unit 11, Pontefract Road, Barnsley, S71 1EZ
01226 296 161

The Brew Company Ltd
Unit C G4 Business Centre, Carlisle Street East, Sheffield, S4 7QN
01142 709 991
www.thebrewcompany.co.uk

The Concertina Brewery
9a Dolcliffe Road, Mexborough, S64 9AZ
01709 580 841

The Sheffield Brewery Company Ltd
Unit 111 The JC Albyn Complex, Burton Road, Sheffield, S3 8BZ
01142 727 256
www.sheffieldbrewery.com

Two Roses Brewery Ltd
Unit 9 Darton Business Park, Barnsley Road, Darton, Barnsley, S75 5QX
01226 388 375
www.tworosesbrewery.co.uk

Wentworth Brewery
The Power House, The Gun Park, Wentworth, Rotherham, S62 7TF
01226 747 070
www.wentworth-brewery.co.uk

ST ANDREWS

Fisher & Co Craft Brewery Limited
Lower Farm, Noke, Oxford, OX3 9TX

STAFFORDSHIRE

Black Hole Brewery Ltd
Unit 63 Imex Business Park, Shobnall Road, Burton-upon-Trent, DE14 2AU
01283 534 060
www.blackholebrewery.co.uk

Blythe Brewery
Blythe House Farm, Lichfield Road, Hamstall Ridware, Rugeley, WS15 3QQ
07773 747 724
www.blythebrewery.co.uk

Eccleshall Brewery
St Albans Road, Common Road Industrial Estate, Stafford, ST16 3DR
01785 257 976
www.slatersales.co.uk

Flash Brewery
Moss Top Farm, Moss Top Lane, Quarnford, Buxton, SK17 0TA
01243 430 020

Gates Burton Brewery
7 Reservoir Road, Burton-upon-Trent, NE31 1NB
01283 532 567

JHS Traditional Brewers
c/o The Fox Inn, Broadheath, Tenbury Wells, WR15 8QS
01886 853 189
www.jhstraditionalbrewery.com

Kinver Brewery
Unit 2 Fairfield Drive, Kinver, Stourbridge, DY7 6EW
07715 842 679
www.kinverbrewery.co.uk

Leek Brewery
Unit 11 & 12 Churnet Court, Churnetside Business Park, Cheddleton, Leek, ST13 7EF
01538 361 919
www.beersandcheese.co.uk

Lymestone Brewery Ltd
Unit 5 Mount Street Industrial Estate, Mount Street, Stone, ST15 8LL
01785 817 796
www.lymestonebrewery.co.uk

Morton Brewery
c/o 96 Brewood Road, Coven, Wolverhampton, WV9 5EF
07988 069 647
www.mortonbrewery.co.uk

Quartz Brewing
Heart of the Country Village, Swinfen, Lichfield, WS14 9QR
01543 483 241
www.quartzbrewing.co.uk

Talke o' Th' Hill Brewery
Mere Lake Road, Talke Pits, Stoke-on-Trent, ST7 1UF
07875 951 399
www.talkeothhill.co.uk

The Backyard Brewhouse
Unit 8a Gatehouse Trading Estate, Lichfield Road, Brownhills, Walsall, WS8 6JZ
07591 923 370
www.thebackyardbrewhouse.com

The Burton Old Cottage Beer Company Ltd
Unit 10 Brian Eccleshall Yard, Eccleshall Business Park, Hawkins Lane, Burton-upon-Trent, DE14 1PT
07909 931 250
www.oldcottagebeer.co.uk

Titanic Brewery
Unit 5 Callender Place, Lingard Street, Burslem, Stoke on Trent, ST6 1JL
01782 823 447
www.titanicbrewery.co.uk

Tower Brewery
The Old Water Tower, Walsitch Maltings, Glensyl Way, Burton-upon-Trent, DE14 1LX
01283 530 695

Townhouse Brewery
Townhouse Farm, Alsager Road, Audley, Stoke-on-Trent, ST7 8JQ
07976 209 437

SUFFOLK

Bartrams Brewery
Rougham Estate, Ipswich Road, Bury St Edmunds, IP30 9LZ
01449 737 655
www.bartramsbrewery.co.uk

Earl Soham Brewery
Earl Soham, Woodbridge, IP13 7RT
01728 684 097
www.earlsohambrewery.co.uk

Elveden Ales
The Courtyard, Elveden Estate, Elveden Estate, Thetford, IP24 3TA
01842 878 922

Engel's Fine Ales
Glencot, Wood Lane, Aldeby, Beccles, NR34 0DA
01502 679 144
www.engelfineales.com

Green Dragon Brewery
29 Broadstreet, Bungay, NR35 1ED
01986 892 681

Green Jack Brewery
Triangle Tavern, 29 St Peters Street, Lowestoft, NR32 1QA
01502 582 711
www.green-jack.co.uk

Kings Head Brewing Co
The Kings Head, 132 High Street, Bildeston, Ipswich, IP7 7ED
01449 741 434
www.bildestonkingshead.co.uk

Shortts Farm Brewery
Shorts Farm, Wetheringsett Road, Thorndon, Eye, IP23 7LS
07900 268 100

The Black Adder Brewery
13 Churchfield Road, Sudbury, CO10 2YA
01787 311 055
www.mauldons.co.uk

The Brandon Brewery
76 High Street, Brandon, IP27 0AU
01842 878 496
www.brandonbrewery.co.uk

The Cliff Quay Brewery
Cliff Road, Ipswich, IP3 0AZ
01473 225 501
www.cliffquay.co.uk

The Mill Green Brewery
White Horse Inn, Mill Green, Edwardstone, Sudbury, CO10 5PX
01787 211 211
www.millgreenbrewery.co.uk

The Old Cannon Brewery
86 Cannon Street, Bury St Edmunds, IP33 1JR
01284 768 769
www.oldcannonbrewery.co.uk

SURREY

Ascot Ales Ltd
Unit 5 Compton Place Business Centre, Surrey Avenue, Camberley, GU15 3DX
0845 257 5575
www.ascot-ales.co.uk

Brightwater Brewery
9 Beaconsfield Road, Claygate, KT10 0PN
01372 462 334

Dorking Brewery
Engine Shed, Dorking West Station Yard, Station Road, Dorking, RH4 1HF
01306 877 988
www.dorkingbrewery.com

Farnham Brewery
104 Upper Hale Road, Farnham, GU9 0PB
01252 735 278

Hogs Back Brewery
Manor Farm, The Street, Tongham, GU10 1DE
01252 783 000
www.hogsback.co.uk

Pilgrim Brewery
West Street, Reigate, RH2 9BL
01737 222 651
www.pilgrim.co.uk

Surrey Hills Brewery Ltd
Old Scotland Farm, Staple Lane, Shere, Guildford, GU5 9TE
01483 212 812
www.surreyhills.co.uk

The Leith Hill Brewery
The Plough Inn, Coldharbour, Dorking, RH5 6HD
01306 711 793
www.ploughinn.com

The Little Beer Corporation Ltd
Unit 16 Enterprise Units, Moorfield Road, Slyfield Industrial Estate, Guildford, GU1 1RB
07919 025 980
www.littlebeer.co.uk

Tillingbourne Brewery Ltd
Old Scotland Farm, Staple Lane, Shere, Guildford, GU5 9TE
01483 222 228
www.tillybeer.co.uk

Wayland's Brewery
6 Marley Close, Addlestone, KT15 1AR
07956 531 618
www.waylandsbrewery.co.uk

SUSSEX

High Weald Brewery
Unit 8 Bulrushes Business Park, Coombe Hill Road, East Grinstead, RH19 4LZ
07836 291 430
www.highwealdbrewery.co.uk

TYNE & WEAR

Big Lamp Brewery
Grange Road, Newburn, Newcastle-Upon-Tyne, NE15 8NL
0191 267 1689
www.biglampbrewers.co.uk

Bull Lane Brewing Company
The Clarendon, 143 High Street East, Sunderland, SR1 2BL
0191 510 3200
www.bull-lane-brewing.co.uk

Darwin Brewery
63 Back Tatham Street, Hendon, Sunderland, SR1 2QE
0191 514 4746
www.darwinbrewery.com

Delavals (Brewers) Ltd
26 Windsor Gardens, Whitley Bay, NE26 3BG
0844 504 2214
www.delavals.com

Mordue Brewery
Units D1 & D2, Narvik Way, Tyne Tunnel Trading Estate, North Shields, NE29 7XJ
0191 296 1879
www.morduebrewery.com

Three Kings Brewery Ltd
14 Prospect Terrace, North Shields, NE30 1DX
07580 004 565

Tyne Bank Brewery Ltd
Unit 11, Hawick Crescent Industrial Estate, Newcastle-Upon-Tyne, NE6 1AS
0191 265 2828
www.tynebankbrewery.co.uk

WARWICKSHIRE

Church End Brewery Ltd
109 Ridge Lane, Nuneaton, CV10 0RD
01827 713 080
www.churchendbrewery.co.uk

Church Farm Brewery
2 Church Cottage, Budbrooke, Warwick, CV35 8QL
01926 411 569

Long Itch Brewery Ltd
Manor Farm, Hunningham Road, Offchurch, Leamington Spa, CV33 9AG
07780 900 699
www.longitchbrewery.co.uk

North Cotswold Brewery
Ditchford Farm, Stretton-on-Fosse, Moreton-in-Marsh, GL56 9RD
01608 663 947
www.northcotswoldbrewery.co.uk

Purity Brewing Co
The Brewery, Upper Spernall Farm, Great Alne, Alcester, B49 6JF
01789 488 007
www.puritybrewing.com

The Atomic Brewery
c/o 1 Lower Hillmorton Rd, Rugby, CV21 3ST
01788 542 170
www.atomicbrewery.com

The Shed Brewery Limited
2218 Stratford Road, Hockley Heath, Solihull, B94 6NU
07910 004 041

The Warwickshire Beer Co
Queen Street, Cubbington, CV32 7NA
01926 450 747
www.warwickshirebeer.co.uk

Tunnel Brewery Ltd
Lord Nelson Inn, Birmingham Road, Ansley, Nuneaton, CV10 9PQ
024 7639 4888
www.tunnelbrewery.co.uk

Whale Ale Ltd
The Brewery, Unit 5B Brailes Industrial Estate, Winderton Road, Banbury, OX15 5JW
01608 686 974
www.whaleale.co.uk

WEST MIDLANDS

ABC Brewery
Aston Brewery Company Ltd, Unit 21 Birch Road, Witton, Birmingham, B6 7DD
0121 328 2655
www.abcbrewery.co.uk

Angel Ales Ltd
62A Furlong Lane, Cradley, Halesowen, B63 2TA
07847300 350
www.angelales.co.uk

Beer Geek Brewery Limited
Unit D3 Seedbed Centre, Avenue Road, Nechells, Birmingham, B7 4NT
0844 272 7207
www.beergeekbrewery.com

Birds Brewery
Ladybird Barn, Old Burcot Lane, Burcot, Bromsgrove, B60 1PH
01527 889 870
www.birdsbrewery.co.uk

Black Country Ales
Unit 4, Tansey Green Road, Brierley Hill, Dudley, DY5 4TL
01384 480 156
www.blackcountryales.co.uk

Enville Ales
Enville Brewery, Cox Green, Hollies Lane, Stourbridge, DY7 5LG
01384 873 728
www.envilleales.com

Olde Swan Brewery
89 Halesowen Road, Netherton, Dudley, DY2 9PY
01384 253 075

Sarah Hughes Brewery
129 Bilston Road, Sedgley, DY3 1JE
01902 883 380

The Beowulf Brewing Company
Unit 3-4 Chasewater Country Park, Pool Road, Brownhills, Walsall, WS8 7NL
01543 454 067
www.beowulfbrewery.co.uk

The Slaughterhouse Brewery Ltd
Bridge Street, Warwick, CV34 5PD
01926 490 986
www.slaughterhousebrewery.com

The Two Towers Brewery Ltd
51 Mott Street, Birmingham, B19 3HE
0121 439 7253
www.twotowersbrewery.co.uk

Toll End Brewery
Waggon & Horses, 131 Toll End Road, Tipton, DY4 0ET
0121 502 6453

Whitworth Brewing Co
34 Dunard Road, Shirley, Solihull, B90 2HR
0121 347 6450
www.whitworthbrewing.co.uk

Windsor Castle Brewery
7 Stourbridge Road, Lye, Stourbridge, DY9 7DG
01384 895 230
www.windsorcastlebrewery.com

WEST SUSSEX

Adur Brewery Ltd
Adur Business Centre, Little High Street, Shoreham-by-sea, BN44 3DG
01273 467 527
www.adurbrewery.com

Arundel Brewery
Unit 7c, Ford Airfield Industrial Estate, Ford, Arundel, BN18 0HY
01903 733 111
www.arundelbrewery.co.uk

Baseline Brewing
Unit 11b Golding Barn Industrial Estate, Henfield Road, Small Dole, Henfield, BN5 9XH
01903 879 111
www.baselinebrewing.co.uk

Bedlam Brewery Ltd
Ibourne Farm, Shaves Wood Lane, Albourne, Hassocks, BN6 9DX
07955 684 041
www.bedlambrewery.co.uk

Dark Star Brewing
22 Star Road, Partridge Green, RH13 8RA
01403 713 085
www.darkstarbrewing.co.uk

Hammerpot Brewery Ltd
Unit 30 The Vinery, Arundel Road, Poling, Arundel, BN18 9PY
01903 883 338
www.hammerpot-brewery.co.uk

Hepworth & Company Brewers Ltd
The Railway Yard, Nightingale Road, Horsham, RH12 2NW
01403 269 696
www.hepworthbrewery.co.uk

Kissingate Brewery
Pole Barn, Church Lane, Lower Beeding, Horsham, RH13 6LU
01403 891 335
www.kissingate.co.uk

The Anchor Springs Brewery Company Ltd
Unit J OYO Business Units, Arndale Road, Wick, Littlehampton, BN17 7GA
01903 719 842
www.anarchybrewco.com

The Baldy Brewery
Parham Park, Parnham, Pulborough, RH20 4HS
07718 641 195
www.thebaldybrewery.co.uk

The Downlands Brewery
Unit Z(2a) Mackley Industrial Estate, Henfield Road, Small Dole, Henfield, BN5 9XR
07775 743 518
www.downlandsbrewery.com

The Langham Brewery llp
Langham Stables, Langham Lane, Lodsworth, Petworth, GU28 9BU
01798 860 861
www.langhambrewery.co.uk

W J King & Co Brewers
3-5 Jubilee Estate, Horsham, RH13 5UE
01403 272 102
www.kingbeer.co.uk

WEST YORKSHIRE

Barearts Brewery
290-292 Rochdale Road, Todmorden, OL14 7PD
01706 839 305
www.barearts.com

Barley Bottom Brewery
Unit 1a Howden Hall Industrial Estate, Howden Road, Silsden, Keighley, BD20 0HJ
01535 656 797
www.barleytrading.co.uk

Bob's Brewing Co Ltd
The Brewers Pride, Low Mill Road, Ossett, WF5 8ND
07789 693 597

Bridgehouse Brewery Ltd
Hawkcliffe Works, Hebden Road, Oxenhope, Keighley, BD22 9SY
01535 642 893
www.bridgehousebrewery.co.uk

Briscoe's Brewery
16 Ash Grove, Otley, LS21 3EL
01943 466 515

Brontë Ales Ltd
Atlas Mill, Atlas Mill Road, Brighouse, HD6 1ES
www.bronteales.co.uk

Cap House Brewery
Victoria Works, 444 Bradford Road, Batley, WF17 5LW
01924 479 909

Collingham Ales Ltd
Leeds Road, Collingham, Wetherby, LS22 5AA
01937 573 096
www.collinghamales.co.uk

Elland Brewery
Units 3-5 Heathfield Industrial Estate, Heathfield Street, Elland, HX5 9AE
01422 377 677
www.eandsbrewery.co.uk

Empire Brewery
Unit 33 The Old Boiler House, Upper Mills, Cannal Side, Huddersfield, HD7 5HA
01484 847 343

Fernandes Brewery
The Old Malthouse, 5 Avison Yard, Wakefield, WF1 1UA
01924 291 709
www.fernandes-brewery.gowyld.com

Five Towns Brewery
651 Leeds Road, Wakefield, WF1 2LU
01924 781 887

Fox Beer Company
Fox & Newt, 9 Burley Street, Leeds, LS3 1LD
01132 454 505

Golcar Brewery
60a Swallow Lane, Golcar, Huddersfield, HD7 4NB
01484 644 241

Goose Eye Brewery
Ingrow Bridge, South Street, Keighley, BD21 5AX
01535 605 807
www.goose-eye-brewery.co.uk

H B Clark
Westgate Brewery, Westgate, Wakefield, WF2 9SW
01924 373 328
www.hbclark.co.uk

Halifax Steam Brewing Company
The Conclave, Southedge Works, Hipperholme, Halifax, HX3 8EF
01484 715 074

Hamelsworde Brewery
16 Longworth Road, Hemsworth, Pontefract, WF9 4SZ
07530 669 332
www.hamelsworde.co.uk

Hand Drawn Monkey Brewing Co
Plover Road Garage, Plover Road, Lindley, Huddersfield, HD3 3PJ
01484 655 262

Haworth Steam Brewing Company
98 Main Street, Haworth, Mold, BD22 8DP
01535 646 212
www.haworthsteambrewery.co.uk

Kirkstall Brewery
Unit 6 Canal Wharf, Wyther Lane, Leeds, LS5 3BT
0113 345 8835
www.kirkstallbrewerycompany.com

Linfit Brewery
The Sair, Lane Top, Linthwaite, Huddersfield, HD7 5SG
01484 842 370

Little Valley Brewery Ltd
Turkey Lodge Farm, New Road, Cragg Vale, Hebden Bridge, HX7 5TT
01422 883 888
www.littlevalleybrewery.co.uk

Magic Rock Brewing Company Ltd
Unit 1 Quarmby Mills, Tanyard Road, Oakes, Huddersfield, HD3 4YP
01484 649 823
www.magicrockbrewing.com

Mallinsons Brewing Company
Unit 1, Waterhouse Mill, 65-71 Lockwood Road, Huddersfield, HD1 3QU
07850 446 571
www.drinkmallinsons.co.uk

Milltown Brewing Co Ltd
The Brewery, The Old Railway Goods Yard, Milnsbridge, HD3 4PE
01422 610 579
www.milltownbrewing.co.uk

Naylor's Brewery
Midland Mill, Station Road, Crosshills, Keighley, BD 20 7DT
01535 637 451
www.naylorsbrewery.com

Oates Brewing Company
4 Ladyship Business Park, Mill Lane, Halifax, HX3 6TA
01422 320 100
www.oatesbrewing.co.uk

Old Bear Brewery
Unit 4b Atlas Works, Pitt Street, Keighley, BD21 4YL
01535 601 222
www.oldbearbrewery.com

Old Spot Brewery Limited
Manor Farm, Station Road, Cullingworth, Bradford, BD13 5HN
01535 691 144
www.oldspotbrewery.co.uk

Ossetts Brewing Co Ltd
Low Mill Road, Ossett, WF5 8ND
01924 261 333
www.ossett-brewery.co.uk

Owenshaw Mill Brewery Ltd
Owenshaw Works, Old Cawsey, Sowerby Bridge, HX6 2AJ
01422 839 010
www.owenshawmillbrewery.co.uk

Partners Brewery Ltd
Unit 12 Savile Bridge Mill, Savile Road, Dewsbury, WF12 9AF
01924 457 772
www.partnersbrewery.co.uk

Pennine Brewery Co.Ltd
Unit 8a Grange Road Industrial Estate, Batley, WF17 6LL
01904 567 567
www.pennine-brewery.co.uk

Salamander Brewing Co Ltd
22 Harry Street, Bradford, BD4 9PH
01274 652 323
www.salamanderbrewing.co.uk

Saltaire Brewery
The Brewery, County Works, Dockfield Road, Shipley, BD17 7AR
01274 594 959
www.saltairebrewery.co.uk

Slightly Foxed Brewing Company
Unit 25 Asquith Bottom Mill, West Street, Sowerby Bridge, HX6 3BS
07412 008 221
www.slightlyfoxedbrewery.co.uk

Summer Wine Brewery Ltd
Unit 15 Honley Business Centre, New Mill Road, Honley, Holmfirth, HD9 6QB
01484 665 466
www.summerwinebrewery.co.uk

Sunbeam Ales
Sunbeam Terrace, Leeds, LS11 6EW
07772 002 437
www.sunbeamales.co.uk

The Bridestones Brewing Company Limited
Smithy Farm, Blackshaw Head, Hebden Bridge, HX7 7JB
01422 847 104
www.bridestonesbrewing.co.uk

The Leeds Brewery Company Ltd
Units 3-4 Sydenham Road, Leeds, LS11 9RU
01132 445 866
www.leedsbrewery.co.uk

The Malthouse Brewery
Unit 4 Parkway House, Ashley Industrial Estate, Wakefield Road, Ossett, WF5 9JD
07980 386 361
www.themalthousebrewery.co.uk

The Revolutions Brewing Company Ltd
Unit B7 Whitwood Enterprise Park, Speedwell Road, Whitwood, Castleford, WF10 5PX
01977 552 649
www.revolutionsbrewing.co.uk

The Ridgeside Brewing Company Ltd
24 Penraevon Street, Penraevon Industrial Estate, Leeds, LS7 2AW
07595 380 568
www.ridgesidebrewery.co.uk

Tigertops Brewery
22 Oakes Street, Wakefield, WF2 9LN
01924 897 728

WharfeBank Brewery Ltd
Unit 4 Pool Business Park, Pool Road, Pool In Wharfedale, Otley, LS21 1FD
0113 284 2392
www.wharfebankbrewery.co.uk

WILTSHIRE

Box Steam Brewery
The Midlands, Holt,
Bradford on Avon, BA14 6RU
01225 782 700
www.boxsteambrewery.com

Braydon Ales Limited
Preston West Farm, Preston,
Chippenham, SN15 4DX
01249 892 900
www.braydonales.co.uk

Hop Back Brewery
Downton Business Park,
Batten Road, Downton,
Salisbury, SP5 3HU
01725 510 986
www.hopback.co.uk

Keystone Brewery
Old Carpenters Workshop,
Berwick St Leonard,
Salisbury, SP3 5SN
01747 820 426
www.keystonbrewery.co.uk

Moles Brewery
5 Merlin Way, Bowerhill,
Melksham, SN12 6TJ
01225 708 842
www.molesbrewery.com

Plain Ales
Bow House, Chitterne,
BA12 0LG
01985 851 105
www.plainales.co.uk

Salisbury Brewery
Unit 1 Oakley Business Park,
Dinton, Salisbury, SP3 5EU
01722 716 440
www.salisburybrewery.com

Shed Ales
Broadfields, Pewsey, SN9 5DT
www.shed-ales.com

Stonehenge Ales
The Old Mill, Netheravon,
Salisbury, SP4 9QB
01980 670 631
www.stonehengeales.com

The Downton Brewery Company Ltd
Unit 11 Downton Industrial
Estate, Batten Road, Downton,
Salisbury, SP5 3HU
01722 322 890
www.downtonbrewery.com

The Ramsbury Brewery
Priory Farm, Axford,
Marlborough, SN8 2HA
01672 541 407
www.ramsburybrewery.com

Three Castles Brewery Ltd
Unit 12 Salisbury Road
Business Park, Salisbury Road,
Pewsey, SN9 5PZ
01672 564 433
www.threecastlesbrewery.co.uk

Three Daggers Brewery
Westbury Road, Edington,
Westbury, BA13 4PG
01380 830 940
www.threedaggersbrewery.com

Wessex Brewery
Rye Hill Farm, Longbridge
Deverill, Warminster, BA12 7DE
01985 844 532

WORCESTERSHIRE

Attwood Ales Ltd
Station Road, Hartlebury,
Kidderminster, DY11 7YJ
01299 253 617
www.attwoodales.com

Bewdley Brewery Ltd
Units 7 & 8 Bewdley Craft
Centre, Lax Lane, Bewdley,
DY12 2DZ
01299 405 148
www.bewdleybrewery.co.uk

Brandy Cask Pub & Brewery
c/o 25 Bridge Street,
Pershore, WR10 1AJ
01386 552 602

Bricktop Brewery
c/o The Gate Hangs Well,
Woodgate Road, Stoke Prior,
Brosgrove, B60 4HG
01527 821 957
www.thegatehangswell.co.uk

Cannon Royall Brewery Limited
Rear Fruiterer's Arms,
Uphampton Lane, Ombersley,
WR9 0JW
01905 621 161
www.cannonroyall.co.uk

Hobsons Brewery & Co
New House Farm, Tenbury
Road, Cleobury Mortimer,
Kidderminster, DY14 8RD
01299 270 837
www.hobsonsbrewery.co.uk

Malvern Hills Brewery
15 West Malvern Road,
Great Malvern, WR14 4ND
01684 560 165
www.malvernhillsbrewery.co.uk

Pope's Brewing Company llp
Unit 73a Brackpole Trading
Estate West, Worcester,
WR3 8TJ
01905 755 016

St. George's Brewery Ltd
The Old Bakehouse, Bush Lane,
Callow End, Worcester,
WR2 4TF
01905 831 316

Teme Valley Brewery
The Talbot, Knightwick,
WR6 5PH
01886 821 235
www.temevalleybrewery.co.uk

Weatheroak Brewery Ltd
25 Withybed Lane, Alvechurch,
B48 7NX
0121 445 4411
www.weatheroakales.co.uk

Weatheroak Hill Brewery Limited
Coach & Horses, Weatheroak
Hill, Alvechurch, Birmingham,
B48 7EA
01564 823 386
www.coachandhorsesinn.co.uk

Wyre Piddle Brewery
Highgrove Farm, Pinvin,
Pershore, WR10 2LF
01905 841 853

NORTHERN IRELAND

CO. ANTRIM

Hilden Brewing Company
Grand Street, Lisburn,
BT27 4TY
028 9266 0800
www.hildenbrewery.co.uk

CO. DOWN

Clanconnel Brewing Co Ltd
PO Box 316, Craigavon,
BT65 9AZ
07711 626 770
www.clanconnelbrewing.com

SCOTLAND

ABERDEENSHIRE

BrewDog Ltd
Unit 1 Kessock Workshops,
Fraserburgh, AB43 8UE
01346 519 009
www.brewdog.com

Burnside Brewery
Unit 3 Laurencekirk Business
Park, Aberdeen Road,
Laurencekirk, AB30 1EY
01561 377 316
www.brewmet.com

Six Degrees North Brewery
The Workshop, Cowgate,
Stonehaven, AB39 2LD
07840 678 243
www.marinehotelstonehaven.co.uk

ANGUS

Mòr Brewing Limited
Old Mill, Kellas, Broughty Ferry,
Dundee, DD5 3PD
07593 245 000

ARGYLL & BUTE

Atlas Brewery
Lab Road, Kinlochleven,
PH50 4SG
01855 831 111
www.sinclairbreweries.co.uk

Fyne Ales
Achadunan, Cairndow,
PA26 8BJ
01499 600 238
www.fyneales.com

Sulwath Brewery
The Brewery, 209 King Street,
Castle Douglas, DG7 1DT
01556 504 525
www.sulwathbrewery.co.uk

The Oyster Bar & Brewery
Ellenabeich, Easdale, Oban,
PA34 4RQ
01852 300 121
www.oysterbrewery.com

CLACKMANNANSHIRE

Devon Ales Ltd
Mansfield Arms, 7 Main Street, Sauchie, Alloa, FK10 3JR
01259 722 020
www.devonales.com

Heather Ales
New Alloa Brewery, Kelliebank, Alloa, FK10 1NU
01259 725 511
www.fraoch.com

DUMFRIES & GALLOWAY

Andrews Ales
c/o 1 Railway Cottage, Cummertrees, Annan, DG12 5QG
01461 700 387

EDINBURGH

Barney's Beer
Summerhall Brewery, 1 Summerhall, Edinburgh, EH9 1PL
07512 253 660
www.barneysbeer.com

Fowler's Ale (Prestoungrange) Limited
The Gothenburg, 227-229 High Street, Prestonpans, EH32 9BE
01875 819 922
www.prestoungrange.org

Stewart Brewing
Unit 5, 42 Dryden Road, Bilston Glen Industrial Estate, Loanhead, EH20 9LZ
0131 440 2442
www.stewartbrewing.co.uk

FIFE

Beeches Brewery
The Beeches, Lochgelly, KY5 9QB
01592 782 474

St Andrews Brewing Co
Unit 4 Fife Food Resource Base, Southfield Industrial Estate, Faraday Road, Glenrothes, KY6 2RU
07879 399 441
www.standrewsbrewingcompany.com

The Eden Brewery
Main Road, Guardbridge, St Andrews, KY16 0UU
07786 060 013
www.edenbrewerystandrews.com

GLASGOW

Kelburn Brewing Company Ltd
10 Muriel Lane, Barrhead, G78 1QB
0141 881 2138
www.kelburnbrewery.com

HIGHLAND

An Teallach Ale Co
1 Camusnagaul, Dundonell, By Garve, IV23 2QT
01854 633 306

Black Isle Brewery Co Ltd
Old Allangrange, Munlochy, IV8 8NZ
01463 811 871
www.blackislebrewery.com

Cromarty Brewing Company Ltd
Canon Street, Davidston, Cromarty, IV11 8XD
07927 037 736
www.cromartybrewing.co.uk

Glenfinnan Brewery Co Ltd
Struth A Lhuilinn, Glenfinnan, PH37 4LT
01397 704 309
www.glenfinnanbrewery.co.uk

River Leven Ales
Lab Road, Kinlochleven, PH50 4SG
01855 831 519
www.riverlevenales.co.uk

The Cairngorm Brewery Co Ltd
Unit 12 Dalfaber Industrial Estate, Dalfaber Drive, Aviemore, PH22 1ST
01479 812 222
www.cairngormbrewery.com

The Plockton Brewery
5 Bank Street, Plockton, IV52 8TP
01599 544 276
www.theplocktonbrewery.com

ISLE OF ARRAN

Arran Brewery
Claddach, KA27 8DE
01770 302 353
www.arranbrewery.com

ISLE OF COLONSAY

Colonsay Brewing Co Ltd
The Brewery, Colonsay, PA61 7YZ
01951 200 190
www.colonsaybrewery.co.uk

ISLE OF ISLAY

Islay Ales
The Brewery, Islay House Square, Bridgend, PA44 7NZ
01496 810 014
www.islayales.com

ISLE OF LEWIS

Hebridean Brewing Company
18a Bells Road, Stornaway, HS1 2RA
01851 700 123
www.hebridean-brewery.co.uk

ISLE OF MULL

Isle of Mull Brewing Co Ltd
Ledaig, Tobermory, PA75 6NR
01688 302 821

ISLE OF SKYE

Isle of Skye Brewing Company
Uig, IV51 9XP
01470 542 477
www.skyebrewery.co.uk

LANARKSHIRE

Loch Lomond Brewery Ltd
Unit 5 Lomond Industrial Estate, Alexandria, G83 0TL
01389 755 698
www.lochlomondbrewery.com

Strathaven Ales
Craigmill Brewery, Strathaven, ML10 6PB
01357 520 419
www.strathavenales.co.uk

MIDLOTHIAN

Top Out Brewery llp
Unit 3 6b Dryden Road, Loanhead, EH20 9TY
0131 440 0270

MORAY

Speyside Craft Brewery
2 Greshop Road, Greshop Industrial Estate, Forres, IV36 2GU
01309 358 052
www.speysidecraftbrewery.com

Windswept Brewing
Unit B13, Coulardbank Industrial Estate, Lossiemouth, IV31 6NG
01343 814 310
www.windsweptbrewing.co.uk

ORKNEY

Highland Brewery Company Ltd
Swannay Brewery, Swannay By Evie, Birsay, KW17 2NP
01856 721 700
www.highlandbrewingcompany.co.uk

Orkney Brewery
Quoyloo, Stromness, KW16 3LT
01667 404 555
www.sinclairbreweries.co.uk

PERTH & KINROSS

Moulin Inn & Brewery
Baledmund Road, Moulin, Pitlochry, PH16 5EH
01796 472 196
www.moulininn.co.uk

The Inveralmond Brewery Ltd
1 Inveralmond Way, Inveralmond, Perth, PH1 3UQ
01738 449 448
www.inveralmond-brewery.co.uk

RENFREWSHIRE

Houston Brewing Company
South Street, Houston, Johnstone, PA6 7EN
01505 614 528
www.houston-brewing.co.uk

SCOTTISH BORDERS

Scottish Borders Brewery Ltd
Lanton Mill, Jedburgh, TD8 6ST
01835 830 387
www.scottishbordersbrewery.co.uk

Traquair House Brewery
Traquair House, Innerleithen, EH44 6PW
01896 831 370
www.traquair.co.uk

SHETLAND

Lerwick Brewing Co
Staney Hill Industrial Estate, Lerwick, ZE1 0QY
www.lerwickbrewery.co.uk

Valhalla Brewery
Baltasound, Unst, ZE2 9DX
01957 711 658
www.valhallabrewery.co.uk

SOUTH LANARKSHIRE

Broughton Ales Ltd
Broughton, Biggar, ML12 6HQ
01899 830 345
www.broughtonales.co.uk

STIRLING

Balmaha Brewery
The Oak Tree Inn, Balmaha, Glasgow, G63 0JQ
01360 870 357
www.oak-tree-inn.co.uk

Bridge of Allan Brewery
Queens Lane, Bridge of Allan, FK9 4NY
01786 834 555
www.bridgeofallan.co.uk

Eglesbrech Brewing Company
Behind The Wall, 14 Melville Street, Falkirk, FK1 1HZ
01324 633 338
www.behindthewall.co.uk

Tryst Brewery
Lorne Road, Larbert, FK5 4AT
01324 554 000
www.trystbrewery.co.uk

WEST LOTHIAN

Alechemy Brewing Ltd
Unit 2C Young Square, Brucefield Industry Estate, Livingston, EH54 9BX
07748 156 937
www.alechemybrewing.com

WALES

CARMARTHENSHIRE

Coles Family Brewery
White Hart Inn, Llanddarog, SA32 8NT
01267 275 395

The Jacobi Brewery of Caio
Penlanwen, Pumsaint, Llanwrda, SA19 8RR
01558 650 605
www.jacobibrewery.co.uk

William Evan-Evans Brewery
The New Brewery, 1 Rhosmaen Street, Llandeilo, SA19 6LU
01558 824 455
www.evan-evans.com

CEREDIGION

Mercian Brewery Ltd
Tyr Ardd, Station Terrace, Llanbydde, SA40 9XX
01570 481 280
ww.mercianbreweryltd.co.uk

Penlon Cottage Brewery
Penlon, Pencae, Llanarth, SA47 0QN
01545 580 022
www.penlon.biz

CLWYD

New Plassey Brewery
Eyton, Wrexham, LL13 0SP
01978 781 111

CONWY

North Wales Brewery
Ty Tan-y-Mynydd, Moelfre, Abergele, LL22 9RF
01745 832 966
www.northwalesbrewery.net

DENBIGHSHIRE

Big Hand Brewing Company Ltd
Unit A1 Abbey Close, Redwither Business Park, Wrexham, LL13 9XG
01978 660 709
www.bighandbrewing.co.uk

Buzzard Brewery
Speddyd Farm, Llandyrnog, Denbigh, LL16 4LE
07972 202 880

Great Orme Brewery Ltd
Y Bracdy, Nant-Y-Cywarch, Glan Conwy, Colwyn Bay, LL28 5PP
01492 580 548
www.greatormebrewery.co.uk

Heavy Industry Brewing Company Ltd
Denbigh Brewery, Old Abbotoir, Denbigh Street, Denbigh, LL16 5AR

Llangollen Brewery
Abbey Grange Brewing Limited, Abbey Grange Hotel, Llantysilio, Llangollen, LL20 8DD
01978 860 753
www.llangollenbrewery.com

Sandstone Brewery llp
Unit 5 Preston Road, Wrexham Industrial Estate, Wrexham, LL13 9JT
07851 001 118
www.sandstonebrewery.co.uk

FLINTSHIRE

Brecon Brewing Ltd
8A Brecon Enterprise Park, Brecon, LD3 8BT
01874 620 800
www.breconbrewing.co.uk

Facer's Brewery
c/o Tan Y Coed, Bryn y Garreg, Flint Mountain, Flint, CH6 5QT
07713 566 370
www.facers.co.uk

Hafod Brewing Company Ltd
Gorwel, Hafod Road, Gwernaffield, Mold, CH7 5ES
www.welshbeer.com

GLAMORGAN

Artisan Brewing Company Ltd
183 Kings Road, Cardiff, CF11 9DF
07505 401 939
www.artisanbeer.co.uk

Bryncelyn Brewery
Unit 303 Ystradgynlais Workshops, Trawsffordd Road, Ystradgynlais, Swansea, SA9 1BS
01639 841 900
www.bryncelynbrewery.org.uk

Bullmastiff Brewery
14 Bessemer Close, Cardiff, CD1 8DL
029 2066 5292

Neath Ales
Unit 37 Endeavour Close, Purcell Avenue Industrial Estate, Neath, Port Talbot, SA12 7PT
www.neathales.co.uk

Otley Brewing Company Ltd
Unit 42 Albion Industrial Estate, Cilfynydd Road, Pontypridd, CF37 4NX
01443 480 555
www.otleybrewing.co.uk

Rhymney Brewery Ltd
Unit A2 Valley Enterprise Centre, Pant Industrial Estate, Dowlais, Merthyr Tydfil, CF48 2SR
01685 722 253
www.rhymneybreweryltd.com

Swansea Brewing Company
74 Hawthorne Avenue, Uplands, Swansea, SA2 0LY
01792 290 197

The Cardiff Brewery
Crawshay Street, Cardiff, CF10 1SP
029 2040 2060
www.sabrain.com

The Hurns Brewing Company Ltd
Phoenix Brewery, Unit 3 Century Park, Swansea Enterprise Park, Swansea, SA6 8RP
01792 797 300
www.hurnsbeer.co.uk

The Vale of Glamorgan Brewery Limited
Unit 8a Atlantic Trading Estate, Barry, CF63 3RF
01446 730 757
www.vogbrewery.co.uk

Zerodegrees Cardiff
27 Westgate Street, Cardiff, CF10 1DD
029 2022 9494
www.zerodegrees.co.uk

GWENT

Tiny Rebel Brewing Company Ltd
Unit 12A Greenwich Road, Maesglas Industrial Estate, Newport, NP20 2NN
01633 547 378
www.tinyrebel.co.uk

Warcop Country Ales
c/o 9 Nellive Park, St. Brides Wentlooge, Newport, NP10 8SE
01633 680 058
www.warcopales.com

GWYNEDD

Cader Ales Ltd
Marian Mawr Industrial Estate, Dolgellau, LL40 1UU
www.caderales.com

Conwy Brewery Limited
Unit 3 Morfa Conwy Enterprise Park, Parc Caer Selon, Conwy, LL32 8FA
01492 585 287
www.conwybrewery.co.uk

Purple Moose Brewery Ltd
Madoc Street, Porth Madog, LL49 9DB
01766 515 571
www.purplemoose.co.uk

Snowdonia Brewery
Snowdonia Park Hotel, Waunfawr, Caernarfon, LL55 4AQ
01286 650 218
www.snowdonia-park.co.uk

MONMOUTHSHIRE

Carter's Brewery
White Hart Inn, White Hart Lane, Machen, Caerphilly, CF83 8QQ
01633 441 005

Cwmbran Brewery
Gorse Cottage, Craig Lane, Upper Cwmbran, Cwmbran, NP44 5AS
01633 485 233
www.cwmbranbrewery.co.uk

Kingstone Brewery
, Tintern, NP16 7NX
01291 680 111
www.kingstonebrewery.co.uk

Tudor Brewery
29 Nevill Street, Abergavenny, NP7 5AA
01873 855 074
www.kingsarmsabergavenny.co.uk

POWYS

Breconshire Brewery Ltd
Ffrwdgrech Industrial Estate, Brecon, LD3 8LA
01874 623 731
www.breconshirebrewery.com

Heart of Wales Brewery
Stable Yard, Zion Street, Llanwrtyd Wells, LD5 4RD
01591 610 236
www.heartofwalesbrewery.co.uk

Monty's Brewery Ltd
Castle Works, Hen-Domen, Montgomery, SY15 6HA
01686 668 933
www.montysbrewery.co.uk

Radnorshire Ales
Timber Works, Brookside Farm, Water Street, Presteigne, LD8 2SY
01544 350 456
www.radnorhillsholidaycottages.com

Redstone Brewery Ltd
The Brewhouse, Tynewydd Farm, Llangors, Brecon, LD3 7UA
07581 878 604
www.redstone-brewery.com

Index

(Page numbers in bold italic refer to case studies)

Mm

Nn

Oo

Pp

The Bayou Graves

A Sophie Quinn FBI Mystery Thriller
Book 2

Georgia Wagner

Publisher: Greenfield Press Ltd

Contents

Prologue

It was a beautiful day to kill himself.

Jack drank in the predawn glow, inhaling deep, breathing in the pale fire from the sky ready to bake the swampland and the yeasty smell of black soil and clay. The steady putter of his motor eased him along familiar river paths.

There were a few homes here and there at first, a few moldering docks where personal rowboats, johnboats, and wide-bodied skiffs like his own lay moored. Soon Jack left all of them behind, along with the distant highway traffic and the wet concrete smell of the city. He whistled to himself, a high and wandering melody from his childhood. His cheeks puffed, projecting the song over the chuckling engine behind him.

He blinked as the sunlight grew stronger, wiping the stray tears in the corners of his eyes, whistling even stronger to spite the sensation.

At the forward helm, his palms lightly leading the wheel like a practiced dancer's hands on his partner's hips, he took narrower and narrower offshoots from the wider rivers. There were paths in the bayou that you could only know by piloting them. Like a city without street signs.

And he knew them all.

He'd come here since childhood. He'd been shown the paths by those who could navigate them by starlight and memory alone. Tougher men than him.

Whenever people shook Jack's hand, they always seemed surprised by the strength in them. The coarse roughness. His greetings were often accompanied by exclamations of "Daggum, but you got some mitts on you, don't ya?" or "Oh my. That's quite a grip!"

He would smile and offer a sheepish apology. But he wasn't fooled. Other people being weak, being soft didn't mean he was strong.

Compared to some men . . . he was weak. Terribly weak.

This was the place.

With a deep rattling breath, Jack cut his skiff's motor and in the slow, dark flow of the mangrove-choked waters, he listened to the plop and ripple of river fish as they took their breakfast from the humming insects dancing on the water's surface.

"Is it today, then?" Jack murmured to himself, idly picking at his nails.

It was, after all, a fine day for suicide.

Bam!

The pounding sound jolted him out of his thoughts with a scowl.

Bam! Bam!

Flat strikes on solid wood from the back of the skiff. Jack looked back, his lips curling as his nose wrinkled in a sneer. He bristled as he laid eyes on the black tarp at the rear of his skiff.

"Help! Let me out!" a woman's voice cried.

Bam! BAM! BAM! BAM!

All thoughts of self-harm disappeared as Jack stepped to the tarp, tearing it back to reveal a wooden coffin, rough, unfinished and nailed at the corners like a prop from a spaghetti Western. He glared at the shaking wood, his thoughts turning red.

"Someone, help me! Get me out! I can't breathe!" The woman's voice choked off the words with hyperventilating sobs.

With a closed fist, Jack pounded on the coffin lid and a frightened yelp answered him.

"You should have spoken up. But you didn't. You never speak up until you're ready to scream. Well, scream all you like, now."

The woman did scream, a primal, wordless shriek of fearful panic.

Jack screamed back, red-faced as he leaned low over the coffin lid.

The knots holding the coffin to the skiff's cargo rails were tight, but his coarse hands were used to working them. He paid little attention to what his body was doing, feeling heat creeping into his face. Anger. Pain.

With one booted foot he kicked the coffin's side, and it pitched over the edge of the skiff, setting the boat wobbling by the sudden change in weight and shooting a geyser of black water sprinkling through the air.

He didn't stay to watch, turning back to the helm and starting the motor once more. He wouldn't look back.

After all, he was a terribly weak man.

Chapter 1

"Expectation is the enemy of observation." Sophie Quinn muttered the phrase like a mantra, rubbing her fingers through the close crop of her auburn hair against her temples. Her eyes were closed as she cleared her mind, the thin fabric of her tank top nightshirt rising and falling with each steadying breath from her athletic frame. The heat of her laptop against her thighs compounded the uncomfortable mugginess of the early morning humidity, last night's dew rising with the sun like sweat all across Louisiana's Gulf Coast.

Without looking, she could imagine the crime scene pictures, witness statements, and forensic reports fed in from the USB stick jutting out from the machine's side like a hangnail. She'd paid a private investigator, Fabian St. Claire, quite the bounty for the information: supposedly, RK's first killings. The infant steps of the River Killer, the man who had murdered her sister and nearly killed Sophie too.

A lump of money and three days of study later, and Sophie could feel her impatience mounting. She rolled her neck, slowly easing out the kinks.

Her breakfast of black coffee and a microwaved cup of ramen was going cold on the sticky, plastic table beside her. Three days ago, she'd moved from the FBI provided suite to a motel outside of New Orleans. Maxwell Meyers, Sophie's boss in the Detroit field office, had called to ask when she'd be back, and Sophie had told him she had reconnected with her parents and needed some time.

The best lies had a nugget of truth. She had reconnected with her parents during her last investigation, by stumbling into them while pursuing a suspect and then berating them like an ill-tempered child for being happy to see her. Sophie's stomach twisted at the memory, and she took another drink of the stale coffee to give it something more tangible to gripe about.

Her parents hadn't tried to call after her sudden departure. Apart from the pain in Sophie's heart, it was as if their six years of silence hadn't been broken at all. Sophie had been tempted once or twice to call them, to be the one to break the silence . . . to maybe even apologize. But each time, she only had to glance at her open laptop and the urge would quiet.

She opened her eyes.

The pixelated grain of the laptop images stared back at her. Many of them were copies of copies, uploaded from scans of physical prints. The originals were no doubt stored in some police server or an FBI database somewhere. An unfortunate side effect of using PIs, they often couldn't cite their sources, pilfered, anonymous, and ill-gotten as they could be.

It would have to be enough.

Sophie looked again, closer, mouthing her mantra once more. "Expectation is the enemy of observation."

Dates on forensic reports put the information at ten years ago, four years before Sophie, her sister Molly, and the other eight victims of the River Killer. Several shots showed the same pair of victims. Two young women, just like the other ten victims, just like Sophie had looked six years ago.

The victims were bound at the wrists by chains, their hands over their heads and looped through the branches of a forlorn willow.

"10 mm low carbon steel with long 40–50 mm links," Sophie muttered to herself. It was difficult to tell from the photographs, but in one close-up shot, the size of the links appeared consistent with the chains RK had used in his string of killings.

The difference that Fabian had noted when handing over the USB was apparent. These were definitely . . . messier. The women had clearly put up a fight; pictures of multiple bruises and cuts were included in a fragment of a pathologist report that indicated they were made before the women were killed.

But the most striking note was in what had killed the women. A single shot, close range, to the head. The report said it was a hunting rifle.

Well, what you think, cher? Not exactly clever, t'just shoot both the girls like that. Then again, I was 'bout to shoot you, you slippery lil' water snake.

Sophie's hands tensed and her mouth drew into a hard line. For a moment, the red light of the motel room smoke alarm seemed to flare like the smoldering end of a cigar. RK's voice, graveled and callous as a country grave, a voice forever imprinted in Sophie's mind, taunted her.

She bit the inside of her cheek—a trick she learned in her training under her magician mentor, the Great Gavriel—and the pain returned her to the present.

Footprints.

RK's crime scenes were infuriatingly clean of boot prints and tread marks, but this scene had multiple sets. Two separate boot treads and then the bare footprints of the two victims. Another difference. Both victims here had their shoes removed, but in the ten victims of RK's infamous spree only one had been missing a sandal from one foot. Why change that?

Maybe it wasn't his idea? If there were two boot prints, then someone else was there. Either during the killings or soon after.

Two people; then only four years later, RK was alone.

Sophie's memory flashed, the sensation of bobbing in the bayou waters, the murky current tugging at her back-length auburn hair, long since cut short. A shadowed face with invisible eyes staring down at her from under a frayed baseball cap. Work boots and blue jeans. A dim flare and drifting clouds of obfuscating cigar smoke. A shotgun pinched in the crook of his elbow.

A shotgun.

Sophie blinked, her mind suddenly racing as she quickly stood and paced the room. Of course, it wasn't a hunting rifle. Another change. Was the rifleman someone else? A partner? Hunching over the laptop, Sophie flipped through the open files, suddenly questioning every detail of the gun.

The casing.

A single bullet casing was found on the scene, labeled and tagged as a hunting rifle. Sophie's fingers rolled across the tabletop. Her mind slipped back to her criminology degree.

After Molly's death, Sophie had abandoned her old life, the sold-out illusion acts and displays of mentalism or hypnotism before astounded crowds. *Mysteries and Phenomena with Sophie Quinn* had been canceled at the height of its popularity, as its star and host disappeared into dual PhD studies in record-smashing time. All in the hope of joining the FBI, being on the team that could bring RK to justice.

She'd perused munitions manuals, trying to familiarize herself with the guns and ammunition she may have to deal with as an FBI agent, preparing for a test the FBI was ultimately unwilling to give her. Psychological evaluation found the frequency and intensity of her flashbacks too high of a risk for a full agent. It was only with great effort that Sophie was even able to convince them to give her a chance as a consultant.

Now, in a stifling motel room in the early hours of the morning, she sifted the pages of the manual in her mind. Rifles, page 189. Casings. Sophie dragged her finger through the empty air in front of her face, her pupils dancing behind her closed eyelids. As a child, her mother said Sophie had an appetite for learning, an *envie* for words and trivia. Her mentalism and hypnosis training under the Great Gavriel added a depth to her ability that drove it to heights she was sure her mother would call miraculous.

There.

Sophie's finger hovered, stopping in midair. She froze for the barest moment, then whispered, "It's the wrong casing."

She hunched over the laptop, double-checking the image from the crime scene. No. She was right. The casing was too long and thin. Someone had mis-logged it or some exhausted forensics tech had simply eyeballed the casing for his report.

This wasn't a bullet casing from a hunting rifle. It was from a military rifle, likely an M16 or similar design.

This fact immediately sprouted limbs in Sophie's mind, her heart racing as she set to pacing once more. The presence of a military weapon returned a fact she'd found the day before to her mind. She picked up her laptop, typing with one hand as she walked the length of the flat, grey-green carpet.

The land the victims were found on, where they were killed, was private property. At the time the investigators hadn't thought anything of it. The owner of the land was cleared. But things change over time. When Sophie had researched the location, she'd found reports about the current owners of the land.

The Mississippi Line Minutemen.

An off-grid militia group.

For a few minutes, Sophie simply stared. She felt stunned, as if all the blood in her veins had gone still. Six years . . . going on seven. Every lead had been a dead end. Even the last case, when the FBI pulled her back to her hometown to finally bring down the River Killer, it had turned out to be a copycat. Was this what she was looking for? She could see it in her mind's eye. A militia member, possibly two, the group still in its infancy. They're not there for any ideals, they're there for violence. They find and kill these women. Somehow . . . it goes wrong. Maybe the group finds out and kicks them out. But one of them can't let it go. He waits, he plans, and then four years later . . . RK appears.

He could be a member, a former member, a friend of the group shielded by them.

Sophie's consultant access to the FBI databases was limited, but she checked it anyway. No follow-up investigation was made to rule out the Mississippi Line Minutemen's involvement in these deaths.

Frowning, Sophie nipped at her cuticle. There was no easy way about this. Militia groups were notoriously protective of their ground and membership. Maybe she could pose as a volunteer? A journalist looking to get their story firsthand? No. That wouldn't work. Sophie sighed, flopping back onto the motel bed. A group that was self-sufficient and off-grid like the MLM would have no reason to let a lone woman snoop around. If Sophie was going to chase this lead, she would need backup, an insider . . . or a couple trucks worth of FBI agents behind her.

She'd need a favor, and she knew just the person to ask.

As Sophie reached for her phone, it abruptly lit up, vibrating with an incoming call. She raised an eyebrow, as it wasn't quite seven in the morning, but after she scanned the name, she accepted the call.

"Quinn," the voice on the line started, a sharp, older woman's voice, a voice used to command. "Good. You're already up. Get dressed. Agent Greer is picking you up in fifteen."

Sophie started. Supervising Agent Lidya Merchant was never one to mince words, but the last time they'd spoken, Sophie had been packing for a flight back to Detroit.

"Ma'am," Sophie began.

"Stow whatever you're about to say," Merchant interrupted. "I know you're still in town because you didn't take your flight. You told Meyer you're here because of your parents, and I know that's bullshit because I know *you*, Sophie Quinn. So that means you stuck around for some other reason. And given our history, I don't think it's unreasonable of me to think you may be sticking your nose where it isn't welcome. Am I getting warm yet?"

Sophie's jaw dropped, fishing around for some rebuttal to the rapid-fire accusations, but for all her skill as a performer, Merchant was the one person Sophie could never seem to outmaneuver. The woman was absolutely unflappable.

"How did you find my motel?" Sophie asked at last, grabbing her slate-grey suit off the back of the closet door.

"Quinn, I've tracked down Serbian arms dealers through the streets of Kosovo, you think I can't track down one wayward magician in my own backyard? Thirteen minutes—besides, you're going to want in on this one." Lidya paused, and Sophie thought she heard the faint rustle of paper across the phone line.

"How many bodies?" Sophie asked softly.

Merchant hesitated a moment, then replied, "We're still finding them. Twelve minutes, Quinn," and the line went dead.

Chapter 2

Talking to kids was not Sophie's strong suit. Intimidating suspected murderers, cold-hearted producers, and the occasional ornery guest star—that she could do. Anyone too young to drink, though? Terrifying.

So it was with some trepidation that Sophie settled into the passenger seat of FBI Agent Jonathan Greer's white Interceptor, the same car she'd "borrowed" during their last investigation together chasing the copycat River Killer. In addition to its tall, maturely handsome, and blue-eyed driver, the sedan held a teenage girl with cherry red hair dressed in black denim and an assortment of silvery steel jewelry. One hand loosely held the strap of a backpack dyed with the smeared grey and black of urban camouflage.

And from the moment Sophie had set foot out her motel door, the young lady had been staring at her in wide-eyed captivation.

"Don't worry," Greer had said, holding up a coffee shop to-go cup to ward off comment. "We're headed to the field office; I've already got someone picking her up there."

"Great," said Sophie, buckling herself in. "Who is she?"

Greer blinked over his cup. "Right, sorry. Miss Quinn, this is my daughter, Kinsley. Kinsley, this is Miss Quinn."

His daughter. Of course. Why else would the straight-laced agent in a pressed, tan suit be driving around a seventeen-year-old girl at seven in the morning? Sophie remembered Greer telling her about his daughter. He'd been drying off after an unexpected plunge into the bayou chasing down a suspect. Sophie's photographic memory pulled the image of Greer's six-foot frame swimming in a borrowed, neon-green shirt, and she suppressed a grin.

"A pleasure to meet you, Kinsley," Sophie said, extending a hand over the backrest.

The girl had looked excited before, but Sophie's offered hand was like a lit match set to fireworks.

"No. Way," Kinsley squeaked. "It's actually *you.*"

And just like that, she was off.

Kinsley's voice had the native Louisiana drag and bounce to its cadence, just like her father, if a little more energetic, and as they pulled away from the parking lot, all the usual questions and comments of an enthusiastic fan started pouring out.

But cruising along the highway, Greer confidently weaving his way through the morning snarl with a practiced grace that would put any cabbie on either coast to shame, Sophie found it hard to pay attention. She was no more comfortable than she had been when stepping into the car.

Her mind was too full.

Merchant's ominous warning about this case she'd been snagged into. The MLM militia and RK's connection to the secretive group. And now, an unexpected meeting with a young fan . . .

Thankfully, Kinsley Greer seemed capable of carrying on the conversation for all three passengers.

"—and that's when my friend Ava found your masterclass on mentalism. She said that it was fake, but I told her 'if you actually practiced, you'd see it's not,' and I showed her your special, *Kaleidoscope,* and she said it's just plants in the audience and not real hypnotism, but—"

"Kinsley," Greer interrupted gently. "I'm sure Miss Quinn is still waking up, like the rest of us."

He took a sip of his coffee, and Sophie was reminded of what the ex-Marine had confided in her only a week before. She wondered if he still had sleeping trouble when he was between cases. Not that being an FBI agent was the only thing he had to worry about. Sophie's glance strayed to Kinsley through the rearview mirror along with her thoughts.

"It's ok," Sophie said, forcing her face to smooth and offering a practiced smile no less genuine in its intention than the real thing. "I've actually been up for hours. I was surprised to get the call, though. Did Merchant tell you anything? She was a bit vague with me."

Kinsley jumped in at the question. "I knew it! You two *are* working another case together. It's got to be something urgent if they need you both there right away. What is it this time? Kidnapping? Grand larceny? Some sort of . . . magician . . . crime? Is that why they called you in, Miss Quinn?"

Sophie shrugged. "Murder. Sounded like multiple-homicide, maybe a serial. I'm not sure why they called me in yet. I guess our boss will tell us when we—" She caught Greer's eye. He wasn't looking at the road anymore, his head tilting and his eyebrow raising in a silent approximation of "what the hell are you doing?"

Sophie's own expression pinched in confusion, then understanding hit her. Greer didn't talk about his job with his daughter, or maybe just didn't talk about the murder cases. Of course. He was a soldier, used to carrying his duty in grim silence, separate from his family. She'd been too distracted and crossed a line.

"—um. I mean—" Sophie started.

Greer suddenly perked up, his eyes returning to the road as he said, "I'm afraid that's classified, baby-girl. Sorry."

"It's alright," Kinsley said, letting the disappointment in her voice show as she slumped back in her seat.

This was part of why Sophie didn't talk to kids. You never knew what topics were landmines for their parents.

The car went silent.

Sophie wasn't at all sure how Greer found this preferable to telling his daughter that he chased murderers.

. . .

No, to Sophie's mind, this was decidedly worse. Insufferably worse.

"Did you like it?" Sophie asked, tilting her chin up to let Kinsley know she was talking to her.

"Huh?"

"You said you saw my special. Did you like it?"

Kinsley smiled again, nodding back. "Yeah. It was really cool."

Sophie reached into the inner pocket of her suit coat. She fished around the brass pocket watch and fingernail clippers, one finger holding aside the leather clutch holding her lockpicks and the foil-lined plastic of an unopened pack of chewing gum alongside a thick, black marker.

There it was.

Pulling out a deck of cards, Sophie began to shuffle them in one hand, her elbow moving out to stabilize her against the jouncing road. "You know, I never get to do close-up magic anymore. Too busy with—my current job." She still wasn't sure what Greer's line was in talking about FBI investigations with his daughter and skirted the topic as best she could.

As she spoke, Sophie fanned the cards, rolling them out and back together, working her thumb along the edges to split the deck in rapid-fire waves of shuffling. She could see the glimmer of sudden attention in Kinsley's eyes again, watching the movement of the playing cards.

"I've seen some old recordings of you on YouTube," Kinsley said. "Doing card tricks with people on the street."

Sophie nodded, cringing a little at the memory of her earliest days after she left Gavriel's show. "Card magic is hard to do with a large audience. If people aren't close, they lose that wonder. Even with small venues, most everyone wanted to see my hypnotism and mentalism acts instead."

Sophie twisted in her seat, trying to keep the deck in Kinsley's line of sight, fighting the constricting seat belt once, twice, then unbuckling and turning around to fully face the back seat.

"Now hold on!" Greer started to protest, but Sophie held up a finger in a hurried shushing motion.

"Bup-bup-bup-bup! Eyes on the road," she said, flashing a quick wink as she leaned her back against the dashboard.

Greer held up a placating palm. "Soph—Miss Quinn, we're on the highway."

"I trust you."

"What if there's an accident?"

"You've driven through war zones. We're fine."

"I'm not worried about *my* driving. There's—"

"Dad!" Kinsley leaned forward, her head bobbing as she rolled her eyes in embarrassed frustration. Sophie was suddenly reminded of her own behavior at that age and felt a pang of sympathy for Greer. Not enough to buckle her seat belt, but a pang nonetheless.

Greer let out an explosive sigh, shaking his head. "Fine, but if you die I'm not—"

"Yeah, yeah, you can say, 'I told you so' when I'm dead." Sophie held out the deck, letting Kinsley pick a card.

"You've got it in mind?" Sophie asked, quickly accepting the card back and shuffling it away. It was a simple trick, one Sophie could do in her sleep, but she'd added her own flare to it over the years.

"So now what?" Kinsley asked. "You make it jump to the top of the deck? Or appear up your sleeve?"

"Which would impress you more?" Sophie replied, continuing to shuffle, her eyes staying with Kinsley.

That seemed to give the teen a moment's pause.

"If it was on the bottom of the deck."

Sophie slowed down a moment, her eyes narrowing just a little as a surprised smile tugged at her mouth. "Why do you say that?"

"Because," Kinsley said, "the card is in your sleeve right now. The left one. It'd be trickier to pass the card to the bottom of the deck with your off-hand."

Sophie stopped shuffling. For a moment they just stared at each other while Sophie gave the young woman a long, appraising look.

Kinsley looked down, her face beginning to turn beet red from the unbroken attention. "I—I mean, I *think* the card is up your sleeve, the left one. It could be up the right. Or—er—not . . ."

The sudden shifting, her eye-line, and quickening in her voice. Sophie clocked all the signs of an improvised fib. Kinsley knew *exactly* where the card was.

She was also a terrible liar. Lucky for her dad, Sophie supposed.

"Well, is she right?" Greer asked, curiously glancing at the pair and the now motionless deck of cards.

Sophie laughed and reached up the left sleeve of her suit coat, pulling the card out with a soft twirl.

"Is this your card?" she said melodramatically.

Kinsley chuckled along, pushing back a stray hair as she nodded yes.

Greer whistled appreciatively, and Sophie gave a small clap. "You're extremely observant. Did you see the pull or have you seen this trick before?"

Dragging her backpack into her lap, Kinsley shrugged. "I didn't *see* the pull. But the way you held your hand . . . There was a moment where you moved the deck and you didn't have to. The only reasons I could think of for you to do that would be to misdirect me or if you were palming a card up your sleeve."

Sophie gave the deck a fresh riffle. "Sound logic. A word to the wise, though, if you catch a trick from another performer, just play along. Sometimes the difference between a pickpocket and a street magician is how polite you've been to them." Sophie winked again, and Kinsley snorted with a quick laugh and a nod.

At that moment, Greer's and Sophie's cell phones both buzzed.

"Hold on," Sophie said, pulling up the text. "It's Merchant."

Greer nodded. "We're still meeting at the field office?"

Sophie scrolled down. "Yes. It's the case file."

Greer shot a quick glance, but Sophie wasn't about to say more.

She'd already opened the link.

The first image was a forensics room at the Louisiana field office. Sophie almost didn't recognize it. The tables, chairs, and equipment had been shoved to the side to clear floor space for tarps. A few forensics techs were still clearing space in the background, but in the center of the frame lay three oblong crates. Broad on one end, skinny on the other, distended hexagons of dark, waterlogged wood.

No. Sophie swept her fingers across the screen, zooming in. Not crates. Coffins. Three dripping, rotting coffins.

Sophie pulled the next picture. She closed her phone abruptly.

"What is it?" Kinsley asked, her voice soft.

"Nothing." Sophie answered hurriedly, blinking a few times and taking a swift breath to steady herself. "I see you've got a backpack. You going to summer school or something?"

Kinsley paused, her former enthusiasm and even her embarrassment erased by concern. But this was a game she seemed used to playing. "Um, yeah. Kinda. I'm actually tutoring. I've got a week before my job at the summer camp starts up again, and my counselor says this sort of thing looks good on college applications."

Sophie nodded as she turned back around in her seat. But her mind was back in the case file. The forensics techs had opened the coffins and taken close-ups of the bodies . . . and of the long fingernail scratches dug into each lid from the inside.

The phones vibrated once more. A single phrase from Merchant.

"Four now."

Chapter 3

The chaos of the FBI's Louisiana field office reminded Sophie of a particular ant hill. Only seven, her father had run over the little, red-dirt mound with the push mower. He'd shouted a word she and Molly weren't supposed to use and ran to hose the little biters off his legs as quick as a jackrabbit. Molly had squealed and Sophie, intrigued and horrified, had crept up to watch the hypnotic spirals and cyclones of tiny ant bodies as they hurriedly tried to get the shape of the catastrophe that had befallen them.

Sophie imagined that's all one can do when their perspective is so small, just look at the pieces they can see and hope to understand what happened.

Looking up at the corkboard dominating the back wall of the meeting room, Sophie's eyes darted from picture to picture, following the red string connecting them to pushpins stabbed into a geographic map of the coastal swamp and bayou that made up the Mississippi Delta south of New Orleans. Trying to find the shape of the disaster.

"We have four coffins. That's four bodies pulled from the water," a rotund man in a blue and gold FBI cap said, punctuating his statement with a fist into his meaty palm. His face was red, and he dabbed at the beads of sweat threatening to escape his forehead and run across his cream-smooth jowls. In spite of this, he kept on the nylon FBI field jacket he'd squeezed over his too-tight office button-down. "That's four too many! And we still don't know how many more are out there or how long it's been happening. We have a BAU unit, for Chrissakes. That're supposed to catch these things, for all the good it's done us." Here the man shot a meaningful glare to Lidya Merchant.

Merchant, standing at the side of the room as upright and imperious as a falcon, didn't reply or give any indication she'd heard the man's comment, her honey-brown eyes staying locked on the corkboard like a stalking predator, drawing a huff of frustration from the large man.

When they'd initially found their seats among the other field agents, search and rescue coordinators, and various lab technicians pulled in for the briefing, Greer had whispered into Sophie's ear that this was Deputy Director Hitchens, and Sophie recalled Merchant's warning from when she'd first touched down in Louisiana.

So this is the man who wants us gone, Sophie thought. The idea that Hitchens was looking for any excuse to remove Merchant and Sophie from their posts didn't shock her. Blustering and heated, the man was antithetical to everything her former mentor or current partner displayed. Where Greer was reserved and polite, Hitchens was thundering and crass. Where Merchant carried her fifty-four years gracefully, Hitchens's body reflected years of neglect and abuse from both the bar and the drive-thru. He may not have been much older than Greer's forty years, but where the former Marine's athletic physique showed discipline, the deputy director's sweating, toad-like frame seemed mainly fit for screaming and finger-pointing, the go-to tactics for men in leadership who lacked the empathy and charisma to actually *lead.*

Even his hasty addition of an FBI cap and jacket seemed designed to convince the assembled agents that he was "part of the team." Sophie would be shocked if Hitchens even left his office before taking his lunch hour.

"The third coffin, pictured here," Hitchens rolled on, vaguely gesturing at the board, "is the oldest. Initial estimates from the forensics team say it's been in there at least a year. A *full* goddam *year!* The decay of the victim inside is consistent with that estimate. Chavez, do we have any updates from the ME yet?"

A young lab tech in the back of the room shook his head. "Medical examiner says she's going to be at it all day and into tomorrow. The bodies are very delicate given their state. It'll probably—"

"Alrightalrightalright." Hitchens waved off the rest of the answer, shaking his head in disgust. "I don't want excuses. I want answers. Let me know the *instant* we get any new information."

A full year. Sophie's brow furrowed at the thought as she traced the locations on the corkboard's map with her gaze. This was definitely a serial killer. The consistent ritual nature of the coffins. The locations they'd been dropped. Her criminology studies returned to mind again as Sophie leaned forward in her seat. It was only one theory, but killers typically fell into one of two categories. Impulsive and compulsive. Impulsive killers had a trigger, some event or situation that sparked the need to kill. They could go months or even years without killing, until they encountered that trigger and the need was on them again. Compulsive killers always felt the need. It would start small, growing stronger, driving them to plan and execute a new kill. They might have a schedule, restraining themselves to only killing once every few years. Inevitably, compulsive killers would shorten that time. Years became months, then weeks. Four within the last year . . . at least.

And then there were the coffins. They were rough, handmade. And people didn't just keep coffins lying around. The killer likely would have made them himself. Another sign he *planned* these killings.

Sophie pulled out her phone, scrolling through the case file to find the close-ups of the coffins.

"I'm sorry, Miss Quinn, am I boring you?" The abrupt accusation pulled Sophie out of her analysis and forced her to meet the deputy director's eyes as his lips curled into a sneer.

Sophie suppressed a sigh. "I was just—"

"Look, Miss Quinn," Hitchens interrupted, folding his toady arms over his flabby chest. "I know your type. I know these meetings aren't *glamorous*."

"That's not—"

"I know you're some hotshot in Detroit now and all this *honest* field-work probably feels pedestrian to you."

Sophie gave up trying to get a word in, sitting back and resisting the urge to fold her arms. There was no need to provoke him further with impatient body language. But there'd be no reasoning or explaining to a man who always assumed he knew best.

"But," Hitchens continued, his voice rising, getting more fired up with every word, "this unit isn't about rock stars and Rambos going off on their own. We're a team here. So you can take your pissy "I don't have to pay attention" attitude back to Hollywood, because we don't need it here!"

There was a pause. Sophie heard Greer's jaw grinding beside her. Merchant shifted in her post against the wall, and a wave of discomfort rippled through the assembled agents in the room. Sophie knew the smart thing would be to let it go. The man was a bag of hot air. Trying to argue with him would only make it worse. This was not a man who took correction or apologized.

"I was just pulling up some details in the case file."

Really, Quinn?

Hitchens scoffed. "Oh! Well, pardon me. You already had a lead and I so rudely interrupted you with my briefing. By all means then, Miss Quinn. Tell us what you have."

Next to her, Sophie heard Greer's chair shift. She realized then how tense the six-foot ex-Marine was sitting. His fists had clenched. In a fraction of a second, she registered what was about to happen. The big man was about to stand, and it didn't take a mentalist to read what his body language was telegraphing.

A distant part of her wondered what Greer's history with the deputy director was. But there was no time for that now.

Sophie popped out of her seat first.

Rounding a few of the other agents, she made her way to the board, glancing back at Greer. He had frozen, still tensed, looking at her in confusion. With a subtle twitch of her palm at her waist, Sophie motioned for the tall agent to stay seated.

In the few steps that carried her to the corkboard, Sophie wondered at Greer's vehement reaction. They'd grown closer over their last case. But they were still effectively strangers. She'd known him less than a week and he was . . . what? Ready to pick a fight with his boss's boss over being rude to her? What did that mean?

Maybe that's just the sort of man he was, she thought. Jonathan Greer's sense of loyalty and morality were clearly another muscle he hadn't neglected since leaving the military.

When Sophie reached the map, she lifted a highlighter from the board's tray, turning to face the group.

"Thank you, Director Hitchens," she started, her voice flowing into the type of confident projection she used on the stage. He wanted to put her on the spot, to make her feel shamed by the sudden attention.

But he'd forgotten just how comfortable Sophie Quinn was under a spotlight.

The rest of the group looked as puzzled as the deputy director when Sophie turned and in crisp, articulate dictation started to highlight the waters around the pushpins where the coffins had been found. "As I'm sure you were about to say, the waters in these areas are noted alligator feeding grounds. Slow-moving bayou, plenty of fish and shore cover. It's the middle of the season, so licensed hunters and poachers alike are in these areas. There's a lot more traffic—"

"I was getting to that," Hitchens snapped. "All four have been known gator fishing territory. We've already marked the other feeding grounds and have search teams coordinating on—"

"Not all of them." Sophie moved the highlighter to other rivers and small lakes, as well as some sections of the map that showed no water at all. "Remember, over time the waters can change course. New rivers form and old ones dry up. The Delta isn't etched in stone. The first creatures that know about the change are the fish, then the gators follow. Next are the poachers. And finally, the government." Sophie capped the highlighter, turning to punctuate her final words with a tap on the outdated map.

The search and rescue coordinators gaped, suddenly flipping through their phones and documentation while Merchant subtly shook her head at the revelation. "She's right," one agent said, quickly rushing to his feet.

"Hold on," Hitchens called out, "this meeting isn't over!" But already half of the search team representatives were headed to the door, their hands at their cell phones as they relayed new search coordinates to their teams and local PD in the field.

Realizing the tide was against him, Deputy Director Hitchens straightened his cap and barked, "Alright. Dismissed!" Then, as he hurried to the door, shooting a last jowl-twitching glare to Sophie, he called out, "Someone get me a new map!"

As the last of the lab-techs and other agents filed out, Merchant and Greer joined Sophie at the main board.

Greer let out a low whistle. "Hitchens is going to be pissed. You might better lay low for a while." Sophie only chuckled in response, and after a moment, Greer joined in. "That was impressive work, Miss Quinn. How'd you know about the other feeding grounds?"

Sophie spun the highlighter between her fingers, rolling it to her thumb and back with a click of her tongue. "Like I said, he's using an outdated map." Sophie briefly considered confessing the actual reason. That for the last six years she'd obsessed over the lay and movement of the bayou, hoping for some pattern or clue that would give her the River Killer. From the knowing looks that Merchant and Greer gave each other, Sophie decided it was something she didn't need to say out loud.

Merchant folded her hands behind her back. "So what else do you have for me, Quinn?"

Sophie and Greer stepped back to view the board as a whole.

Greer was the first to speak, though. "All four victims were women." He frowned as he said it. "Unfortunately, that's the case with most serial killers," he finished. The tragic assumption was that most killers would be men and most victims would be women in these sorts of serial murders. It certainly helped to begin crafting a profile but wouldn't tell them much. Not unless they knew more about the women.

"No IDs yet?" Greer asked.

Merchant shook her head. "Four Jane Does. The ME sent samples for DNA testing, but I'm not holding my breath."

Sophie closed her eyes. She'd already been staring for most of the morning. She needed a fresh perspective. Taking a breath, she fell back on old techniques, opening her mind, letting her perception widen. Her senses sharpened as they reached out to embrace the room around her. The squeak of the tile under Greer's shoes. The shadows of the agents in the pit outside the meeting room cutting past fluorescent lights, leaving flickering shadows against her closed eyelids. The frosty touch of the overzealous air-conditioner and the crackle of pages catching the vent's breeze, drifting up from the board and back.

Autohypnosis. In the hands of a skilled hypnotist, the technique was a tool to remove preconceptions. To let your subconscious mind take over and show you what was right in front of your face. The same technique she'd used all those years ago . . . chained to a boat anchor . . . looking for a key on the bottom of a river . . . the night she'd lost her sister . . . Every sight and sound held in equal importance. Trusting that if something was there, her senses would find it.

Sophie breathed in.

Exhaled.

. . .

"Quinn?" Snapping fingers.

Sophie's eyes shot open. Her hand was up, her finger pressed to a victim photo on the board.

Greer was looking on, concern carving out lines under his eyes. Merchant snapped a finger beside Sophie's head again.

"You ok there, Quinn?" Merchant asked. The older woman's head tilted, showing off the tight silver bun tying up her hair. "Are you having a flashback?"

Sophie cleared her throat, shaking her head. She was about to explain when her attention was suddenly drawn to the picture her finger was pointing to. It took her conscious mind a moment to catch up, and when it did, it was as if she was remembering the details of some long-ago memory.

"Look at how they're dressed."

Merchant and Greer turned back to the board. The clothing on three of the bodies was molded and in various degrees of rot, but one was more recent, perhaps only a week old.

"What am I looking for?" Greer said, squinting.

Sophie tapped a nail against the board. "No blue jeans. No T-shirts or casual clothes between them. Every one of them is dressed like they were going out. These women weren't taken from their homes or jogging some trail. They were out on the town, maybe celebrating something?"

"Or on a date?" Greer supplied. "Our last case, the copycat. One of his inspirations was Norman Gummer, the Bayou Bluebeard."

Sophie chewed her lip, stepping back. "I remember. Gummer would take women out on dates. Charm them. Woo them. Then when he had them alone . . ."

"Nothing about coffins, though," Greer quickly added. "I don't think we have a copycat here."

"I agree," Sophie said. "I can't think of any famous killers who were Casanovas *and* into ritual burial. Whatever this killer's doing, it's all his own invention."

Folding his arms, Greer grunted his assent. "Maybe we can find something else at the scene. It sounded like we'll be waiting a while on those autopsies."

Merchant's phone chirped, and the three all turned to face each other as the supervising agent let out a sigh. "You have your pick of crime scenes," she said, a note of simmering fury rising behind the words. "Search and rescue just pulled two more coffins."

Sophie froze. Six victims.

And who knew how many still hidden.

Chapter 4

They say that it was a lonely sailor staring into the glare of morning light that started the legend of the mermaid. Heartbreak and harsh sun that turned the shadow of a passing manatee into the illusion of a beautiful, welcoming woman. Jack knew too well that loneliness had a way of warping one's perception. He closed his eyes and pressed his palms into the lids until red and white starbursts replaced the darkness, rolling like kaleidoscopes.

His hands were still shaking; agitation hung onto him from his morning on the river like cloying sweat. He had to clear his mind.

Walking the streets of New Orleans, the summer sun beginning to bake the old cobbles and sear the dew off the sprawling oaks, Jack saw the city was already awake with its midmorning dance. The usual traffic and birdsong of coffee shop addicts adding pep to the mix among the sleepy trudging of those sleepwalkers shuffling to their respective places of employment.

All the ordinary people around him, just *existing*.

Jack said his good mornings to passers-by just as he did most every morning. He ducked into a coffee house. He waited in line. He smiled to the barista. Anyone could be polite and decent for five seconds. Or a minute. Or ten years . . .

His cheek twitched, and Jack could feel his smile nearly slip at the thought. But he held on. He could hold on for five seconds. Just five seconds while he left the coffee shop.

Cup in hand, Jack hurried to the door, back to the street.

New Orleans City Park spread out in front of him. Much too busy for this time of day. Jack squinted at the tiny figures rushing between trucks and flatbeds. Huge signs and stages seemed to have miraculously appeared between his walk into the city and his first sip of coffee.

At the edge, he saw banners with pictures advertising bands, many he didn't recognize. A concert? No, some sort of festival. Jack's vague stare latched onto one of the pictured performers. One he *did* recognize. A local musician, a lone girl with a guitar. He'd seen her posters around the restaurants and coffee houses. What was her name? He took another sip of his coffee.

"Sir!"

Jack flinched. He froze, instinct telling him if he just held still, the danger would pass.

"Excuse me, sir! Hold up!"

Danger? What danger? You coward. You're weak, Jack. Weak. No wonder you crumble at the slightest confrontation like a startled jackrabbit. That's what you are, Jack. A scared, little jackrabbit.

Jack blinked. His hand was trembling, and he clenched it into a fist.

Steady, he told himself. Footsteps were scuffing up the sidewalk behind him. Right behind him.

Jack turned, and an affable smile broke across his face as he held up a now-steady palm.

It was the barista from the coffee shop. A young woman with a freckled face and a brown smock. She'd been jogging after him and took a quick moment to catch her breath as she held up a dark-brown fold of cracked leather.

"I'm glad I caught you. You left your wallet on the counter, sir." She smiled to him reassuringly. Kindly. The dimples in her cheeks and her soft brown eyes supporting the unspoken good humor of the gesture.

Jack grinned in return, his palm going to his forehead with a comic slap.

"Aww, geez, ain't that something. I swear, I'd forget my own head if it wasn't screwed on."

The barista laughed, a bright, honest ripple.

"Oh, don't worry about it. Happens to the best of us."

"Thank you very much, sweetheart," Jack said as he took back his wallet. "That's very generous of you."

"I've seen you around, the coffee shop, I mean."

Jack swallowed, his smile faltering for just a heartbeat. "You have?"

The girl nodded; she was fishing something from her smock's pockets. "I know you come in every weekend, sometimes during the week."

His heart was racing. What was this? Hurriedly, Jack took a sip of his coffee, nearly dropping his wallet as he fumbled it into the back pocket of his jeans. He made a noise with his throat that could have been agreement or disagreement, looking over his shoulder at the stage being assembled in the park like it was a train he needed to catch.

The barista pulled out a flyer and a thick marker. "So I'm sure you've noticed the music festival's starting tonight. A bunch of us from the shop are going. How about you?"

Was she blushing? Jack blinked, swallowing his coffee as realization settled on him.

"I . . . well, I hadn't given it much thought."

"You should come." The young barista handed him the flyer, using both hands to make sure the morning breeze didn't pull it away. Her smooth, soft fingers pressed against his callouses. "My, you do have some strong hands, don't you?"

Jack's heart stopped. His lip quivered as the smile on his face ticked in a momentary grimace. But the girl's eyes were on his hands.

"What do you do, for a living, I mean? Some of us at the shop have theories, but you never stick around."

"I'm sorry." Jack's voice croaked. His throat felt paralyzed. The trembling urge was moving from his heart to his hands, and he knew he wouldn't be able to stop it long. "I have to go."

The girl's eyes fell, though her smile stayed. "No, I'm sorry. I'm holding you up. Just, you know, keep it in mind." The girl tapped the flyer in Jack's hand.

In one motion, Jack nodded and turned. He did not look back. His hand shook, crinkling the flyer. He heard the exasperated sigh from the barista behind him as he hustled toward the park.

Strong hands.

Jack looked at his hands. He remembered holding the wood planks down. Driving the nails in. *Bam!* He remembered holding the women, refusing to let them go, his strong hands keeping them from escape. *Bam! Bam!* He remembered another hand in his, slipping away. Pulling out of his strong grip.

Because you're not *really* strong, Jack. You're weak.

She'd made a fool of him. Trembling like a delicate flower. He was a *man!* He was *supposed* to be strong. Men don't shake. Men don't get frightened by soft, weak women.

A snarl formed on Jack's lips. He threw the half-empty coffee cup into the street. The flyer curled in his grip, and he nearly threw it too. But he saw it now. The barista had scribbled her name on one side with a telephone number.

Jack's breath caught, his step stumbled.

Annie?

No. He shook his head, clearing his thoughts. *Anna*. The barista's name must be Anna.

Something must be wrong with the girl, he thought. He was likely ten years older than her. What would she want with a man like him? His money, likely. His dignity. Just another toy for her to play with until she gets bored. Sure, it's all smiles and kisses at the start. Then it's venom and screaming. And the lies.

Jack's jaw cracked, his teeth grinding as he entered the park. He was seeing them around him now. Young women. Drawn by the festival. The flyer's picture of shirtless men and scantily clad women holding cups of whatever liquor you wanted to imagine. They'd be here, baiting the hook, ready to trap some man.

Now he was beside the banner he'd seen from a distance. The local girl with the guitar. He still couldn't think of her name. It could be Annie . . . It might as well *be* Annie. She had the same hollow, lying smile.

There would be people around. But he was used to that. Maybe even security. That was fine. The more Jack thought of it, the more the idea excited him. Safe, secure, wrapped up in her own little world.

Jack was a weak man, but he could feel it. He was getting stronger.

Chapter 5

"Why here?"

Sophie squinted across the flat bayou. Six coffins had been dredged from the turbid water over the last week. Four within the last forty-eight hours. Her showdown with Deputy Director Hitchens in the field office hadn't been all bluster. The bodies *were* turning up in gator hunting grounds, but that was a thin thread to follow.

Beside her, Greer took in the scene, his crystal blue eyes following the water's edge. He gave no answer and Sophie didn't expect one. But that didn't mean an answer didn't exist. The killer had built those coffins with care. Selected his victims. And dropped them in *these* spots. There had to be a reason.

But they were all different. This was only one location out of six . . . No, out of however many they still hadn't found.

Creaking mudbugs and the sporadic ripples of catfish or bass were hushed by the chop of police search and rescue boats assisting the FBI and the shout of the officers dragging the bottom of the bayou for other victims they may have missed. The afternoon sun baked the lush green canopy over the water's edge and gave the algae an almost neon clarity.

When she was a child, Sophie loved places like this. Her sister, Molly, and her friends would fish with homemade poles not likely to catch anything, but they didn't mind. The point wasn't to come home with a fish. It was to feel like Tom Sawyer and imagine the whole world behind them didn't exist.

That purity had lasted until she was in her twenties. Until the River Killer took her and her sister. Now, between her childhood memories and painful study, Sophie knew these waters as if she'd dug them out herself and filled them one bucket at a time.

So why couldn't she see the pattern?

"You were right," Greer sighed. "This is another bend known for gators." Brushing some sedge off the pant leg of his tan suit, Greer gestured out into the water. "Maybe that's the point? I remember what you said in our last case. Gators don't eat their food all at once. They stow it. Stuff it up under a log or a little dugout in the river bottom so they can come back and finish later. Maybe that's what our killer is doing?"

"You think he planned to come back and fetch them?" Sophie asked.

Greer shrugged. "I remember hearing about the bog-bodies in Scotland. People that were buried in the peat bogs and preserved for hundreds or even thousands of years. Could be he's trying something like that. Dropping them, coming back to get them later, like a gator. Companies find them occasionally, the bog bodies, I mean. Still in one piece, almost like they've been pickled."

"Mummified," Sophie murmured.

"How's that?" Greer said, turning to meet Sophie's eye.

"Mummified," she said more clearly. "Peat bogs are acidic, like vinegar, and they don't have moving water like the bayou. That means there's no oxygen. No oxygen means almost no decomposition. The bayou moves slowly, but it moves. It's also alkaline. You'd be lucky to find bones after a few years, let alone full bodies."

"Could the coffins help?"

This was something Sophie had grown to appreciate about the tall agent. She was used to men getting flustered by her corrections, hurriedly trying to assert some other irrelevant fact to plaster over their wounded pride. Not Greer. He heard her, took in what she said, and used it. If he was put off by a woman who knew things he didn't, he certainly didn't show it.

Sophie tilted her head, considering his point. A high-quality casket *could* keep a body preserved for an extremely long time. "Depends on how well sealed they are," she said after a pause. "These are handmade, and according to the report, they've all been found flooded. Maybe they'd hold up for a few hours, but I doubt they'd last much longer. Particularly if the women were scratching and banging from the inside. They wouldn't get the lid off, but they'd make cracks that let the water in."

"God Almighty," Greer whispered.

Sophie pointed to a lip of earth overhanging the water, not letting the horrible image linger in her mind. "There's a perfect spot to stow the body if he was emulating them. The kind of spot an alligator would choose." Her finger trailed to the center of the bayou, where the police boat still slowly circled. "That location tells me he just wanted to dump them. He may not have even stopped the boat. Just trusting the weights will hold them down."

The tall grass scraped behind them, and Sophie turned with Greer to see the local detective on scene approaching. His suit was the same slate-grey as Sophie's, though it was soaked through with sweat. As he approached, he pulled a handkerchief from his pocket and dabbed his forehead, a caterpillar mustache weaving over his lips as he stretched his jaw and the moldy bread smell of swamp water squelching out with every step.

Their initial introductions had been perfunctory, but Sophie's memory had no trouble pulling the name. "Detective Miller, are we ready to proceed?"

The caterpillar flapped as Detective Miller nodded. "Thank you for your patience, agents. We wanted to make sure we got all the details we could from Mr. Green before we got him off the water."

Sophie didn't correct the detective. She was a consultant, not an agent. But it certainly didn't hurt if he thought of her as one for the moment. "I understand. And Mr. Green—"

"Found the body. That's right," Miller interrupted, coughing into his hand and then waving for the pair to follow him.

Trudging up the shoreline, they reached a small fishing skiff. Two police officers flanked a tall man who looked to be in his thirties, likely no more than a couple of years older than Sophie. His lean build wasn't too dissimilar from Greer's, and he had a striking pair of green eyes, shaded from the afternoon sun by a weather-beaten baseball cap. In some ways, he was the ideal bayou man. Handsome, outdoor strong, and tall. Just the sort Molly, or even a much younger Sophie, would have crushed on.

Sophie took him in, the same way she assessed audience members at her shows, looking for those who would cooperate and those looking to cause mischief. His hands were in his pockets, his shoulders were slumped but not weak. He seemed to be trying to keep his eyes on the two officers beside him, only shooting quick glances at Sophie and Greer while they approached. She heard a nervous chuckle in answer to something the righthand officer had said, but his voice was otherwise low . . . soft. Almost like he was trying to disappear. If he'd found the coffin, he certainly had reason to be rattled, but Sophie found her brow instinctively knitting. His body language didn't say shock. Not completely anyway.

He was scared. But of what?

"Mr. Green," Detective Miller said, when they were within a dozen steps of the man. "These are the agents that I told you about."

Mr. Green nodded, his attention now going to his boots. "I don't know what else I can tell you what I haven't told these officers already, but I'm happy to cooperate."

Sophie noted the way Green articulated the word "cooperate." She doubted he was doing it consciously, but she picked up on the simple trick. If you tell someone you want to help them, that you're cooperating, and say it often enough, sometimes they don't realize that you haven't answered their questions.

Sophie cleared her throat and offered a hand. "Thank you, Mr. Green. I'm sure we won't take up much more of your time."

The witness glanced at Sophie's hand, then gave a hurried, limp handshake. She felt callouses and the dry crack of working hands, but there was no pressure. The man's fingers barely even curled. Greer swept in with a quick handshake of his own once Sophie's hand was free, clasping the outside of the man's knuckles with his other hand and giving it a firm shake.

"That's quite the grip you have there," Greer chuckled. "What do you do for a living?"

Another difference, barely touching Sophie's hand but giving Greer a hearty handshake. It could simply be the man's upbringing. Some people were raised to treat women like they were made of glass. She wondered in that moment how the killer had been raised.

Mr. Green wiped his palm against his shirt, giving Greer the faintest tick of a smile. "I'm a fisherman. I work the boats out on the Gulf when I can. When I can't, I've got my skiff and make do with catfishin' or whatever else I can get."

Green pointed to the water and a wide, flat-bottomed skiff. Sophie gave the boat a quick perusal. Hooks. Steel rods. A couple boxes that could theoretically hold bait and tackle. Sophie's gaze followed the wide open bottom of the boat. Lingering a moment as her lips pursed in bemused focus.

"You clean your catch in the boat too?" Sophie asked.

"Sometimes," Green replied, his eye squinting as if this was a difficult question.

Sophie nodded. "And you were out fishing today?"

Green nodded. "That's right. I was fishing out across that ways when my line caught up on something. I messed around with it a bit, just trying to get my gear untangled, and then I find this box. It kinda—"

"Where are your poles?"

"Sorry?" Green blinked, stumbling over his words.

"You said your line got caught. Where's your pole?"

"Oh—well . . . probably the bottom of the bayou now. Dropped it in all the mess there. See when I was trying to get myself untangled, I found I was hooked on this box. It sorta twisted under the boat and then thumped against me. Nearly turned over the skiff."

Detective Miller coughed and added, "We think there was probably an air pocket or maybe gas from the body. When it came loose from the weights holding it to the river bottom, it came up."

Green shrugged. "I told them, I don't know anything about any gas. It just banged up against my boat and knocked my catch right back into the water. Lost a morning's work."

For the first time since they'd started talking, Sophie caught the distant look of genuine memory. Chin down. Eyes trailing to his off-hand. Whatever Green was really saying here, he seemed to mean it. But it wasn't what the police assumed he meant.

"And then what happened?" Greer asked.

Green bobbed his head, gesturing to the skiff again. "I hooked the box up, dragged it ashore. I thought at first it was some cargo trunk or something, mighta fallen off some old riverboat back in the day. That's when I noticed what it was. I—I almost left it, but I called the police instead. I didn't open it. Didn't touch it. I'd seen the news reports on those others . . . those poor women . . ."

"Did you say you lost your catch?" Sophie asked, her eyes returning to the fishing boat.

Greer and Detective Miller both turned to her, a single eyebrow apiece raised in a silent question. Green seems equally confused.

"Uh . . . yeah. When the coffin hit the skiff," he repeated.

"So you didn't have a chance to clean any of your catch today?"

Green paused, his train of thought once more derailed.

"No. I—well, maybe."

"So you might have cleaned your catch? Do you clean each fish as you catch it?"

Detective Miller adjusted his belt, clearing his throat. "Look, agents, I'm sure Mr. Green here has had a long day. I'm not sure I understand how—" But before the sentence was finished, Greer had raised a hand to cut off the detective.

"We're just trying to get the full idea of how this happened, detective," Greer said amiably. Again, Sophie was glad for her partner's cooperation. It was clear from his expression he didn't know what she was after, but he'd sided with her nonetheless.

"It's a straightforward question, and now I'm curious too." The tall agent leveled his blue eyes on the lanky fisherman. "So do you clean each fish as you catch it?"

Mr. Green picked at his pockets and shook his head. "No. I, uh, well, I usually store them on ice and clean them at home."

"And you lost your cooler when the coffin hit the skiff?" Sophie supplied.

"Yeah, yeah that's right."

Sophie chewed her lip for a moment. "And you didn't cut yourself when the boat rocked?"

"No, like I said, I wasn't cleaning anything yet."

Sophie took a deep breath, letting her eyes hang closed for a moment.

"If you didn't clean anything, and you didn't cut yourself, then why do I see blood in the bottom of your skiff?"

Chapter 6

"I told you back on the river. I was fishing! I called *you*, why would I do that if I did somethin', huh? What kinda sense does that make? *Couyon.*"

Sophie had let Joseph Green talk all the way back to the station. Sure, the detective had read the man his rights, but Green had jabbered over that too. Like a switch had been flipped, the shy man had erupted with a never-ending string of excuses and complaints. Much of what he said was nonsense, asserting and contradicting himself that he must have cleaned a fish in the boat and forgot about it or that he had a nosebleed he forgot to mention.

She and Greer would have been happy to continue the conversation by the river, but as soon as the holes in Green's story started appearing, Detective Miller had decided they'd better record the man's ramblings.

Now, sitting across an interrogation table in a red brick police station south of New Orleans, it was clear that Greer had reached the end of his patience with the stream of color commentary.

"Right," the tall agent said, dropping a flat palm on the table. The stern look and sudden sound was all it took for Joseph Green to button up. "We've been over this a couple times already, but why don't we try once more, just on the off chance you don't want to spend the night in a cell. Why is there blood in your boat?"

"I was fishing. I cut myself cleaning a fish." The words were flat, lifeless, like he was reading lines from a script.

Which is likely what he is *doing*, Sophie thought. He hadn't said anything that Sophie hadn't said first. Whatever the truth was, even now, Green was trying to sweep it under the rug.

Sophie checked her phone. Nothing new from the local PD. They were testing the blood and doing a few other checks that Sophie requested. But right now, all they had was Joseph's timely discovery of the coffin and the blood in his skiff. Hardly an airtight case, but the man was clearly hiding something.

"Bullshit," Greer snapped. "We both know that's a lie, so why bother with the act?" Greer leaned back, shaking his head in disappointment. "Joseph, we're not playing a game here. Alright? That wasn't a box of nails or someone's garbage they threw in the river. It was a woman. A human woman. And you know that. You know a lot more than that. A lot more than the lines you've been feeding us so far. So why don't we start again? How did you know she was there, Joseph?"

"I didn't," Joseph said, bluntly. "I was fishing. I was just trying to do the right thing."

Greer tensed beside her, but Sophie held up a hand. She looked over the hunched man cuffed to the table in front of them. His head was so low, it nearly touched the table. His hands were folded over his lap. His whole posture reminded her of a leaky faucet, each slow blink another drip. It was the look and sound of a man who felt completely trapped.

He said he was trying to do the right thing. But he was lying. Could he be the killer? It didn't feel right to Sophie. His body language was all wrong. It was more like . . .

And then it clicked.

Suddenly, Sophie understood. But if she was going to get more from him, she needed to make a way.

"I believe you," Sophie said.

Joseph Green's eyes flicked up to meet hers, a similar look of shock also appearing on Greer's face. "You do?"

Nodding, Sophie reached into her suit coat's inner pocket. She pulled out the pack of gum beside her deck of cards and lockpicks, slipping out a piece for herself and offering another to Joseph. She held it close to him; she didn't want to remind him of the cuffs around his wrists if he couldn't reach it. With a grunt of thanks, he took it, and Sophie smiled to him, pouring out all the reassurance she could muster. She needed him calm. Together they unwrapped the gum. Joseph popped his piece into his mouth whole, while Sophie folded hers first. For a moment, they sat in silence, softly chewing.

Greer looked on as if he were the only sane person remaining in the room. *Thankfully*, thought Sophie, *he did so silently*.

Sophie tapped a finger on her pant leg, softly. So soft it almost couldn't be heard at all. Within a few seconds, her chewing was in sync with the tapping. Then, unconsciously, Joseph's chewing did the same. She could see his nostrils flare with a sedate breath. His pupils dilating. He was falling into her rhythm. The first, basic layer of hypnosis. Really just relaxation, openness.

And hopefully, all she'd need to get the truth.

"You really were fishing after a sort, right, Joseph?"

Without thinking, the man nodded.

Greer turned to Sophie, about to ask a question, but she hurriedly shook her head and Greer closed his mouth.

"Those waters are good for gator, aren't they?" she prompted.

Joseph nodded again.

"You know. All we care about is the girl. We don't care if you don't have a license for gator." As much as Sophie tried to keep the Cajun roots out of her voice, she needed Green to trust her. So she let a little back in. Unaffected, native. She was like him. More like him than he had realized. That's what she needed him to realize, at least for the moment.

Joseph Green shook his head, letting out a long sigh. Sophie kept the steady tapping on her thigh, letting him find his own time to speak. After another breath, he did.

"You know it's impossible to get those licenses, right? They set up a lottery to keep it fair. Give everyone an equal chance. Like we don't all know it's rigged."

"You couldn't get a license this year," Sophie said, softly.

Green nodded. "Not for the last five. There's money in gators. Meat. Leather. Teeth. But only if you got a license. Otherwise, you're just a damn poacher." With a rattling breath, the man leaned down and wiped at his eyes with his cuffed hands. "I wasn't even going out today. But my momma, her back is in pain again. She needs the surgery, been putting it off, but she *needs* it now. And there's no money. I know it ain't right, and if it was just me—I wasn't even going out. But she needs it. When I—when I found that box, I wanted to leave it. Wanted to pretend I never even saw it. I knew I was gonna get in trouble. That's always what happens to folk who try an' do the right thing. No good deed goes unpunished."

He was spiraling. Sophie interrupted, but quietly. Using her voice to bring his focus back. "But," she started. "You couldn't just walk away. You had to do the right thing."

Closing his eyes, Joseph took a long breath through his nose, his breathing slowing down, falling back toward Sophie's steady rhythm.

"You lost your catch," Sophie prompted.

"Tossed the whole damn gator back in the water," Green murmured. "No time to dress it or get it stowed in the truck."

That explained the blood in the skiff. Judging by Greer's glance at her, he was having the same thought. No doubt the police labs would come back with alligator blood from the sample.

But did that explain it all, really? Gator poaching carried a hefty fine, and if money was at the heart of things, it might be enough for Green to lie. But enough to get arrested? Brought to the station and interrogated? No, there was something more to Green's story.

Sophie checked her phone again. No new messages.

Leaning close, Greer whispered into Sophie's ear, the sound a smooth, velvet rasp like the voice a warm cup of coffee might have if it came to life. Not unpleasant to Sophie's mind. No, not at all.

"I don't think he's our killer," Greer said.

Tapping him on the shoulder, Sophie and Greer retreated to the hallway where they wouldn't be overheard.

"He's still hiding something," Sophie said, once the door had latched behind them.

"What? He's already admitted to poaching."

"I don't know yet. But he's antsy and he's latched onto the idea that he wasn't supposed to be there. I don't think it's just about poaching. I think he was already nervous to be on those waters, and until we know why, we could be missing a key piece to finding the real killer.

Besides"—Sophie looked back through the door's slim, one-way glass window—"I can't rule out that he is the killer and using this as a cover to remove himself as a suspect."

Greer scratched at his neck. "So—what? He dumps the body, calls it in himself, and makes an excuse about poaching gator so we stop looking into him? Miss Quinn, that's a bit of a stretch even for me."

"Would you let yourself be brought in for a murder investigation before you admit to poaching first? That aside, this is one of the freshest coffins we've found," Sophie rebutted. "The victim asphyxiated, but it had enough of a seal to bob to the surface. Unless we have another secret he's keeping—"

Sophie's phone buzzed, and she gave it a hurried glance.

"What is it?" Greer asked.

Turning the phone around, Sophie showed him the message. Greer squinted, then his eyes shot wide.

"There's our secret," Sophie said. Then quickly shooting her spent gum into a nearby trash bin, she turned and burst back into the interrogation room.

Green jumped in his seat, recoiling from the sudden charge. Both hands slapping the table, Sophie snapped, "Drugs. That's what this is about."

Green paled. "W-what? What drugs? I don't know about—"

"There's a compartment on *your* boat stuffed with dozens of baggies filled with crystal meth, ready for sale."

Stammering, Green tried to wave his cuffed hands in protest. "Whoa, whoa! I don't know anything about any—"

Greer's voice suddenly filled the room with a military bark that made Sophie jump in her skin and nearly sent Joseph Green out of his chair. "Enough. We are *way* past that, my friend. Now you're looking at prison. So your options have come down to this." Greer leaned across the table, his palms pressing into the metal top and his brow bunching in focused fury. "You either cut the stalling and tell us exactly what happened on that water, or we're going to make it our personal mission to ensure you serve the rest of your natural life in a bright orange suit. Do you understand?"

Joseph looked from Greer to Sophie. Stowing all her former sympathy, she only gave back a hard stare. He'd get nothing else from them. It was his turn.

"Talk!" Greer barked, smacking the table for emphasis.

"Alright!" Green cringed back, his fingers spread in surrender. "I really was out there poaching, but!" he hurried to cut off Greer before he could be accused of stalling. "But, I was also doing a run. I was going to make the delivery after, but then I found the box, and it all got so complicated. After I called the cops, I got scared. Scared they were going to find out."

"Who?" Sophie shot back.

"The Dream."

Sophie gave Greer a quick glance, but his face was impassive. After a moment, the former Marine folded his arms. "What do they have to do with this?"

Joseph swallowed, his eyes darting around nervously. When he answered, his voice was low, as if he was trying to avoid being recorded. "They've got claim on that stretch of river now. They let me pass through because I cut them in on . . . on the meth. Like a toll. But if they find out I'm poaching their waters, I'm a dead man." Abruptly, more urgent now, Joseph stood, his voice tumbling as panic took over. "You can't let them know I said that. Please! I'll do anything else you want. You have to protect me. That's your job, right? You're the law!"

Greer shook his head. "You're in police custody now. You want protection? I suggest you extend the same cooperation to them." Turning to leave, Sophie quickly followed. Joseph Green's final pleas were cut off by the interrogation room door.

After several steps, when it was clear Greer would not offer any explanation, Sophie asked, "Who are the Dream?"

"Freedom's Dream Fishing Club," Greer replied, his voice tight. "A private group with a clubhouse not too far upriver. Publicly, they are what they say, a fishing club. Privately, they're a splinter from a much more infamous group . . . one that was fond of white hoods and burning crosses."

"Sound charming," Sophie murmured.

"They're not. They've been staking more and more river as their 'turf' over the last year. Even shot a man for 'trespassing.' Never mind that his truck was parked a full ten miles from their clubhouse property."

"Then if someone was dropping coffins in their turf, they'd know. Unless—"

"Unless they're the ones doing the killings," Greer finished.

Sophie chewed her lip. Armed and secretive groups with a history of violence could be extremely difficult to crack. But if that's where the evidence pointed, what little evidence they had . . . "That's our next stop," Sophie said. It was not a question.

"No," Greer sighed. "Our next stop is get you a Kevlar vest and a sidearm. Knowing the Dream, it may end up being *that* kind of visit."

Chapter 7

Every crack and rustle from the old oak and sprawling mangroves sent a spike of anxiety through Sophie's spine. The crackling undergrowth and glimmering eyes of bullfrogs hiding in the rushes took on a more sinister tone in the setting sun. Soon it would be too dark to see and they would break out the flashlights Greer had insisted on.

Still, if all went to plan, the Dream wouldn't see them coming.

There was a dirt road that led from a small town off the river branch to the Freedom's Dream Fishing Club, but it was barred with a chain link gate, and from their initial drive-by, it looked to be watched with at least one security camera.

Of course, the clubhouse had a dock on the river, but they didn't need to scout it out to know it would also be watched. People who dabbled in drug running and stockading themselves against the possibility of a "race war" in America were nothing if not security minded.

Instead, the pair had opted to infiltrate the property from the side, cutting a hole in the fence and making a beeline over the marshy woodland. Given Joseph Green's testimony, they had secured an emergency warrant, but Deputy Director Hitchens had made it clear the local police's main priority should stay on the coffin search. So unless they *needed* backup, she and Agent Greer would be on their own.

She might have been inclined to argue against Hitchens, but then there were seven.

Seven.

They were up to *seven* coffins now. Another had been called in while Sophie was getting her pistol from the field office's armory. It made Sophie shiver to think how long they might have been lost, left under the bayou to rot, if that first coffin hadn't been found. Seven women who needed justice. Nearly as many as RK had killed. But no, now there were two more she knew about. Sophie thought again about the USB drive sitting in her laptop bag back in Greer's white sedan, about the militia group the River Killer may have belonged to. Or not . . .

Still too many questions.

Greer held up a hand, and Sophie stopped short, nearly colliding with him as she started out of her thoughts.

The athletic FBI agent had exchanged his tan suit for a dark jacket and bulletproof vest. He'd insisted Sophie wear the same, the dark coloring to keep them from being seen, the Kevlar in case they were.

"Wires," Greer whispered, pointing ahead and to their left.

Sophie squinted, but even her keen eyes took a moment to register the glimmer of fishing line catching the rose light of the setting sun.

"They've set booby-traps?" she asked, noticing her own voice peaking with disbelief. "That can't be legal."

"It's not," Greer marked grimly. "That doesn't stop a lot of folk around here, though. Watch your step."

Carefully, they crept on. Sophie's athletic shoes squelched in the soft soil, and she swatted at a mosquito as it landed on her ear. The whole place was alive with movement and sounds, and it was making it difficult for her to focus on the danger they were walking into.

Sophie saw now that Greer had his own pistol out, holding it in a two-handed grip and pointing the barrel to the ground as he eased himself through the brush, careful to make as little noise as possible.

"Are you really expecting a firefight?" Sophie asked.

Greer glanced back at her. "I don't want one, but you have to be ready. There's more than one killing that the DA would love to tie back to the Dream. If they're behind these killings, they aren't likely to come along quietly. Even if they're not, I don't expect them to be gracious hosts."

Sophie glanced at the dull metal. Even over the earthy smell of the bayou, she could scent the oiled gunmetal from her own pistol at her hip.

When she'd been training in hypnotism under the Great Gavriel, the old magician had taught her the importance of presence.

People hear what they see, he'd told her. She could see him in her mind's eye, propped up on the hollow table he used in his stage show. *If you turn around, they will not hear you, y'understand? You can shout all you want, but if the audience don't see those lips flap, they are hearing jack-all, right?* The fifteen-year-old Sophie in her memories nodded firmly. Arms folded, flannel shirt and jean shorts already holding a workday's worth of sweat while her long, auburn hair frizzed out in the humid air. *So,* continued the old magician, *If you come out charging, arms waving, it doesn't matter what you say, because all the audience* sees *is some psycho woman comin' right for 'em! If you want them to think you done said something mysterious and* wise, *then walk like you're—like you're . . .*

Socrates?

I was going to say Oprah, but sure. Him too.

Sophie cracked a smile at the memory. Then her eyes returned to Greer's drawn pistol.

"People hear what they see," she murmured.

"What's that?" Greer halted, tilting an ear to Sophie.

"Your gun," she said, gesturing. "We're already sneaking in, are you worried we might send the wrong message if we show up with drawn pistols too?"

"We might not get a chance to draw them when the moment comes," Greer replied darkly.

"Maybe," Sophie answered, weighing the reply. "But as much as it is in our hands, we want this to go peacefully, right?"

"Miss Quinn, I promise, we'll be peaceful. We're just going to protect ourselves too. Unfortunately, some folk are more polite once they see the other party has a ready gun."

Sophie put a hand on Greer's arm. For a moment, she was distracted by the toned contour of the forearm underneath. Tense. Ready. Despite being a forty-year-old man, it was clear Greer's military discipline had not failed to follow him into his FBI post.

"Miss Quinn?"

Sophie looked up. The setting sun caught the sky blue of Jonathan Greer's bright, intelligent eyes. She saw the resolve, the determination driving him . . . but also a sadness. Sophie's years of experience cold-reading people for mentalism and hypnotism immediately broke out what she was seeing.

A man who did not want to hold a gun, but who would never put it down while people were in danger.

The sun set. Greer's eyes disappeared into shadow.

With a shuffle and a click, Sophie pulled out her flashlight.

"I'm sorry, Agent Greer," Sophie said at last, releasing his arm and casting a quick beam into the sedge grass and snarling roots around them. "I just want to make sure you get home to Kinsley in one piece."

Greer sighed, pulling out his own flashlight. "Me too, Miss Quinn."

As they shuffled forward, Sophie came up alongside him, each of them sweeping the ground in front for signs of other traps.

An awkward, uncomfortable feeling was squirming inside her. And Sophie found herself chewing at her lip.

It was the touch.

She'd grabbed his arm and looked him in the eye. If she was reading another person, her well-honed checklist would be ticking boxes left and right. Interest, obviously. Attraction, yep. Only, no, she was Sophie Quinn. She was past that part of her life. Sophie Quinn doesn't get . . . twitterpated.

It had to be something else.

Yes. The embarrassment she could feel coloring her cheeks was just the tension from her brief moment of physical contact with her partner. Greer was kind and easy to get along with, but he was a traditional sort of man. She had ditched Louisiana before she was even twenty and had been the star of her own show. She was a certified globe-trotter who had her own mansion in Cali. Used to. *Used to* have her own mansion in Cali. The point was, they weren't anything alike, and she didn't want him getting the wrong idea by grabbing his arm in the middle of the woods. And it *was* the wrong idea.

An image flashed in Sophie's mind of a motel in Peillon during their last case. Greer had been outside as she pulled in, just a thin white undershirt over his well-muscled chest. The silver in his five-o'clock shadow and close-cropped hair catching the moonlight.

Sophie shook her head, clearing the thought by sheer force of will.

"She's nice, by the way," Sophie said conversationally.

Greer glanced her way. "Who?"

"Kinsley. Your daughter. I remember you mentioning her. She's quick too."

Greer chuckled, a heartfelt smile spreading on his face. "She is. Though, she still surprises me. Looks like she surprised you too, catching out that trick of yours."

"She must get that from her mother, because I know I had you fooled."

Greer's grin faltered; for a moment his eyes were distant, and he let out a small cough as he replied, "Yes. She did have a keen eye."

There was a tightness behind the words. Sophie had seen that Greer didn't wear a wedding ring, but beyond that . . . Sophie didn't know if Kinsley's mother was even alive, and despite the time they'd spent together, personal details about her impromptu partner were still fairly sparse. Strong. Reserved. He hadn't talked about his family beyond his daughter, and he likely wouldn't. Men like Greer didn't talk about themselves unless prompted.

But with the anxiety of creeping onto the fishing club's property driving her mad, Sophie needed the distraction.

"So, what's the story with Kinsley's mama?"

"What do you mean?"

Sophie rolled her eyes. "You don't wear a ring. You have a daughter. I know sometimes agents and police take off their rings at work, it's the same in show business, but you don't even have a tan line. Add to this the faraway look you just gave me, and—well, there's a story there."

"There is," Greer replied flatly.

"So," Sophie prompted. "Indulge me."

For a moment, the choir of frog song and clicking insects filled the silence between each rustling footstep, and Sophie was afraid she'd crossed another line in the former soldier's personal boundaries.

"When I came back from my last tour in the Marines, I was dealing with a lot of things," Greer said at last.

Sophie clocked the regret in his voice, and despite the pause, waited for Greer to continue. She glanced his way, but his eyes stayed focused on the woods in front of them.

"Guilt. Anxiety. In war, there's a lot you're expected to do, and they tell you it's for the greater good. You can't really think about it at the time. You do your job. You do it well and people live. Your squad. The people you're there to protect. But eventually you start to think about the others. The people who died *because* you did your job well."

Greer shook his head. "Everyone has to make peace in their own way. I was still making mine, but I had a wife at home and a seven-year-old daughter too." Taking a deep breath, his broad shoulders rose and fell as Greer seemed to dispel some jabbing thought that had crossed his mind. "I didn't make it easy. It took a toll on our marriage."

Sophie felt a pang of sympathy for the former Marine engineer. She'd known a few veterans around Peillon when she and Molly were growing up. Men she'd thought of as old-timers who'd done their time in Vietnam. Young men back from the Gulf War. Combat always leaves scars, though some carry it easier than others.

"Margo was a good mother," Greer said at last. His tone brooked no argument, as if it was a point he'd debated to the point of frustration. He took another breath and his cadence was gentle once again. "Just like I had to find my way of dealing with what I'd been through, what I'd done, she had to find her own way of coping with me. Her way was taking more and more time away from the house. Away from me and Kinsley. Her escape nights turned from a few drinks with friends to partying with strangers, drinking alone. 'Cutting loose' is how she described it. And I don't fault her for needing that. I wasn't much fun to be around."

Sophie nodded. She'd never had many close friends, but the few she'd had vanished after Molly's death. Her trauma was of a different sort, but Sophie doubted that she had been much fun to be around either. Though, if she hadn't decided to investigate RK's case on her own, there's nothing to say she couldn't have taken a similar path to Greer's wife, Margo.

“Drugs?” Sophie asked when silence began to stretch.

Greer shook his head, then stopped himself. “Not at first,” he said reluctantly. “But by the end, she must have been getting something from her party friends. I told myself, if I could get it together, if I could just get myself totally ‘fixed,’ then she’d come home. I was in therapy. I worked on myself. Wrestled my demons. But it seemed the better I got, the worse Margo became. Then, one day, she just didn’t come bac k.”

Sophie’s heart twisted. “What do you mean? Is she . . .”

“No,” Greer waved a hand, brushing aside the implication. “No, she’s still alive. As far as I know. The last I heard, she lives in Vegas now.”

“She just left you and Kinsley?” Sophie felt a small boil of indignation swelling in her chest. Then a burnt voice in the back of her mind whispered, *Tragedy strikes and the young fille run away, lose herself. Now who dat sound like, cher?* RK’s charcoal black laugh tugged at her thoughts, and Sophie bit the inside of her cheek, wincing at the sore flesh but staying in the present, hurrying another question before her own thoughts could take her away. “Does that mean you’re still married? Technically, I mean?”

Greer held up his bare ring finger. “No,” he replied. “I couldn’t reach her by phone, but I finally got a letter to her, years ago now. No reply. The only thing I got from her was a set of divorce papers. No call. No note. Nothing. I sent them back two weeks later.”

"That must have been hard," Sophie said, quietly. "On you and Kinsley."

Greer sighed, pausing to inspect the path between the trees in front of them. "It was. She loved her momma. I think a part of her still does."

Sophie stopped alongside the tall agent, catching his eye in the castoff from their flashlights. "Losing someone you love is maybe the hardest thing a person can endure. I'm glad she had you. I'm sure she's glad about it too."

Her chest felt tight, and Sophie stepped a little ahead of Greer, turning her face away. She was thinking of Molly again. It wasn't exactly the same, losing a mother and losing a sister, but at least Kinsley had someone beside her willing to admit that it had happened. Willing to work through it with her.

The trees were parting ahead. In the moonlight, Sophie could make out the outline of a building. The Freedom's Dream Fishing Club clubhouse. From the vague, shadowed shape, it looked more to Sophie like a military bunker.

"Sophie!"

Greer's hand slapped down on Sophie's shoulder. She froze, her feet locking to the ground and her shoulders and hips twitching in a brief moment of fight or flight as she instinctively started the motions of a wristlock on her partner. Thankfully, she managed to stifle the reaction.

"Don't. Move," Greer said softly.

Holding perfectly still, she watched as the former Marine crouched behind her, holstering his pistol and inspecting the ground around her feet. Sophie looked down. And saw the glint of fishing wire bending around her shin. Less than a centimeter of pressure arcing the nearly invisible line.

Greer drew a pocket knife as he followed the line to a nearby tree.

"Hold perfectly still," he warned.

"What is it?" Sophie asked, unable to repress her anxious curiosity.

"It's a mine. A claymore."

Sophie's mind again raced through her knowledge of military arms. Claymore mines were directional traps designed to explode when triggered and send shrapnel in a cone pointed toward whoever set off the trap, leaving the area behind the mine intact. They were decidedly *not* for civilian use.

"Can you disarm it?" Sophie asked.

"I've disarmed these before."

Sophie frowned. "That wasn't a yes."

"It's been almost a decade. But I remember how to do this."

"That's still not *yes*," Sophie shot back, a nervous laugh chasing out the words.

"I can. Just give me a moment," Greer replied.

Sophie closed her eyes. "Great. Good. I'm not quite ready to die yet."

"You're not going to die," Greer said, his voice starting to strain. "Many mines are set up to maim or injure. To make the enemy spend time and resources caring for wounded troops."

"Jonathan, I hate to break it to you, but that might not be the reassuring bit of trivia you think it is."

Greer chuckled and Sophie found herself laughing along with him.

The tripwire across her shin went slack.

No explosion.

Sophie let out a sigh of relief, turning to face Greer as he pocketed his knife. She swallowed the words of thanks she'd intended, though, as a floodlight kicked on with an electric bang, blinding them. The last thing Sophie heard before diving into the underbrush was the crack of a rifle shot.

Chapter 8

Carly Monroe slapped a palm across the butt of her guitar. The resonance of the acoustic-electric's hollow body against her chest was reassuring. Vibration. Like when her momma would wrap her arms around her as a little girl and hum her favorite songs. It was comforting, reassuring. Which was fantastic since Carly felt herself dancing on the edge of raw panic.

"It's ok, Carly. You've played stages before."

Not this big.

"You've been making your bread with this guitar for five years now."

Sharing an apartment with five roommates and eating rice and beans every day.

"This is your chance. You were made for this."

Listen to them. You really think this crowd turned out to hear Carly Monroe?

The roar of the festival crowd rose again as the current band, a folk rock group out of Colorado, if Carly remembered right, broke out in a fresh reel of amplified fiddle, drum, and the singer's warm, buttery lyrics. Even though Carly's trailer was cordoned off with the rest of the backstage area by fabric walls and metal fencing, she had no trouble hearing their raw number. It was like all of Louisiana had turned out.

Behind closed eyes, Carly bobbed her head, humming melodies she'd written herself, fighting her doubts. It was a miracle of networking and a freak ski accident that she was performing instead of the original singer the event booked. But she was here. The festival was being live streamed, live recordings would go out.

This was it.

The moon was out, the summer air was stifling but just beginning to cool. It was her, then the headliner. And the crowd was ready for her.

"Miss Monroe?"

A knock on her trailer door accompanied the meek call through the thin wall.

"Yes?" Carly called back.

"It's tech. Listen, we need you out here. There's some conflicting info about your setup, and we need you to confirm some details?"

Carly's heart seized, and she took two quick gulps of air before she could answer.

"What do you mean? We already did soundcheck and the board operator has the details. I'm on after this song."

"I know, I know," the technician called back. "We've got most of it, but we can't find the tech sheet for your set. We can't ask the boss because he's in the medical tent puking his guts out."

"What?" Carly stood now, laying her guitar on the half-flattened couch she'd been sitting on and rushing to the trailer door.

"Bad tacos or something. We've got craft services on it. But look, we just need you to confirm a couple things and you'll be back in time for curtain."

This couldn't be happening.

Carly threw open the door. The man on the other side pulled his cap in what could have been a nod to her and spun, immediately waving her to follow him through the fabric corridors of the park's hastily constructed backstage.

The pair shouldered past other bands, stumbling roadies, and men and women in the same dark "Festival Crew" T-shirt as her guide.

"So, what's the issue?" Carly asked, calling up ahead as the technician rounded another bend, sidling past a row of locked cables and rig wheels for the stage lights.

He didn't answer. Or if he did, it was swallowed by the sound around them.

Backup speakers and crates stamped with the names of bands and the production company hosting the event piled up on each side of the narrow space. There was barely room for the two of them to squeeze by, let alone any other tech crew or performers.

Carly's heart raced. The man's silence was making her imagination run wild. If they had to rebalance the audio before she could go on, the crowd could get ugly. They'd been drinking for hours, feeling the vibe of the bands and the hot summer night. Any sound errors or lighting glitches could turn those singing voices to heckling. Delays could mean boos and shouts for her to get off the stage and make room for the headliner.

Blinking away a rising tear, Carly nearly ran into her guide. He'd stopped.

"Listen, we know you were excited for this effect, but we couldn't remember which part of the stage it's supposed to be set on." He gestured ahead and Carly leaned forward, squinting into the dark storage area.

That was it? Carly swallowed a quick bubble of indignation, resisting the urge to turn and snap at the crewman. Her setup wasn't complicated. A few inflatable gimmicks that would come up toward the end of her set, during the ballad. She just had to point out which went where and then she'd be on.

The crowd swelled. The guitar was rising with them. The song was almost over, and she had to be on stage.

Between the mixers, backup electronics, and pyrotechnic canisters, Carly spotted a plain pinewood box. Shallow and long as a person.

"What is that?" she murmured.

A firm hand clapped a damp rag to her face. Carly screamed, the sound swallowed in the heavy paw as another arm wrapped around her shoulders and pinned her flailing arms to her body.

She took a panicked breath, the rag pulling into her mouth and nostrils. A chemical smell rushed across her brain.

And then Carly was floating away.

Chapter 9

"This is private property," shouted a gruff voice from behind the spotlight.

Sophie crouched among the roots of an old oak tree. She had turned off her flashlight, her pistol in her hand as she leaned a shoulder against the bark.

"I'd be perfectly in my rights to shoot you. Why don't you make this easy on all of us and come on out."

Across the band of light flooding between the trees, Sophie saw Greer in a similar position, posted up against a tree of his own. At the sound of the rifle shot, all other noise had gone out of the bayou, all except the trickling of the dark river and the rustle of wind through willow vines.

Greer suddenly shouted back, his voice strong and clear. "FBI. We're here under legal warrant to investigate the area. Put down your guns and come out."

"Oh, I don't think we will," the coarse voice called back. "But if you want to step into the light and show your badge and warrant, I'll be happy to inspect it."

We. The speaker had given away there were multiple people at the clubhouse. That was bad for them. A single person could be reasoned with, but with a group behind them, it would be harder to get the gunman to cooperate.

A dim light briefly blinked in Greer's hand, his cell phone. Sophie heard him speaking under his breath, catching the words "backup" and "fishing club." Hitchens had said the local PD weren't to get involved unless they *needed* backup. Even now, Sophie suspected the deputy director wouldn't be rushing to see that they got it. And even if the cops did show up, Sophie had the sinking feeling it would only make it harder for them to get any information. No, she had to figure out a way to get the Dream to talk before things escalated.

Sophie thought back over what she knew about the Dream, what Joseph Green had said and what the FBI had in their database, and a plan began to form in her mind. It was a gamble, but they didn't have time. Either the Dream would come out to hunt them or the cops would show up and turn this into a siege. She needed to act *now.*

Sophie stepped out.

Greer made a harsh hissing noise with his teeth, his eyes wide in shock as he beckoned her back. But Sophie simply holstered her gun, twisting her palm in a quick gesture to "stay back."

Clearing her throat, Sophie suppressed her nerves, taking the fact that they hadn't shot her immediately as a good sign. She let her native accent slip back into her words. "I'm Sophie Quinn," she called out. "I'm not an FBI agent. Not a cop neither." She was a consultant for the FBI, but there was no reason for the Dream to know that. "I was born in Louisiana, raised here. I'm here because something's not right on the river. Something that you can help us put right."

In the moment of silence that followed, Sophie felt she could sense the Freedom's Dream Fishing Club considering this new complication.

"If you're not FBI, what are you doing with your friend there, and why are you trespassing on our land?" the anonymous voice called back.

Sophie licked her lips. It was the same voice. The tone was clear, commanding. Someone used to being in control. An older man's voice too. Maybe in his fifties. The leader? Sophie scanned the FBI dossier in her mind, plucking a name out of memory.

"William Landry. I didn't come here to shout into the void. My hands are empty. Now come out and talk to me, man-to-man."

Sophie chose her words carefully. Equally authoritative, almost condescending. It didn't have to be mentioned, but would no doubt be noticed, that she was a woman. A short-haired woman in a suit coat and athletic shoes. The sort of woman who didn't mind using an expression like "man-to-man" and taking a tone with him. Just the sort of woman a man like Landry would chafe at being disrespected by. The sort of person he wouldn't dare risk looking frightened of.

Half of mentalism was simply knowing the sort of person you were talking to, the sort of phrases that would push them one way or another. Sophie could only hope she had the right idea about the hidden rifleman.

In the glare of the spotlight, Sophie heard the crunch of gravel and the scrape of grass from approaching footsteps. Then William Landry, wide-shouldered, pot-bellied, and with a face sporting a cleanly trimmed beard, propped his rifle in the crook of his arm and stepped into the light. He stopped a foot from Sophie, clearly expecting her to take a step back. But Sophie had used this trick herself and stood her ground, forcing the large man into an awkward stutter-step as he held short of running into her.

Sophie grinned, drawing a scowl from the man as he glanced into the dark of the tree line behind her.

"There now. Face-to-face," he said, his voice low with taut anger. "Now, Miss Quinn, I'm not even going to ask how you know my name. I'm known around these parts. My club is known around these parts. You claim you're born and raised around here, but I *don't* know you. So why don't you tell me what on God's green earth you're doing on my land before the boys and I give you a real reason to run."

The smell of tobacco and the pomade in his beard rode the man's breath, making Sophie flinch from the sudden odor, but she held her ground.

"Someone has been murdering women on this river," she said, slowly, giving each word as much gravity as she could muster. "For at least a year, likely more."

Landry snorted. "And you think one of us had something to do with it."

Us again. How many more were waiting in the clubhouse? Perhaps with rifles of their own trained on her as they spoke. Sophie pushed the thought to the back of her mind. She had a job to do.

Sophie shrugged. "You control the river."

Landry looked as if he might protest, and Sophie held up a palm as she rushed to get ahead of the complaint. "Some people *say*. I'm aware of what your club's official stance is, just as I'm sure you know what people say about the Dream."

At that Landry smirked. He didn't just accept the reputation the Dream had, he loved it. The expression sent an invisible shiver down Sophie's spine as she continued. "Either way. Your club, as you say, is well known. So who would be dropping bodies in your river? Who *could* be dropping bodies in stretches of river that we know you have . . . an *interest* in?"

At this William Landry gave Sophie an appraising look. He turned, starting to pace the grass between the clubhouse and her. "We've been disturbed to discover these killings too," he said at last. "Half a dozen women sealed in coffins and sunk into the water. It's enough to make a man's blood boil." He bit off the words with a dramatic grimace.

"But then what do you expect when society indulges just any kind of person? When it demonizes those of us who just want to protect our people."

"I'm sure you'd go to any lengths to protect the people you care about," Sophie said.

Looking her in the eye, Landry grimly said, "Not just my people. My culture." She was playing a delicate game, provoking and sympathizing at intervals, trying to coax Landry into speaking more. If he was involved, he'd give himself away. All she needed was time for him to talk.

The dossier had given plenty of speculation and association between the club and racial violence in the area. And the motives were often linked. Psychologically, those who bought into the dogma of white supremacy often tied in other ideas about the inherent inequality of men and women, tight strictures about what a "moral" woman should look like, or the obligation to violently protect whatever pseudo form of Christianity or paganism allowed them to justify their hate.

Would it be a stretch for violent men to switch from attacking people of the "wrong" skin color to women they saw as morally corrupt? Women dressed for a night on the town? Sophie's stomach turned. These were dangerous men. But were they the men she was looking for?

"William," Sophie said at last. "What have you done?" The words were innocent enough, but Sophie wielded them with the precision of a fencer's foil. Her eyes squinted slightly. Her lips parted. Her tilted chin supported the look, and a slight breathiness to her voice bound it all in an expression that subconsciously told the large man, "I know what you've done. I'm horrified. But you can explain it." That was the most important part. The head tilt, the subtle body language that told him she was listening.

Landry sneered. "I don't know what you've heard. But anything we've done, we've done to protect people. Real *people,* not a broken, racist system bent on putting white men on the endangered species list."

"Is that why you killed them?" Sophie prompted.

Landry's eyes bugged. His face flushed. "Why, you bitch!" Turning on his heel, Landry's jowls trembled as he took two rushing steps toward Sophie.

This time she did step back, bumping into what she thought was a tree behind her. Until a pair of hands took her shoulders and steadied her.

Greer.

Silently, the former Marine had come up to support Sophie. His pistol holstered, he patted her shoulders and stepped around her, standing forward and just a little to the left of Sophie, like a shield.

In the moment, five other men came tumbling out of the clubhouse. Each carried a pistol or rifle of their own, but only one, a scrawny twenty-something in a green T-shirt, had the barrel raised. Landry glared between Sophie and Greer, stepping back to stand beside his cadre.

"I knew it. That's what this is about. You're trying to pin this on us," the club leader said, his voice sharp and poisonous.

Greer spoke up. "Calm down. We're here to investigate. No one is accused of anything . . . yet."

"Bullshit," one of the club members spat. A lanky man, though with the same trim and oiled beard as the leader despite looking a decade younger.

"Someone has been killing these women. Someone comfortable on the river," Greer replied.

Another club member, a thickly muscled man with a shaved head and close-cut chin strap, shouted back, "We protect our women! You take us for some kind of yellow-bellied dog that would kill innocent, young, white girls? We've had to chase a half-a-dozen lowlifes off this run in the last six months. Cops aren't doin' shit, and now you're all surprised these women are showing up dead? But you don't want to talk about that. No, let's harass some good old boys at the fishing club." The man spat into the dirt, giving Greer an evil eye as he finished.

"We can keep this civil," Greer said, an edge of threat rising in his voice. "Let us look around. Consent to be questioned—"

"Why?" Landry barked. "So you can twist our words and force us to fit your narrative? Like hell. We do not consent to this unlawful trespass."

Greer patted the lapel of his field jacket. "As I said, we have a warrant. Your consent is unnecessary."

Sophie grimaced. It was true, but it was absolutely the wrong thing to say. One by one, she watched the nervous club members' expressions curdle into black anger. Fingers tightened on firearms. Shoulders bunched and feet braced. Sophie couldn't help but notice Greer doing the same, his hand surreptitiously slipping over his pistol grip.

Then her eyes snapped on the lanky young man with the raised gun.

His eyes twitched, looking between the men beside him and Sophie. He licked his lips, and she could see the moonlight catching beads of sweat on his forehead. The night was hot, but she could tell a nervous sweat when she saw one.

Landry's and Greer's voices were rising. Soon the other club members were shouting as well. All except the wild-eyed young man still looking frantically between the groups. There was something there. Sophie could sense it. She just needed to cut through the anger boiling over around her.

Greer's hand settled on his pistol.

A sharp, shrill whistle cut through all the noise.

Sophie lowered her fingers from her lips, every eye now on her.

She had to seize the moment now or lose it for good.

"That's enough. Agent Greer, stand down." Sophie had taken a half step back from the cluster, catching a field stone with her heel that let her stand a bit taller. Now that she was eye to eye with the larger men around her, she adjusted her posture to match. Upright, commanding, the same sort of pose she'd used to orchestrate her shows in front of thousands of speculative eyes. Sophie threw her arms out like spreading wings.

People hear what they see.

"I did not come here to be stopped by your indignation. And I couldn't give a rat's ass about your feelings." A bit cruder than she cared to speak, but she needed to speak their language. These were hard men, and sympathy had gotten her as far as it would. "Right now, I don't even care about your *club* or its *beliefs*. I care that there are seven dead women and you'd rather have a pissing contest than help us bring their killer to justice. You know something. You've seen something. You might not like it. It may grate against every fiber of your being to admit it. But someone here knows who is behind these deaths. And the blood of every woman who dies while they stay silent will be on their hands."

Silence.

The men looked between each other, a ripple of uncertainty in their posture that only seemed to exaggerate the red-faced rage that set William Landry vibrating where he stood. Sophie looked at them each, one by one. Then her gaze landed on the thin young man with the rifle. The barrel had now drooped toward the ground, and he was looking at her in wide-eyed astonishment. His mouth was open, his head bobbed ever so slightly.

"You," Sophie said, bringing her hand forward in an open-palmed gesture that urged the young man forward. "You know something. You've had it on your face since we arrived. Tell me."

Landry spoke first, though. "George, you don't say nothing to her, she—"

"Don't look at him, George. Look at me," Sophie interrupted. With two quick steps, she closed the distance, stepping so close their noses nearly touched. George instinctively stepped back, meeting Sophie's stare then looking away as if the momentary eye contact had burned him. "You know something. Something the others know too." It was a guess, but given Landry's haste to silence the boy, she took the chance. "Something they don't want you to say."

Now the young man, George apparently, looked terrified. He clutched the rifle like it was the rung of a ladder holding him over a hundred-foot drop.

"I swear to God, boy. If you open that fool mouth of yours, we are going to have a problem." Landry tried to step between Sophie and George, but before he'd taken his second step, Greer was already between them.

Sophie only saw this out of the edge of her field of view; she kept her eyes on George.

"George," she said, softly. The young man looked back to her. "Tell me."

There was no time. No time for hypnotism or anything sophisticated. She'd done everything she could up to this point, breaking out what it meant for them, for the victims. Now it was up to George.

"You—" George's voice was uncertain, a slight tremble behind the vowels. But he swallowed, white-knuckling the rifle, and tried again. "You said the killings started a year ago?"

Sophie nodded. "That's right. That's the earliest victim we've found. Did something happen a year ago?"

George looked back into the group, and he found the tall man with the same beard as Landry. Now that Sophie was looking closer, the man could have been the beer-bellied leader's kin. Maybe a brother or nephew? Just like the leader, this man's face was red with rage. His fists clenched and veins on his neck stood out as he mouthed a silent threat.

The young man closed his eyes. "Miles beats women," the young man said.

"That's a lie!" Landry shouted. "Don't you slander my brother, George Hillock! I'll put you in the ground, you lie about my family!" One arm trying to force Greer out of his way, the large man made a wild swipe for George, but Greer stood impassive as a streetlamp, holding him back with a warning glare of his own.

George blinked, his words now tumbling. "He beat his wife. We all knew it. We've all seen it! She left him a year ago." He was hyperventilating now. Tears beginning to stream at the corners of his eyes, the young man shook the rifle in front of him. "Did you kill them, Miles? Was it you, you woman-beating pieceashit!"

Suddenly a shadow flashed by Sophie's head. She narrowly ducked, just as Miles Landry's meaty fist slammed into George's jaw.

Then all hell broke loose.

Chapter 10

A heavy body slammed into Sophie. With one hand reaching out for the lanky George Hillock and the other going for her pistol, she couldn't catch herself before belly flopping onto the hard ground at her feet. Sophie saw stars. The wind completely knocked out of her, she croaked as her lungs failed to draw air.

Beside her, she saw the lithely muscled Miles Landry throwing the much younger man's rifle into the tall grass, another wild haymaker knocking George to the earth. He wasted no time straddling the skinnier man across the chest and bringing up his elbows for another volley of punches.

Sophie rolled on her side, her ribs stitching as she retched and shuddered.

Bad luck. Caught right in the stomach and chest, her lungs were seizing. Her vision swam and her ears rang.

Above her, William Landry, the leader of the Dream, was trying to bring around his rifle. Agent Greer had the barrel gripped in one hand, forcing it skyward as he shouted in the pot-bellied man's face.

The three other men who had come out of the clubhouse with Miles and George stood apart, circling the twin fights with the anxious energy of men trying to corral poorly trained dogs back into the house.

Fear, anxiety. Sophie picked up the string of micro-expressions and contracting limbs that told her these men were torn. Loyalty at war with guilt. Was George right? Did they all know about Miles beating his wife? Attacking other women? Would they decide to come clean like George or seize the moment to "tie up loose ends"? Sophie fought to get onto her hands and knees, her lungs stuttering to life once more and pulling in a gasping breath. She couldn't afford to wait and see what the club would do.

And judging by the beating he was taking, neither could George.

Greer snapped his head forward, catching William Landry with a headbutt. The man's hands came off the rifle, and Greer snatched it away as the big man tumbled to his back, his hands caught between trying to cover his now-bleeding nose and attempting to stop his fall. Even in Sophie's dazed state, she noted the way Greer had slipped an ankle behind Landry's feet in the same moment he struck, guaranteeing the man would fall.

However, Greer was now standing over their leader with a rifle in hand. And the three circling men turned their split attention back to the FBI agent.

Not good.

"He's killing George!" Sophie cried.

Her voice was still in tatters from her hard fall. Blinking, swallowing to try and strengthen her throat, she screamed with all the pity and fear she could muster. "Stop him! He's going to kill George!"

There are certain hardwired instincts that apply to most people. Hit someone in the nose and they see red. Show someone a crying baby and anxiety follows, usually enough to prompt them to try and comfort the child. For a skilled mentalist and hypnotist, these instincts can be a shortcut. Hard to control, but reliable to trigger. There was one trigger that might help her here, the only one she could use before the men closed on Greer.

So as much as it irked her, Sophie forced herself to cry.

Focusing on the pain of her ribs and the genuine fear that she kept suppressed to this point, Sophie's eyes teared over and her voice rose again. "Miles is killing George! Stop him!"

Immediately, the three club members, Greer, and even Landry from where he lay on his back all turned toward Miles. Two of the men lowered their guns and sprinted forward. They grabbed at Miles's shoulders, but he threw back an elbow to shake them off, raising his fist for another blow.

"Git off me! Gonna talk about beatings, George? I'll show you a beating!" Miles shouted.

A crack bit off the rest of Miles's ranting as one of the club men swung the butt of his rifle into the back of Miles's head.

He fell like a bag of flour, and Sophie's stomach twisted at the flat thud his head made as it hit the dirt.

A hand grabbed her by the elbow and Sophie started before she saw it was Greer pulling her to her feet. Similarly, the two club members who had rescued George dragged him out from under Miles. One took off his overshirt, handing it to George and helping him wipe the blood from his swelling eyes.

William Landry and the club member with the shaved head rushed to Miles's side, slapping his cheek and shaking him to rouse him again.

Without really intending to, three groups formed the points of a triangle, eight paces to a side. On one point, the two unnamed club members propping up a battered George Hillock. On another, William and Miles Landry with the cue ball–headed man, all once again on their feet. On the third point, Sophie and Greer, still holding William's rifle.

"Fine!" William barked, his bearded face shaking as spittle flew from his lips. "You want to side with the goddamned government over your kin, you go right ahead, George. You and your FBI chums get the hell outta here! I never want to see any of you traitors again, y'hear?" With an accusing finger, William swept up the other two points of the triangle and pointed to the clubhouse driveway leading back to the main road.

"That's not how this works," Greer said in return, his voice firm but calm.

"You're pressing your luck, mister," the bald-headed club man growled.

Sophie couldn't help but agree. Greer had taken William's rifle, and George's gun was somewhere in the grass nearby, but there were still plenty of pistols and shotguns to go around between the other men.

Greer spoke again, his voice implacable. "You are all coming in with us. Drop your weapons and get your hands where I can see them."

William gave a derisive and decisively hog-like snort. "How about this? You put down *your* weapons and walk away. Otherwise . . . well, people get lost in these swamps all the time. Accidents are known to happen. Maybe someone will find *your* body in a year or so."

Sophie's chest tightened. There was authenticity to the threat. A blackhearted certainty spoken with the voice of a man used to getting his way no matter the cost.

There was a strong chance this could all still end in blood.

A dim light bloomed inside Greer's coat. His phone was silently ringing. If he noticed, he made no move to reach for it. Sophie could tell by the ready way he held the rifle that Greer wasn't about to let himself be distracted.

Sophie glanced up, doing some quick calculations in her head. The moon wasn't contributing much under the glare of the clubhouse spotlight. She couldn't even see the stars for all the ambient glare around her. The wind howled through the trees behind her with the scraping rustle of serrated leaves sawing against old bark.

And it gave Sophie an idea.

Turning her head slightly, Sophie raised two fingers to her ear, raising her voice a bit more than was necessary to be heard over the breeze. "Copy that. Situation is under control for now. No, keep sighting. Wilco."

George's eyes went as wide as their swollen and battered state allowed. "Hey! Who is she talking to?"

William's own eyes narrowed into a deeper glare as he added, "What then? Backup? Good luck, we haven't done anything but defend ourselves on our own land. As for this little scrap with George"— he spat in the younger man's direction—"a misunderstanding between friends. I'm sure it'll all be settled before the week is out."

Sophie shook her head, pointing a finger skyward. "You hear that, right?" The group shared a confused look, and a few risked a quick sidelong glance at the sky. "That's an FBI surveillance drone. Between the video that's been taken and what we've recorded on our body mics, we've already got more than enough to make your lives extremely uncomfortable. But if anything *were* to happen to us . . ."

Sophie let the words hang.

Miles Landry scoffed. "She's lying, Will. They probably don't even have that warrant they were jawin' about. Let's put 'em in the ground and be done with it. We got other business tonight." Miles flashed a menacing look to George, then pointed his pistol at Sophie.

Greer's rifle snapped to ready position, locking onto Miles. Half a second behind, the other club members' firearms came up, some pointed to the FBI, some pointed to Miles.

"Just know," Sophie said, not bothering with her own pistol, as more bullets wouldn't solve this, "whatever you do, it'll be the nine o'clock news tomorrow, video and all. Think hard, because you're going to see this footage over and over at your trial."

Another breeze kicked up, and nervous eyes darted skyward, anxiously trying to spy the threatened drone.

Greer's phone buzzed once more, the dim light visible inside the shadow of his dark FBI jacket.

Miles looked to his brother, but the larger man gave no signal. He was too lost in his own thoughts, considering the rifle in Greer's hand, the dark sky over them. Sophie could see she'd rattled him.

"To hell with this." Miles's off-hand steadied the pistol. He lined up the barrel with one eye, sighting it directly at Sophie.

Her pupils dilated; her stomach dropped.

Suddenly, bright light flared over the clubhouse in red and blue. Headlights rounded the driveway through the woods, and sirens kicked on, only a few dozen yards from the standoff.

Three police cruisers skidded to a halt. Voices over bullhorns called out, "Drop your weapons and put your hands in the air."

After the barest moment of hesitation, Sophie heard the rattle of rifles and pistols being lowered or tossed to the ground. Local PD with their own sidearms in hand rushed forward. Greer lowered the rifle and held up his ID, receiving a quick nod as one of the officers directed them back to the safety of the cars.

"Nice trick with the drone, Miss Quinn," Greer said, the relief in his voice pulling out the words with a long sigh.

Sophie chuckled, her own voice light as the tension left her body. "Thanks. I figured that was our backup calling you. I'm glad they decided to push ahead without confirmation. That was about to get nasty."

Greer nodded grimly, pulling his phone from his jacket to check the missed calls. "It certainly kept their attention divided between us and the sky instead of on the road up to the clubhouse."

"So, I suppose we're in for a long night at the station," Sophie said. "Though, I suppose it will take some time to get them processed, and if any call for a lawyer, we may get to sleep before interrogations tomor—" Sophie froze.

Greer's brow had furrowed. His face drained of color as he put the phone to his ear.

"What is it?" Sophie asked, nervously.

"It's Kinsley. She was the one calling. She's at that music festival in New Orleans tonight. She sent a text that—" Greer's head cocked as the call connected. "Kinsley? Kinsley, baby, where are you?"

Sophie couldn't make out what Greer's daughter was saying.

But she heard the scream from the phone's tiny speakers just before the call died.

Chapter 11

The police cruiser tore across highway 90 toward New Orleans.

Greer flashed the siren and lights with sniper-like precision, breaking holes whenever the flow of traffic appeared it might slow them down from their hundred-mile-an-hour pace.

"Police claim the unrest began with the delay at tonight's concert," proclaimed a carefully neutral voice from Sophie's phone. She turned up the volume, trying to listen in on the streamed broadcast over the thunder of the straining engine and the friction of tires on asphalt. "Local singer-songwriter Carly Monroe was seen leaving her trailer shortly before her performance and has not been seen since. Tensions have been high in The Big Easy since the discovery of what is now seven bodies in the Delta beginning late last night, and witnesses claim rumors spread like wildfire that the killer was *in* the park.

"Festival goers spilled over Bayou St. John into the streets of New Orleans with local police called in to contain the rowdy crowds. At least one fire has been confirmed, but local firefighters are having difficulty breaking through to reach—"

Sophie snatched the overhead grab handle, barely stopping her head from cracking against the window glass as the cruiser whipped around a clot of slower vans and family SUVs, momentarily turned to catch the flare of headlights from the oncoming lanes before Greer found the fast lane again.

Greer's face was a hardened mask, his eyes flitting between the mirrors and traffic around him as he wrestled the wheel, forcing the high-speed vehicle to bend to his whims.

After his call with Kinsley dropped, Greer hadn't wasted time going back to his own car. Instead, he commandeered one of the cruisers there to round up the Dream. Not waiting for Sophie nor commenting as she jumped into the passenger seat, he drove with a single-minded focus, his clipped answers to her questions leaving Sophie to figure out what had happened on her own.

From what she gathered, Kinsley had gone with a friend to the music festival. Her calls and texts had been to tell her dad that things had gotten rough, that it felt like a riot brewing around her. She and her friend were trying to get out but couldn't get back to their car. If they knew where they were, they hadn't been able to tell her dad, and none of Greer's attempts to call his daughter back had gone through.

"The cell towers must be overwhelmed with all the chaos right now," Sophie said as Greer slammed his phone back into the cupholder beside him. He didn't reply, instead tapping his turn signal as he pulled another hair-raising turn onto an off-ramp into the city, flowing in and out of the emergency lane as he dodged other cars with the same idea.

"Hey," Sophie said as they broke out over the Mississippi River and caught sight of neighborhood rooftops over the highway barricade walls. "We are going to find her. Ok?"

Again, there was no reply, but Sophie could swear the former Marine's grizzled jaw clenched just a hair tighter.

They were forced to slow down. First by the ordinary constraints of city roads, then by the groups of people, many in the multicolored glow-in-the-dark accessories you'd expect from a nighttime music festival crowd, wading through the streets. Wide-eyed, some in anxious fear, others in excitement.

"Hot summer night, alcohol, and the threat of murder," Sophie said, more to herself than to Greer. No wonder things had gotten so out of hand. Violence always rises along with the heat; add to it the surreal, sleep-deprived quality of a night of drinking and the festival became a recipe for disaster.

The car was nearly stopped, but Greer forced it further, cutting through any break in the traffic he could find like a worm through soft soil.

"Damn it!" Greer pounded the steering wheel. It would be useless to add his own horn to the cacophony of blaring from the jammed traffic around them.

They stopped. Caught in the knotted traffic.

Sophie felt an uneasy chill rising in her chest and hands. She stared ahead, into the streetlights past the traffic snarl. Something wasn't right.

More people filled the street. Whoops and screams now filled the space between the horns.

Sophie heard shattering glass somewhere nearby.

Now she saw that some of the people in the street weren't carrying beer bottles or energy drinks. Some had bricks, bits of pipe that looked like they had been wrenched from one of the festival scaffolds.

Greer shook his head. "The park is just another block that way." He gestured. "We just need to get there and we can load up Kinsley and—"

"Jonathan," Sophie interrupted, putting a hand on Greer's arm.

"What is it?" Greer paused, then followed Sophie's eyeline. She felt him stiffen under her palm.

"Get out of the car," Sophie said, quickly, unbuckling herself in the same breath.

"What?"

"We have to leave it. Come on, we've only got a minute or so before this street becomes a crush." Sophie's eyes flicked to the crowd working its way between the cars, the shouts and car horns growing louder by the second.

"Are you kidding?" Greer grabbed Sophie's shoulder. "How are we supposed to get out of here once we find Kinsley?"

Sophie locked eyes with Greer, one hand tightening on his forearm, adrenaline making her feel flush and cold all at once. "We do *not* have time, Greer. Trust me. I've been in some nasty crowds before. This is past the breaking point and you do *not* want to be trapped in a car when it gets here."

And with that, Sophie turned back to her door. They were too close to the nearest car, so she rolled down her window, squeezing out and over the cruiser's hood.

"Cop car! Oh, we gotta cop car over here!" Wild jackal-like laughter joined in with the tumult.

Sophie didn't look. Didn't try to pick out the voices out of the thousands of people she could now see milling their way up the avenue.

Distantly a police bullhorn was shouting, "Return to your homes. Clear the streets and return to your homes," as weak and ephemeral as an invisible angel sitting on the city's shoulder.

A bottle smashed against the cruiser hood beside her, and Sophie recoiled from the splash of broken glass. A rough hand grabbed her and Sophie turned, thinking it was Greer again. Instead, a shirtless man wearing dozens of plastic-bead necklaces had her wrist.

He flashed a smile that promised a bad time and gave Sophie's wrist a wild jerk. Her body already flooded with adrenaline, Sophie stepped forward, not allowing herself to be yanked off-balance, planted a foot, and drove her knee into the man's groin.

Letting out a howl, the man's grip withered and his hands flew to protect his crotch as he slumped between the cars—

And revealed a half dozen other men behind him, laughing maliciously at Sophie's attack.

A hand clapped her shoulder from behind and Sophie spun, shooting a palm up into the second attacker's jaw.

It wasn't until she pulled her hand back that she saw who her second attacker was.

"Greer!"

The tall FBI agent waved away the comment with one hand while the other grabbed his jaw. Shaking off the blow, he pointed off to his left, and Sophie rushed to follow.

"Hey, where you goin'?" laughed anonymous voices behind her, but Sophie didn't look back. She saw the gap Greer had indicated. A side alley leading to the next street over. She didn't stop until she stood in the shelter of the building's alleyway, looking back to see the group they'd left hard at work bashing the police cruiser with whatever blunt objects they could scavenge or just heavily stomping across the roof.

Greer pulled Sophie further into cover. "You were right. Let them have it. We have other problems." Greer stopped suddenly, his eyes downcast. "Look. This is bad. Much worse than I thought. I have to find Kinsley, but if you want to—"

"Leave?" Sophie threw up an eyebrow. "Greer, I know we don't know each other well . . . but I have to think you know better than that by now."

Greer shook his head. "I can't ask you to put your life at risk to help me find my daughter. Even if it wasn't my daughter, I'm an agent. This is what I do. You're a consultant. I can't ask you to follow me into a war zone."

Sophie's mouth drew into a tight line. In the chaos around them, she didn't have the focus to analyze why, but she felt—hurt. Insulted even. Folding her arms, Sophie stepped up to her six-foot partner, practically standing under his chin. She tilted her head up to meet his gaze. "What part of my history makes you think I'd be ok with leaving a young woman in this?" She held his eyes, those sharp, crystal blue eyes.

And a moment later, Greer snorted, a rueful grin breaking out as he nodded.

"Fine. But stay close. We won't see Kinsley if we're constantly looking for each other."

The next avenue was only slightly better. The same mischievous spirit hung over the people blocking traffic, but there weren't nearly as many. Stopping on the sidewalk, Sophie could see the distant glimmer of the park and the festival architecture. She could also make out the orange-red glow of distant flames. The news report had said at least one fire had broken out.

Slowly. Fighting the current of people moving away from City Park, Sophie and Greer reached the edge of the festival grounds. On one side of Bayou St. John, the slow-moving river dividing the park from the east side of New Orleans. It was like watching an ant hill consume a child's Tinkertoy set. Small fires here and there. Shadowed people climbing scaffolding, car roofs, and nearby buildings.

"Kinsley?"

Sophie halted. Greer had his phone to one ear, his hand covering the other.

"Where are you, baby-girl?" Greer hunched, closing his eyes to focus on the sound of his lost daughter's voice, his head softly shaking in frustration. "Baby, I can't hear you. You have to speak up. Tombs? Kinsley? Kinsley!"

Greer pulled the phone away, frantically trying to redial. It was no good, though. The towers were down again.

"What did she say?" Sophie asked, pulling on Greer's arm, keeping them moving.

"She said she and Camille are in a tomb?" Greer's eyebrow furrowed. "But that doesn't make any sense . . ."

But Sophie was already adjusting their course. "Come on. We have to get across the bridge."

"Wait, the festival grounds are back that way," Greer protested.

"She said 'tombs,' right?" Sophie replied, not pausing to take in Greer's confused expression. "There's a city cemetery right across the St. John."

As they both looked across the bridge, Sophie's eyes went wide. With a volcanic gout of flame, the orange glow they'd seen earlier erupted into a tower of fire.

"A gas line," Sophie said weakly.

And suddenly, Sophie had to jog to keep up with Greer as the tall man pressed ahead, driving a path through the crowd across the bridge.

Straight toward the fire.

Chapter 12

"Kinsley!" Greer shouted into the rows of mausoleums, his voice echoing strangely amid the din of chaos in the streets behind them. The orange flare of the gas line fire set the air ablaze less than a block ahead of them.

If Sophie were back in Detroit, the city's cemetery would be filled with knee-high headstones and they'd have no problem finding Kinsley as long as she *wanted* to be found. But New Orleans was below sea level. Heavy rain and saturated soil meant caskets placed in the ground had a way of moving, shifting, or even bubbling to the surface. Consequently, stone mausoleums sat in tight rows like English houses, creating a granite and marble maze for the pair to search, block by block.

"Kinsley!" Sophie joined in. It wasn't crowded here. It seemed only a few other souls had come here to escape, and they seemed more interested in taking shelter than making mischief.

Sophie bit her lip. If Kinsley and her friend were hiding or, God forbid, *in* one of these crypts, it could be a long time before they found them. And if one of them was injured or needed help?

Spinning on Greer, Sophie threw up a hand. "We can't keep searching like this."

"I'm not leaving my little girl," Greer bit back. "Kinsley, baby! Where are you?"

Sophie slapped a granite statue beside them, giving a sharp shake of her head. "No, I mean we can't keep searching *like this*. Is there anything else she told you? Anything else you heard on the line?"

Greer bit back a sharp retort, his jaw clenching. Then he closed his eyes, taking a deep breath. Focusing. "Sounded indoors. The connection was bad but there was a slight echo. Could be inside one of these crypts."

"Or any building around here," Sophie replied.

"She *said* 'tombs.' That was clear enough."

"And you said earlier she and Camille were trying to get to Camille's car?"

Greer nodded.

With each passing minute, Sophie could see the former Marine Corp engineer fraying. He was holding it together, but anxiety was tearing him up inside. Of course, what else would she expect? His only daughter was missing. In danger. She might die if he didn't find her soon. There were a hundred horrors ready to prey on his mind, likely already playing across his thoughts. Sophie felt an upwelling of sympathy, and she put a hand on the big man's shoulder, feeling him tense. For a moment it felt as though he might pull away, but instead, he relaxed.

"What do we do?" Greer asked, softly.

Sophie put a knuckle to her teeth, tapping her mouth as she puzzled over the situation. It didn't make sense. Sure, "any port in a storm," but why go to a cemetery if you were trying to get to your car? Had they been caught in a snarl like their car had? No, they'd have said so. They were trying to reach it.

"Tombs," Sophie said.

Greer swept a palm around them. "Yes. Likely a thousand of them or more."

"No," Sophie rolled her neck, her brow beetling. She replayed every time she'd been to New Orleans, every faint memory of driving the streets. A memory began to spark. "She said 'tombs' plural. Not 'tomb.' You're sure?"

Greer nodded.

Abruptly Sophie turned, racing for the street.

"What? What is it?" Greer asked, sprinting after her.

"Not 'tomb.' Tombes. Tombes Pay & Park. They were trying to reach her car."

Silhouetted against the gas fire, a squat, solid-walled parking garage dominated the adjacent intersection. The only light was the queasy mix of the blaze beginning to consume the back half of the structure and the emergency lights adding a hellish red to what could be seen.

"Help!" a young, feminine voice cried from within.

"Kinsley!" Greer put on a fresh burst of speed, his long legs carrying him to the wide-mouthed entrance.

A security gate was pulled down, steel cages covering the only visible paths in or out. And on the other side, two teenage girls were rattling the metal.

Kinsley looked as if she'd been crying, her dark makeup streaking and mingling with her sweat in waving streaks across her chin. Her friend, Camille, Sophie presumed, was a well toned girl of the same age, wearing a black T-shirt that read FESTIVAL CREW in block letters.

"We're trapped!" shouted Kinsley, bouncing on her heels as she glanced anxiously to the fire spreading behind them.

The first thing Sophie noticed was the heat. The burst of gas fire had set some of the cars ablaze on the far side of the prison-like garage, and now it was making its way through. It might take a few minutes or an hour, but if help didn't arrive, the car fires would spread, garage's main lines would go up, and the whole building would become an inferno.

Greer grabbed the gate, straining as he jerked up on it. The tendons on his neck stood out and a curve of tricep strained the previously loose nylon of his dark FBI jacket. But the metal seemed locked in place and only gave a shimmering rattle in response.

Kinsley shook the gate again, standing directly across from Greer as she said, "The power went out and the gates came down. They're locked in place."

"Do you have any bolt cutters or-or anything?" Camille asked anxiously.

Greer shook his head, then stepped back, his eyes narrowed as he scanned the edges of the gate. "There has to be a control panel," he muttered. "Emergency services have to be able to get in."

Taking another step back, Greer held up a hand. "It's ok, baby. Everything is going to be ok. Just hang tight. I'm going to get this gate up."

If Kinsley was going to reply, it was swallowed up by her shriek as the gas tank of a sedan exploded. A palm-sized section of bumper shot against the gate with a shotgun *bang*, and Sophie's stomach fell as she witnessed the splash of flammable petrol and shimmering glass spread the flames to the cars around it.

Greer rushed back to the gate, but Sophie stopped him, putting both hands over his heart. "I'll stay with them. Get the gate open. Now, soldier!"

Blinking once, Greer nodded and jogged away, following the edge of the building.

Camille and Kinsley were crouched now, but slowly regaining their feet.

"Hi, Kinsley. It's good to see you again." Sophie gave the girls a gentle smile, keeping her voice controlled and calm. It was an odd juxtaposition with the situation, but she hoped it would help keep them all calm.

Kinsley smirked, giving a nervous chuckle. "Yeah. Good to see you too, Miss Quinn."

"I don't suppose you two found your car?"

Camille twisted and pointed to a little two-door parked close to the gate. "We almost got out before the gate dropped."

Sophie nodded. It would be cramped, but they had a ride out at least.

"Ok, get it ready to go. Once the gate's up, I want you to pull it out right away."

The girls nodded and turned to start the car. Before she left, Kinsley met Sophie's eyes. "Thank you, for coming for us."

Sophie's throat tightened. For the barest moment, she could see her sister, Molly, in the seventeen-year-old's eyes. She murmured a hurried "you're welcome," then waved for her to get to the car.

Sophie looked to Greer, no more than twenty yards away. He seemed to have found some sort of panel in the wall.

The gate rattled. It hopped an inch, bouncing and closing again. Greer glanced her way and Sophie shook her head.

Returning his attention to the panel, the steel shook again, this time moving up a few scant centimeters before stopping. Sophie frowned, taking the gate in her hands and trying to force it up. Despite the inch it had given, it refused to budge any further.

Sophie waved an open hand to Kinsley in the car, then jogged to Greer's side.

"It's started but isn't going more than an inch or two."

Greer swore, shaking his head as a spark flew from the open panel in front of him. "I almost have it. I just need some way to bridge this circuit. I can't just hold it in place with a live current."

Sophie leaned in. "Show me."

As he did, she reached into her suit coat. Within a moment she had her leather clasp of lockpicks in hand, looking at the steel instruments with a small sigh of regret. "Here. I needed to replace these anyway." It wasn't true, but the white lie made the sacrifice a bit more palatable.

Greer gingerly placed the bridging piece of steel into the circuit, grunting as a few errant sparks flashed in his face.

They couldn't see the gate, but in the space of two heartbeats, headlights flared and a rust-bottomed two-seater swung out into the street beside them.

From the passenger seat, Kinsley called out, "Need a lift?"

***

It was midnight, but Sophie accepted the iced tea anyway. She didn't think the caffeine stood a chance against the exhaustion settling into her limbs, and the cool drink was helping her to forget the fire and the silent drive back from the city.

Kinsley and Camille hunched over drinks of their own, a faraway look in their eyes, sinking into the couch while Greer returned the pitcher of tea to the refrigerator.

Greer's house was west of New Orleans. A nice suburban two-story with a frame that hinted it was once what the realtors would call a "fixer-upper." But the air conditioner worked and the tap water was clean.

And it was quiet.

Crickets and the whisper of trees settling in the cooler night air were the only sounds. There was no traffic at this hour. Barely any lights apart from the porches of the generously spaced neighbors.

"They said there was a killer loose in the festival."

It took Sophie a moment to register what Kinsley had said, then she remembered the brief blurb she'd heard as she and Greer barreled into the city. That people had panicked, gone crazy as rumors spread about the missing performer. Monroe, that was her name. Carly Monroe.

“Were strange things happening?” Sophie prompted gently. “Before the singer disappeared, I mean.”

Greer gave a small cough, glancing anxiously between Sophie and the girls. He didn’t have to speak to let her know he’d rather give them space. But Sophie hadn’t started the conversation. Kinsley had, and if she saw something or knew something . . . Well, at bare minimum, it might be good for her to get it off her mind.

Kinsley either hadn’t heard her dad or didn’t care. Still staring into her glass, she said, “Yeah. Weird things. A bad feeling.” For a moment, the teen girl bit her lip, and Sophie noticed Camille stiffen beside her. Her fingers curling as they clutched the cold drink in her hands.

“Where did it come from, the bad feeling?” Sophie asked, her voice soft and reassuring. “What did you see?”

The girls both blinked. Too quickly. Too rapidly. For a moment, they seemed to stop breathing. Kinsley took a sip of her drink. “I-I don’t know. Probably nothing. Just my imagination after everything that happened . . .”

There was something here. Something they were reaching for. Sophie chewed her lip. It was late. Any minute and the girls might want to get to sleep or Greer might shut down the conversation and insist they rest anyway. The rumors might just be idle talk. Alcohol, summer heat, and a missing singer were enough on their own. But what if it was their killer?

Sophie took a sugar packet from the table beside her and a teaspoon. Softly, she stirred it in. Letting the gentle clink of the ice and spoon in the glass set a rhythm. Just enough to relax, to draw them into the subconscious cadence. To open the girls up and let them share what they were seeing in their mind's eye. Barely even enough to count as hypnotism. Sophie tilted her head, opening her throat into the warm, inviting tones that best suited her powers of suggestion.

"Tell me what you saw," she said, calmly, expectantly.

Camille's pupils were wide now. Her hands were trembling, but she raised her head, and Sophie again saw the white block letters on her shirt, FESTIVAL CREW.

"I thought it was just a weird gig box," Camille said. "They make guitar cases shaped like . . . like *that*. I thought it was some goth roadie's gig box. But it was wood. Just regular wood. Like you could get a splinter from it." Her voice was shaking now.

Sophie kept the steady rhythm of the ice in her glass, suppressing the sudden spike of adrenaline in her own heart. She didn't say anything, too worried to disrupt Camille's thoughts.

"It was too big to be a guitar case . . ." the girl continued. "Too big."

She was beginning to loop in on herself. Sophie exhaled, letting out her question with practiced calm. "What was it?"

Camille blinked, tears forming in her eyes. "It was a coffin. Like the ones they found."

Kinsley's shoulders shuddered, and Sophie noticed Greer's daughter was as upset as her friend.

Sophie turned her attention to the curtain of curly red hair draped over Kinsley's eyes.

"Kinsley, what did you—"

A dish clattered into the sink. Sophie and the girls all started, and Sophie nearly tipped over her glass, quickly catching it as she looked to the source of the noise. Greer was standing in the adjacent room over the kitchen sink.

"Sorry," he called out, just a bit too loudly. "Clumsy." As he said this last bit, his eyes met Sophie's. His lips were drawn in a hard line, his eyes completely unamused and simmering with the hard edge of anger held in check. The look evaporated as he turned to the girls.

"Camille, we have a guest room if you'd like. Or—"

"You can sleep in my room, Ce," Kinsley interjected, getting to her feet. "We haven't had a sleepover in too long."

Camille gave a grateful smile. "Ok, Kins."

Sophie didn't blame the girls. Not many people would want to spend the night alone after the sort of day they'd had. Going back to happier memories, just being teens at a sleepover for even a few hours might be just what they needed.

As the girls went upstairs, Sophie brought her glass to the sink. Greer took it without looking, rinsing it under the hot water as the silence stretched between them.

Sophie sighed. "Look, I'm sorry if—"

"Don't." Greer turned the washed glasses over, setting them on the drying rack a bit more forcefully than necessary. "I understand this is your skill set. I still don't understand how you do what you do, but I've seen it work. I know it works." Knocking the faucet to "off," Greer turned and sat back against the countertop, his arms folding over his chest. "But my family is off-limits."

Sophie looked away, nervously tonguing at her teeth behind her lips. He had seen her use her unique brand of hypnotism before. At times, he'd even seemed impressed by the results. Sophie kept her voice gentle as she replied, "I'm not doing this for my own amusement, you know?" She met her partner's eyes once again.

Greer's clear blue eyes were cold and implacable, making no attempt to hide his feelings but giving no answer. The look simultaneously told her his mind would not be changed and all that was left was for her to agree or get out.

Sophie folded her arms as well, giving a quick bob of her head to indicate she got his message. It was too late to argue. Too exhausting to fight. But she didn't back away; instead, she rotated and leaned against the kitchen counter beside her partner. He might take issue with her techniques, but they still had a case to solve.

“You heard what she said?” Sophie asked.

Greer nodded. “I did.”

“Do you think he’s escalating? Now that his first coffins have been found, he may be more aggressive. More open and taking more risks.”

Greer let out a long sigh through his nostrils. “Maybe. Or she’s just a frightened teen who saw a pine crew box and was freaked out by it.”

“Maybe,” Sophie admitted, but she didn’t feel it. And judging by the distant look in Greer’s eyes, neither did he.

Greer tilted his wrist, checking his watch.

After midnight now. Sophie shifted uncomfortably. They wouldn’t be able to get Greer’s car or her laptop until the morning; it had been left outside the fishing club where they’d begun their infiltration. Thankfully, she still had her motel key on her at least.

“I’ll get an Uber to pick me up, if you want to go to bed. We can meet up in the morning.”

“No,” Greer said gruffly, pushing up off the counter. “That’s silly. Camille isn’t using the guest room. Stay here. We’ll go get the car and our things together tomorrow.”

Greer had made it to the stairway, not waiting for Sophie and leaving her no choice but to hurry after him.

"I don’t want to be trouble.”

"You're already trouble," Greer said with a soft laugh. "But you're not a burden. I just need to figure out where I put those spare blankets . . ."

As Greer pushed the door open, they both saw that the queen bed was already made up and a linen box was propped against one wall.

"Huh," he said softly. "I guess Kinsley already took care of it . . ."

"Wow," Sophie said with a smile. "Thoughtful girl."

Greer grinned at the compliment, but the expression didn't reach his eyes. Too much pain. Too much anxiety. Sophie guessed her partner would be up at least once to make sure his little girl was still safe and sound in her room tonight. Seventeen . . . Sophie could imagine the strong, older man, already frosted at the temples and in the haze of day-old growth on his strong jawline, going white with worry when she finally left for college.

With a quick exchange of goodnights, Greer left, closing the door behind him.

Sophie was correct. The caffeine in the tea did nothing to stop her collapsing into sleep.

. . .

But she felt far from refreshed when her cell phone's ringing ripped her from slumber.

"Merchant?" Sophie asked, her throat dry and sticky.

"6 a.m., Quinn. The killer is definitely on the move."

Sophie blinked at her phone, confirming the time. "What do you mean?"

"We found coffin number eight," the commanding, older woman said. "Coroner says she's been dead a day at most. Probably much, *much* less."

Chapter 13

"It's not her."

Sophie felt cold dread settle in her stomach. The morning sun was harsh on the bayou and the buzz of insects was like static in her ear, drowning out the voices of the search and rescue team that had dredged the coffin, of the forensics tech who had opened the waterlogged box giving his initial impression.

The girl was young, early twenties. She looked like any number of drowning victims, her florid party skirt and shoulder-length hair clinging to her like the wiry ridges on a stucco wall. She matched the profile of the other victims. A woman dressed for a night on the town.

What she was not, was the missing singer from last night's concert.

Greer grunted in agreement, crouching over the open coffin, angling to look into the dark corners. "Definitely not Miss Monroe."

"The body found by Joseph Green was recent too," Sophie said, circling to the other side of the coffin and bending down to join Greer.

"If we assume the killer took Carly Monroe from the concert, he's up to three girls in the last week. Two within the last forty-eight hours."

"He's accelerating," Greer agreed, reaching forward with a pen and gently turning the victim's head.

Suddenly, Sophie wrinkled her nose, blinking her eyes against an odor rising from the coffin. Raising her hand, Sophie blocked her nostrils for a moment as she fought the sudden urge to sneeze.

"What is it? Do you smell something?"

"You don't?" Sophie asked, genuine surprise in her voice.

Leaning close and taking a whiff, Greer's eyes narrowed as he tried to home in on the scent. Shrugging, he leaned back again.

"There's something," he admitted. "It's subtle. Or maybe it's the bayou covering the odor. I can't quite . . . Maybe a little . . . sweet?"

Sophie nodded. "It's semisweet. I got a good whiff. It feels . . . familiar somehow. Some might have been trapped in her clothing. It irritated my nose. Maybe paint thinner?"

"Some kind of cleaner?" Greer offered.

Sophie closed her eyes, focusing on the scent. The sense of smell was the most powerfully tied to memory. If she could only trace back where she'd smelled it before . . . Her eyes flew open.

"Ether."

"Ether?" Greer echoed. "So it is paint thinner?"

Sophie shook her head, running her hands over her knees as she mulled over the possibilities. "Not necessarily. It's used in solvents. Plenty of industrial processes use ether. Even taxidermy to an extent. It also has a long history as an anesthetic, though nobody in the US uses it that way anymore."

"Why not?"

Sophie shrugged. "Better options were developed. It's harder to control in some ways, and there is a fine line between knocking someone unconscious and killing them."

Greer gave a low, impressed whistle. "Wow. Good catch. We'll get the lab to run some tests and confirm. Are you thinking that's our cause of death?"

Shaking her head, Sophie sighed. "No. I don't think the victims were that lucky." She pointed to the upended lid of the coffin and the long, jagged scratch marks carved in by human nails. The same marks found in every coffin so far. "Whoever our perp is, he seems to know his way around ether enough to knock them out but leave them alive to wake up in the coffins later . . ."

Sophie blinked, a sympathetic tear threatening to spill from the corner of her eye. For a moment, she allowed herself to imagine those last horrific moments. Trapped. Desperately trying to claw her way out. Every strike loosening the tightly sealed planks that held the dark waters out and bringing her closer to drowning. Knowing if she didn't fight, she'd suffocate anyway.

But you, you a fighter. Aren't you, cher? That why you live to suffer another day. Are you jealous? She die terrible . . . but her *sufferin' is over.*

Sophie bit the inside of her cheek, driving the coal-voiced demon from her mind, returning to the present. She wondered how long she'd drifted, glancing anxiously to Greer.

Their eyes met.

"You alright there, Miss Quinn?" he asked, his voice soft, just them and the dead girl between them.

Sophie looked away, rubbing the bridge of her nose with her fingers. "I'll be ok. It's just . . . I'm thinking. About the case."

Greer nodded, his gaze lingering just a moment, drawing a blushing heat to Sophie's cheeks. Then, mercifully, he was talking to the forensics tech. "Excuse me. Can we get some gloves?"

Moments later, the pair took turns gingerly lifting the limbs, the creases of soaked clothing, looking for any clue.

"She hasn't been in long," said Greer. "We might have something usable."

"No purse or clutch," Sophie added. "Check if she has a pocket on her skirt."

Greer shook his head. "No pockets. Wait . . . hold on, we have one of those little inseam deals."

Gently probing at one of the skirt's vertical seams, Greer used a pen tip to gingerly hook the pocket, turning it out. "There's a card here. Looks like a business card."

"What's it say?" Sophie sidled around the coffin.

The big man delicately held the wet cardstock between two fingers, holding it close to his eye. After a moment of back and forth adjustment, he blew out a frustrated sigh. "It's water damaged. The ink is blooming, and I can't make out the words. There's a symbol, but I can't tell if it's a sun or a moon or . . . I don't know, some other circle-shaped motif?"

Sophie leaned over Greer's back, putting her chin over his shoulder as she squinted close.

Gradually a smile bloomed on Sophie's face. "I know that symbol."

Greer's head jerked, half turning to face her. "Wait. Really?"

Sophie nodded, putting a firm hand on Greer's shoulder and giving it a playful squeeze. "That's not a sun or moon. The ink bleed is warping it, but that's a footprint. Barefoot T & C."

"How on earth do you know *that*?" Greer asked, squinting at the blurry image again.

"One, you're holding it upside down," Sophie said, quietly. "Also, I used to work for them."

"Right," Greer murmured, fetching up an evidence bag for the card. " *Mysteries and Phenomena with Sophie Quinn.* I almost forgot."

Sophie gave a light laugh, fetching her phone. The fact that this card was on their victim gave her a new thought, and there was something she needed to confirm before she brought it to Greer. "No, Barefoot was way before I hit it big. They're a smaller operation. Local actors, models, and . . ." Bingo. "Singers."

Turning her phone around, Sophie showed the screen to Greer. "Barefoot were the ones who organized the festival. They were the ones who brought on Carly Monroe at the last minute." Sophie's grin began to fade, implications beginning to spiral in her mind. "And maybe that means that these girls aren't dressed up to just have a night on the town."

Greer stood, realization dawning for him in the same moment. "They're dressed up for an audition?"

Sophie's mouth drew into a thin line. "Nine times out of ten, the advice young actresses get about auditions is to go like you're taking the casting director on a third date. Dress sexy. Even if the role doesn't call for it, go over the top."

Greer's fists clenched. With a grim shake of his head, he turned, snapping off the latex gloves as he stalked for the car.

Sophie hurried after, balling up her own gloves. "Wait. Hold on, we can't just go in there guns blazing."

"Why not?" Greer said flatly, not bothering to stop or look back.

"Because companies like this are insular. They don't risk their reputations by cooperating with police. They go to ground. Bury problems and send victimizers out of state. If you go in there now, you're just giving them warning that the jig is up and they have time to prepare." Sophie jogged up ahead of Greer, turning around to face him head on, walking backwards when he didn't slow down. "I have industry connections. Let me make some calls. We'll get them while their guard is down."

Greer snorted. "Miss Quinn, with all due respect, the time for subtlety is long gone. We have two victims in the last week, eight total *found*, another still missing. If we don't get there *now*—"

"—if we go in heavy, we may *never* find who did this," Sophie protested.

Greer stopped, his eyes locking onto Sophie's with cold determination, a deathly focus that sent a shiver down her spine. "They put my daughter in danger. Do you understand what that means?" Greer paused, but Sophie found she couldn't speak. A hundred rhetorical tactics scrolled by in her mind. A hundred instincts on how to manipulate a mark or bend a crowd to her will. All of it withered before what she saw now.

The fury of an honest man.

When Greer spoke again, his voice was slower but no less grave. "Miss Quinn. I don't expect you to understand. But don't ask me to wait while more young women are buried in the bayou."

Sophie took a steadying breath, clearing the tremor of anxiety racing through her. "We need to be smart. Please."

Willow branches rustled as a breeze still cold with night air escaped into the sunlight. Greer sighed, straightening the sleeve of his tan suit. Taking the final few steps to the car, his personal white Interceptor, he opened the door and handed Sophie her bag. "You can use the spare car. Merchant won't mind if it's returned a bit late. If Hitchens throws a fit, you can blame it all on me. But I can't wait."

The engine roared to life, and Sophie met Greer's eyes through the driver's window one last time. A brief pause as if offering her a chance to join him.

Instead, she turned. Frustration simmering in her chest.

Fine. He wants to be a cowboy? Let him. He'll still need to secure an emergency warrant. As determined as he was, Greer played by the rules. That should buy Sophie some time. In the back of her head, Sophie remembered catching Greer with his own set of lockpicks at the trailer of a suspect on their last case, and suddenly she wasn't so sure she would have that extra time. But she didn't have a choice. If she was going to get anything, she had to act now. Running a hand through the close crop of her auburn hair, Sophie opened her cell phone with the other hand, searching the internet for a phone number.

"Hello, Barefoot T & C? Hi, my name is Sophie Quinn." As she spoke, Sophie checked her suit coat pockets and her underarm holster. Consultant ID. Pistol. Playing cards, marker, gum, pocket watch, lockpicks—sans one steel pick. "Do you have an agent available? I'd like to discuss working with Barefoot. Today. Excellent! And who . . . ?"

Sophie froze. She must have misheard. Glancing over her shoulder, she saw Greer's taillights as he turned onto the road back toward the city.

"Sorry. Could you repeat that name?"

But it was the same. A name from her past. Out of a memory from fourteen years before.

Chapter 14

"Dinner is delicious, hon. Honestly, I don't know how you do it." Jack accompanied the compliment with a manly belch and an open hand swat to his stomach as he pushed his grease-stained plate away. He'd worn his nice button-down and a clean set of pants. After all, when someone you love goes out of her way to prepare you a meal, the least you can do is look nice while you eat it.

The sole of his shoe made a sticky peeling sound as he lifted it from the damp concrete floor and crossed it over his knee. He frowned at the sound. Glancing across the table, he lit an after-dinner cigarette and muttered, "Wouldn't kill you to pick up around here once in a-while, though."

The woman across the simple oak table shuddered. She smiled so prettily in her posters back in the city. She hadn't smiled for him. Not once. No, for him her lovely shoulders slumped. Her bright eyes were wet and puffy.

"Y-yes, dear," she said softly, blinking. Whether from tears or the glare of the sunlamp Jack had propped against the wall, he couldn't tell. He'd done what he could. A window frame and curtains hung from the grey ceiling by a chain, partially obscuring the artificial sun in his concrete kingdom, but he had to admit it *was* a touch intense at times.

Pinewood cabinets, handmade of course, lined some of the walls. A workbench sat on another, as well as a twin bed on its opposite. He'd have liked to have brought in a queen bed, but the steep twist in the steps leading down to his sanctuary didn't allow large furniture. He tried his best though. A lamp here. An end table there. A part of him knew it looked like a third-grade play version of a home, but it was *his* home. His sanctuary. Still, no matter what he did, the concrete walls would show. The dark lines of black mold seeping in from the damp ground all around them. Basements were a terrible idea in the Delta, but of course he'd insisted. They always held a mystery to Jack, a sort of dark appeal.

A ten-inch television idly chatted beside the workbench. The volume was low, tuned to a local news station. More to add a flicker of living light and sound to the crypt-like room than anything else. But then again, all happy homes had a tv going in them. Didn't they? Jack thought so.

Maybe the gator was frightening her?

Glancing over his shoulder, Jack confirmed that the dried and preserved gator head on his workbench was staring straight at the table. He'd been careful to get the meat packed away in the freezer, and the hides were tanning on the racks elsewhere, so all that was left were the bones and head for some kitschy tourist shop souvenir.

"Sorry, dear," Jack laughed, hurrying to the bench and turning the gator's glass eyes to face away. "Didn't realize. You know how I get after I've been out fishing."

The girl at the table looked at him in confusion, her lip trembling. But she said nothing.

Jack frowned. "No, I suppose you don't. You never notice anything. You're selfish, you know that?"

"I'm sorry . . . dear. Thank you."

Jack grimaced. He could hear the fear in her voice. Like he was some ogre. He wasn't a monster. He was a human being. A person who needed love and respect just like anyone. Respect.

Jack pounded a fist on the workbench, and the girl jumped. Under the table the chain binding her ankle to the floor rattled on the concrete.

"Goddammit! It's not just a hobby, you know?" Jack barked. "I bring in money with these gators. Real money. Not that you care. Throwing it away on another goddamn night of drinking. Another dress you'll never wear!"

It felt good to shout. To get it off his chest. When he was done saying all the things he should have said, maybe he'd be the *man* he'd always wanted to be. He had to make her understand. Annie had never listened. Well, now she would. She'd hear him if he had to spend all night.

All day? It was difficult to tell in the concrete. A small voice told him this dinner probably should have been breakfast . . . but that wasn't really the point.

Jack looked to the singer's plate. She'd barely touched her food. Picking at it from some sense of obligation. Spinning it around her plate. Ungrateful. He'd been willing to pretend she'd cooked for him . . . but there was only so far he could lie, and it hurt to see food he'd made for her go to waste.

"I know," Jack laughed darkly, his callous hand slapping his forehead. "I forgot the wine. No wonder you don't have any appetite." He stalked around the table and put both his enormous hands over her shoulders. She flinched, stiffening and curling down under his touch, but he held her in place, working his fingers into the muscles around her neck in a deep, insistent massage. She let out a small, pained yip, but real massages always had a little pain to them. It helped to release the tension. He'd heard that . . . somewhere.

Jack leaned down, planting a delicate kiss on her cheek and tasting the salt of dried tears. "No need to cry about it," he whispered. "I forgot the wine, but maybe if you didn't drink the whole fucking bottle every night, we'd have some around, Annie."

His heavy hands pinched down on her shoulders in a vice grip. The girl screamed. Grabbing her plate, she swung it up into Jack's face. Mushy meat and potatoes slopped over him, followed by the sharp slap of plastic. Jack let go, pulling back and wiping his eyes, his cigarette extinguished and lost amid the mess.

The chain rattled, and he heard the slap of skin on concrete.

His face flushed, hot red flashing behind his closed eyes. Once he'd blinked them clear, the girl was on the ground, her dress getting damp from the floor as her fingers pulled at the chain on her ankle. Those fingers were tougher than most, all that guitar playing he supposed. But she'd never get the chain loose.

Staying where he was, Jack glowered down at her.

Other men wouldn't let themselves be treated this way. What's wrong with you, Jack? Why are you so weak? Why aren't you the one calling the shots?

Jack's hands began to shake. They trembled until all he could do was clench them into a white fist. It still wasn't enough.

Then he threw the table over with a crash.

"Other husbands don't have to put up with this shit, Annie! Other husbands' wives give them *respect*!"

"I'm not Annie, you psycho!" the singer shouted back, tears streaming from her eyes again. "I'm Carly! Car-ly! Let me go! Let me go!"

Jack shouted back, a wordless roar that hurt his own ears in the claustrophobic room. The girl curled up on the floor, her shouts disappearing into sobs. "You never speak up," Jack said, his blood boiling as he circled the room with determined stomps. "You never say a thing until you're ready to scream! Well, what else is there? Huh? Are the-Are the curtains not the right color?" He ripped the fabric from his fake window, throwing it at the crying woman. "Is it too cold? Too hot?" He kicked the sunlamp, sending it crashing down with a sparking fla sh.

Now the room was only lit by the flickering white of the ten-inch tv on the workbench. Jack stalked toward it. "I suppose this has to go too, huh? I forgot, it's not a show *you* chose, so off it goes!"

He reached for the tv, ready to throw it to the ground, but then Jack stopped. His breathing ragged, he tried to hold his breath. His eyes locked on the screen. He turned up the volume until it shouted over the blood pumping in his ears and the woman's ragged crying behind him.

"*Local police have partnered with the FBI as yet another body was discovered early this morning. The latest work of the so-called 'Coffin Killer.'*" Footage zoomed, getting a distant shot of a group of officers around a waterlogged coffin. A man in a tan suit was pacing, but the camera moved away from him to focus on a woman in a slate-grey ensemble. Her auburn hair was cut short around the ears, but Jack still recognized her. He couldn't believe it. It had to be a woman who just looked like her, but no . . . as the woman in the shot looked up, he saw the same face. A face he saw every week . . . maybe seven, eight years ago?

"You used to love her show," Jack muttered. "Couldn't get enough. *Mysteries and Phenomena with Sophie Quinn.*" He shook his head, gently setting the television back on the bench. He turned, stepping back into the dark, walking slowly to let his eyes adjust as he followed the fearful whimper and scrape of chain on concrete. He crouched low. "I used to wonder what you saw in her. I have a guess now." Like a rattrap, his arm swung out and snatched the back of the singer's head, slapping her while grabbing a fistful of her hair and drawing out a fresh cry of panic. "I bet she's just like you. A liar who doesn't know how to treat a man."

Yes. The more Jack thought about it, the more sense it made. Sophie. At least he *thought* her name was Sophie.

It could be Annie.

It may as well be Annie.

Chapter 15

Sophie walked like a defiant prisoner being led to her execution. Head high. Hands at her side. Eyes clear with purpose. It was strange being back in these halls. It almost felt like . . . starting over.

The summer sun through the floor to ceiling windowpanes in Barefoot T & C's main lobby chased her as far as the back hallway before pale fluorescents took over. The carpet, as flat as a baking tray, was crisscrossed with patterns that would have been called "quirky" and "exciting" in the mid-1990s, probably a decade before the stick-thin girl in high heels and a pencil skirt that was leading her to the back offices had been born.

They stopped at a solid wood door with a black nameplate etched in gold lettering that said, "Charles Randy: Talent Acquisition." A door somewhat more worn at the edges than Sophie remembered it, but otherwise just as it was when she'd seen it thirteen years ago. Sophie felt her chest tighten and forced herself to breathe, her face staying implacable.

With a pleasant smile that told Sophie "stay put for a second," the girl knocked and was told to come in by an animated voice, half cigar smoke, half undisguised annoyance. She opened the door to its bare minimum that allowed her to slip inside, leaving Sophie standing in the hall.

Nervous.

Sophie clocked the emotion, closing her eyes, returning to techniques she'd mastered over the years to calm herself. The same sort of techniques that let her hold her breath during a water tank escape or stay in complete calm while she commanded audiences of thousands. Why should she be nervous? She wasn't the same girl she was at eighteen.

And yet, before coming here, she'd stopped off at a convenience store. Picked out the same dollar-aisle blush and eyeshadow that she'd made do with back when she could only pay with what was left after gas money and her daily ramen.

But people hear what they see. The Great Gavriel had taught her that. And she knew exactly what she needed to show Charles Randy, agent to the desperate, if she was going to get any information out of him.

Partly to ease her nerves, partly to pass the moment, Sophie reviewed the case file in her mind. DNA from the first few bodies had come back, and forensics had some of the girls' names now. Evie Shipman. Layla Kelley. Jewel French. The most recent victim had been ID'd as well. Addison Garnier. Apparently, her parents had submitted a missing person report just after she was found, and the local PD had matched her picture immediately.

Once I'm in, I'll turn the conversation. Get him talking about new girls, Sophie thought. Someone gave Addison a card. Carly was under Barefoot contract when she vanished. If the other girls had contact with Barefoot, that's a connection.

Maybe it's him . . .

Sophie pushed the thought from her mind. It was too early to jump to conclusions, and it was dangerous to let her biases have free reign here. Expectation is the enemy of observation, she told herself. Closing her eyes, repeating the mantra.

"Mr. Randy is just finishing a call and will be with you in a moment. But you can go on in." The office girl gave another perky smile as she hustled out of the doorway, leaving it open for Sophie as she retreated up the hall and returned to the front desk.

Stepping inside, Sophie took in the office. Pictures of tropical beaches served in place of windows. Stock images, no friends or family to be found. An oak desk dominated one side of the room, polished enough to give the illusion of fine craftsmanship under piles of loose documents, an ancient-looking desktop PC, and an overfull cigar tray.

On the other side, beside the door she was standing in, sat a wide, sagging couch that looked as if it were half disintegrating under years of harsh cleanings. The room smelled of stale sweat and cheap cologne with a distinct bite of spilled liquor. A singularly uncomfortable-looking wooden chair sat on her side of the desk. And on the other side sat a pudgy man in his early fifties, hair slicked back and dyed charcoal black, wearing a maroon suit that looked off-the-rack fresh.

He gave a sharp, blindingly white smile and waved Sophie to the chair. His other hand held the receiver of his desk phone to his ear, and he grunted something in his too-loud, rusty-nails voice. But his eyes never left Sophie, and after a halfhearted farewell, he hung up the phone.

Sophie knew there was no one on the other end. It was a power play to make him seem busy and important. To make her feel like she was intruding on his time and give him the power in the conversation.

She also didn't care.

There was no telling how long she had until Greer kicked in the door with a warrant and any hope of finding out if Barefoot was involved vanished.

"Sophie Quinn, as I live and breathe," the flabby man said, giving his fleshy chin a quick shake. "So good to see you again, doll. It's not often we get visits from our alumni after they hit the stratosphere." His eyes narrowed and he leaned back, his chair reclining with a small groan as he did. "It's a shame we parted ways. But all good things must come to an end. One of the great *mysteries* of life I suppose. A real *phenomenon*." Charles gave a wicked little grin at his jibe over Sophie's canceled show, and Sophie returned the look with a small smirk of her own.

"It's good to see you again, Mr. Randy," she lied. "I'm sorry to hear about your own misfortunes. Losing two full days of festival after a riot on the first night?" She shook her head sympathetically. "That has to hurt."

Charles barked an ear-shocking laugh, slapping the desk and pointing to Sophie. "Ha! Haven't lost that bite of yours have you, kiddo. Good for you. Tough girl. Strong female lead and all that shit. And still got that figure of yours . . . most of it anyways. Good for you."

Sophie gave a short, mock bow from her seat, gritting her teeth while her face was turned away. "I do my best."

Giving her an appraising look, Charles rolled his thumb across his fingertips and licked his lips. "So," he said at length. "I gather this isn't purely for the pleasure of my company. What can I do for you, kid?"

Here it goes.

Sophie straightened up, folding her hands over her lap. "As you know, I've been out of the game for a bit. Six years now. After I canceled *Mysteries and Phenomena*, I haven't been on the stage. I was grieving my sister and, well, I let it get to me."

"I've heard," Charles said cautiously. "Burned every bridge from here to Hollywood. Is it true you burned down your mansion too?"

Glad to see the rumor mill is still in full swing, she thought. "No. Sold it. Liquidated everything I could."

"Ah," Charles said, flashing a sharklike smile. "I see. Needed cash. Don't tell me you were chasing the dragon? I never took you for the type." Sophie never got into drugs the way other performers so often did, but she could see in the wicked gleam of Charles Randy's eye that he truly hoped she had. Regardless of the truth, that would likely be his contribution to "insider gossip" before the day was out.

"No," she confirmed. "But I've done what I needed to. Got it out of my system." She gestured with two upward palms to emphasize the thought. "I'm ready for a fresh start. Ready to get back in the game."

"Back in the game, eh . . ." Charles chewed on the thought, rocking his chair side to side as he looked her up and down, his eyes invading every crevice of her slate-grey suit, lingering a moment on the athletic shoes on her feet and even more apparently on the curve of her breasts through her blouse.

A familiar crawling feeling sent a ripple up her spine, and Sophie resisted the urge to pull her jacket closed.

"Six years is a long time, sweetheart," Charles continued. "And I ain't tracked how much bad blood you left behind. This could be a risky endeavor for us."

Sophie stood, walking confidently to the desk as she spoke. "Come on, Charles. I'm a known quantity. How many of these new girls show up to work? How many ghost you after a week?" Sophie gestured back to the closed door behind her to emphasize the point.

She froze, just for a moment.

Above the door hung a bleached alligator skull.

Not exactly uncommon in Louisiana, but distinct. Had Charles gotten into gator hunting over the years? Poaching? The killer left the bodies in known gator hunting grounds . . .

Sophie cleared her throat, turning back. "I have experience, Charles. Tell me about the new girls. What did they have?"

A bit clumsy, perhaps, but the goal wasn't a direct answer. Not right away. If she could get him thinking in that direction, he might give himself away.

Charles laughed. "Experience? Ha! Experience she says." With a grunt, the big man levered himself up, his chair's springs gasping with relief as he shifted around the desk to stand face-to-face with Sophie. "You know what experience is, Soph?" He poked a pudgy finger against her chest. "It's another word for *used up*. You've been off playing FBI for half a decade. The only way you stay sharp in this game is to stay *in the game*, sweetheart. Do you even remember how to perform? How to sacrifice? What makes you even think anyone would *want* to see Sophie Quinn again? No. You're expired goods now, doll."

It was a bluff. Sophie knew it was. Bluster. Negging her to get the upper hand, maybe even revenge on an eighteen-year-old Sophie who had gotten as far away from Charles Randy as she could once she saw what a creep the man truly was.

But it still hurt.

Sophie's stomach clenched. In the back of her mind, she supposed she'd always thought once this was done, once *all this* was done and

RK was found, once he was behind bars or dead, she'd go back. She'd pick up her specials. Go on a new tour. *Sophie Quinn, Back from the Dead*. Charles was cruel and manipulative, but he was right about one thing. The show must go on. If she wasn't there, it was somebody else. Every day that went by, she was forgotten more and more. When she was ready to go back, would anyone want her to?

As he spoke, the big man walked a circle around Sophie, stopping against the front of his desk with one hand as the other came to rest on her chin. "It's the hard facts o' life, Sophie girl. But don't worry. I'm sure I can find something for you."

Sophie jerked her chin away, anger flaring. A small part of her told her to calm down. Smile. Play the game and get what you came for.

But for another part of her, enough was enough.

"I wasn't playing," Sophie said, letting a hard edge back into her voice.

"What's that now?" Charles looked confused.

"I wasn't playing FBI. I'm a consultant. I'm investigating the murders of several young women. One of which was found dead with your business card in her pocket."

This was exactly what Sophie hadn't wanted, what she'd warned Greer against. But she was in too deep. She needed answers, and playing coy had only put Charles in a place to play his own games. Suddenly, she couldn't have stopped herself if she wanted.

"How many others?" she insisted. "Missing girls from the last year. How many came to Barefoot first? Evie Shipman. Layla Kelley. Jewel French. Addison Garnier. Tell me about them. Did you speak with them? Meet with them? Talk."

Charles leaned back, his double chin folding as his brow shot wide with surprise.

"Tell me what you know," Sophie demanded.

And he laughed.

Sophie's stomach clenched, and she felt a cold feeling in her fingertips.

"Oh, I get it now," Charles said, wagging a finger. "I had to admit I couldn't believe it when I heard. Sophie Quinn, looking to start over. And she comes back to Barefoot? I should have known." Stepping back to his chair, Charles collapsed into the faux leather like a king onto his throne. Or a Mafia don about to grant a favor . . . "Now with our sweet Carly gone missing, you think we've got the deets that'll put the whole puzzle together. What was that last one? Garnier?" Charles chewed on the name a moment. "Yeah, I remember her. Sweet young thing in here a week or so ago. Really *enthusiastic* about her prospects as a model. You remember how it goes. Don't you, Soph. Well, maybe I know something more. The names ring a bell, but I may need something to help me jog my memory." A shark-tooth grin split over his mouth as he gave her another indecent look. "Haven't had my exercise today. Maybe you scratch my back—and a few other places—and I scratch yours."

Sophie hesitated. This wasn't what she expected. No nerves. No protest.

This was not a man on the verge of being caught. Not a killer trying to deflect suspicion. The posture, the look in his eye. It was Charles Randy as she'd always known him. A conman looking to see what he could *get* out of situation. Ready to exploit the murders of eight young woman if it could get him a bit of action on his office couch.

He wasn't the murderer. She could spend precious hours proving it, but everything in his behavior told Sophie what she needed to know. Then again, Charles Randy wasn't the only agent at Barefoot . . . What if he did know more? Sophie glanced to the desktop computer.

"What'll it be, darlin?" Charles cooed. "Think you can swallow that self-righteous attitude long enough to play ball?"

Sophie's heart was racing. It was now or never. If she walked out that door, the information she needed could be shredded, deleted . . . disappeared any number of ways. Now wasn't the time to be cautious. She had to make a choice.

Turning slowly, Sophie rolled back her shoulders. Years on the stage had taught her how to carry her body to direct attention where she wanted it. She gave a coy smile over her shoulder as she stepped to the door and locked it. Suppressing a grimace, she let her hips sashay with each step as she walked back to Charles's desk, slipping off her suit coat as she did, laying it over his desk as Charles gave her a wicked, hungry grin.

"I always knew you were the same, sweetheart." Charles's knees were spread apart in his chair, and Sophie stepped between them, tracing a fingertip over his tie; she took it in one hand, pulling it softly as she met his eyes. "For all your posing," he said, licking his lips, his pale flesh already perspiring, "I knew you were just another—"

Sophie's fist clenched down on the tie. Yanking down, she slammed the big man's face against his desk.

Charles Randy let out a shriek, his hands coming up to cover his now-bleeding nose as he flopped back into his chair. As he recoiled, Sophie shot a kick straight into his chest. His arms flew out, wheeling for balance, but too late. The chair overturned and the agent hit the ground with a thud that set the monitor on his desk wobbling.

"Bitch! You broke my nose!"

He squirmed, trying to regain his feet, but Sophie had stepped through on her kick and now stood with one heel pressing him to the floor. She took one shifting step, moving the athletic shoe to his throat. A coughing yip turned to a gurgle, and he started to swat at Sophie's leg, pushing and pulling as his face turned red. But Sophie only leaned forward, applying pressure until the soft man let fly a chunk of spittle over his own cheek and gagged, "What the hell is thi s?"

"Answers, Charles. Evie Shipman. Layla Kelley. Jewel French. Who worked with them? When did they meet with Barefoot?" Sophie kept her shoulders back, slipping her hands into her pockets in a show of nonchalance. A posture designed to reinforce her control over the situation.

The flabby agent's eyes goggled. "What? You're crazy! You'll never work again! I'll have you arrested for this, you—*aggck*!"

Giving his throat a fresh press with her heel, Sophie clicked her tongue disapprovingly. "I can't believe I was ever scared of you." Reaching to the desk, Sophie pulled the drawers one by one. "Do you still keep that bottle of tequila? You know, the one you use to *congratulate* teen girls on their contracts. The one you use to make it clear how *grown up* you see them? There we are." Sophie pulled the high-proof bottle from the desk, tossing the loosely screwed cap onto the floor and grimacing at the acrid scent.

It wasn't sweet, but it made her think of the ether coming off of Addison Garnier's coffin, and Sophie's heart went cold. She turned it over, pouring the liquor onto the man's spluttering face, soaking his greasy hair, his chest and his pudgy hands thrown out in protest.

"Wh-what are you—" He choked on his words as Sophie reached to his desktop, pushing aside one of the half-smoked cigars and fetching up a lighter. "NO! You're insane! HELP!"

Charles was shouting, trying to wrestle free from his pinned position, but he was caught by the three-way pinch of his chair, his desk, and Sophie's foot on his neck. She didn't worry about being heard. She had seen the doorframe on her way in, the odd stillness in the room that told her it was soundproofed. And, of course, there were no cameras. Charles Randy wouldn't want "evidence" of what he did in this office.

Sophie's face was a mask of nonchalance as she calmly flicked the lighter over and over, letting Charles see the flame between her face and his. Fresh sweat mingling with the flammable alcohol on his forehead.

"Talk," she said simply.

Charles swallowed, his cries for help vanishing. For a moment, she could see his mind spinning, looking for a way to get the upper hand again, but Sophie merely flicked the lighter back to life, and he withered, curling back like a caveman recoiling from lightning.

"Talk," Sophie repeated, more insistently this time.

"Look," Charles whined. "I was just playing, Sophie. Just playing. Of course, I'll get you whatever you want, anything you want."

Sophie let out a long, angry sigh. "Evie Shipman. Layla Kelley. Jewel French. Addison Garnier."

"Garnier. She was a new contract. I'll get her file if you want. The others—We're a small operation, Sophie—sweetheart. You know that. I don't recognize them. I—" His eyes locked on the waving flame from Sophie's pilfered lighter. "I mean, I'll double-check. If they're in the system, it'll show. Even if it was just a 'follow up later' type meet. Just let me at the computer. I'll show you."

Taking her foot off, Sophie stepped back. It took a few moments of rolling back and forth, propping himself up by his elbows first, but Charles Randy soon righted his chair, smoothing out his hair, black dye coming out with the alcohol to stain his hands, and wiping his face with a soft towel from his desk.

Sophie watched in open disdain, careful to keep his screen in sight so that he couldn't hide or delete any files while she was watching. Soon the printer at his desk clicked to life and began to noisily vomit out a few select pages.

Addison Garnier. Carly Monroe.

"See," Charles said nervously. "No Shipman. Only Jewel is a Mavis. And no Layla Kelley either. Barefoot had nothing to do with them. *I* had *nothing* to do with them? Right? You see that now, right?"

A dead end. Sophie felt her jaw tightening. She'd expected *something*. Between Monroe and Garnier . . . it felt like too much of a coincidence. But she'd seen the files with her own eyes. There was nothing here. Nothing that she could see anyway.

Sophie's phone buzzed in her pocket as she reclaimed her suit coat. Her hands were shaking, and she tensed her arms to silence the tremble. She didn't spare a look at the man from her past, leaving him to squirm in his chair and worry when she'd be back.

"Greer?" she said as she snatched up the printed files and turned for the door.

"Meet me at the field office." Greer's voice was warped by the sound of commotion behind him. "We have a suspect. He's being prepped for interrogation now."

Sophie blinked. Surprise put a spring in her step as she cut a beeline to the front lobby of Barefoot T & C. "That was fast. I'm leaving Barefoot now. It was a dead end."

"No, it wasn't," Greer responded, a muted moment as he put the phone to his chest and shouted something Sophie couldn't hear. When he came back, he sounded breathless. "Barefoot had another contract at the festival. A side stage from earlier in the day. The performer is something of a drifter, but he has roots in the area."

"That lines up," Sophie admitted. "What's the name?"

"Ron Gavriel. Goes by a stage name. 'The Great Gavriel.'"

Chapter 16

"You have the wrong man," Sophie said for the third time since she entered the field office.

Through the dim room's one-way glass, Sophie could see her mentor, Ron Gavriel, lying over a metal interrogation table, his fingers rhythmically tapping the surface.

He looked older than she remembered. The bald spot at the back of his head had claimed several more inches and left the grey fringe looking desperately thin. His paunch had similarly expanded and now folded over his lap as he hunched in his seat. Black socks, black slacks, and a white button-down turned yellow with summer heat and neglect. The man seemed almost a caricature of a tired old magician.

Greer stepped beside Sophie, his arms folded. "He's a craftsman. We found a pinewood coffin in with his other props."

"He's a magician. It's an escape trick. I can show you the false back." Sophie shook her head vehemently. "It's not the same as what we've been finding."

"He has connections with Barefoot," Greer insisted. "That's a connection he shares with Monroe and Garnier."

"I've ruled out that connection. Barefoot doesn't have a link to the other victims," Sophie replied adamantly.

"Unless he killed them. Then he *is* the link," Greer said grimly. "We found ether in his kit. A partially used bottle."

Sophie had no reply to that. She couldn't imagine it. The man who had taken her in, had shown her the ropes and taught her those first crucial skills. Mentalism. Hypnotism. Stage presence . . .

"He's not a murderer," Sophie replied softly, turning to face Greer.

The tall agent looked down at her. "The evidence says he is."

But even as he said it, Greer looked away. A small twinge in his eye. The set of his jaw. Uncertainty? Doubt?

Sophie put a palm on the cold glass. "I used to work with him."

Greer started. She could feel his eyes on her, but he said nothing. Sophie reached into her bag, pulling free a charm bracelet. The haphazard collection of dissonant beads and clay miniatures clicked and tapped like a sprinkling of rain over glass. Each one a gift from Ron Gavriel. Each one a memory, a postcard or letter from whatever city Gavriel had set his top hat in that week. Sophie slipped the bracelet over her wrist.

"I used to work backstage. I was a teenager, but he saw something in me. Taught me the ropes. Made me believe I could get out of Peillon. He's not a murderer. He's seventy years old."

Greer shifted on his feet. Discomfort. Unease.

"He's got grey in his hair, but look at his forearms . . . He's strong enough. If ether is knocking the victims out, that limp of his is less of an issue too."

Anger flared in her chest, and Sophie looked his way, his blue eyes meeting hers. Each of them appraising the other in that moment.

"What about the Dream? We arrested a whole fishing club of suspects and haven't even talked with them."

Greer's eyes darkened then, shaking his head in disgust. "Hitchens cut them loose."

"What?" Sophie's breath caught. That couldn't be right. After all they went through. "Why? Why would he do that?" she demanded.

Greer threw up a palm, slapping his leg as he paced the room. "The festival singer went missing while they were all in custody."

"We don't even know for sure that the killer took her," Sophie shot back.

"He didn't care. Said it was that and the fact that the search of the clubhouse showed zero evidence of the Dream's involvement," said Greer.

"They found *nothing*?" Sophie couldn't believe it. After everything they'd seen, there was no way the clubhouse came back clean. "We found a military mine on their property. And there was 'nothing' in the clubhouse?"

"Oh, they're being charged over the illegal weapons. Posted bail, had some guns seized. But with regard to any missing persons cases? With regard to murder? Nothing to tie them to our case. Nothing to keep them in a cell."

"So what then? They're just walking free?" Sophie asked, biting the words off in frustration.

"They've been warned to stay out of the way. To not get involved in the case or leave town," Greer added. "I told Hitchens what they were saying. But the deputy director doesn't seem to think the Dream's threats of vigilante justice against this killer warrant any immediate action . . ."

This was insanity.

Sophie's mind was spiraling. Their prime suspects were walking free and her oldest friend was cuffed to a table in the next room. Tilting her head back, Sophie took several deep breaths. One thing at a time. First, she had to get Gavriel out of here. Next, she had to find the killer before the Dream found a scapegoat and put some other innocent man in the ground.

"I'm going to talk with him," Sophie said. It was not a question. Stepping to the door, she shot a look to Greer, but he wasn't looking her way. He was focused on the glass and the slumped form of Ron Gavriel.

A moment later, an officer closed the door of the interrogation room behind her.

The old magician looked up at the sound, squinting for a moment in confusion.

"Sophie?"

"Hi, Ron."

He stood and Sophie met him on his side of the table, wrapping her arms around him in a bear hug. At first the old man was too shocked to reciprocate, then he tried, only to realize the cuffs on his wrists prevented him from offering more than a warm pat to Sophie's forearm as she released him.

Smirking, he let out a mirthless chuckle, and Sophie offered him an apologetic grin in return.

"I was wondering if you were ok," Gavriel said, taking his seat as Sophie returned to her side of the table, grabbing her chair and bringing it around to sit on the same side of the table as Gavriel. "There's a postcard waiting for you in Detroit," he said.

"Where from?" Sophie asked.

"Barcelona, if you can believe it." Relaxing a bit, Sophie heard some of the Boston accent beginning to pop in the old man's words.

"Who paid for you to work Barcelona?" Sophie said incredulously. "Do you even speak Spanish?"

"A working vacation, Soph. I have a friend there, Martin," he said, exaggerating the long E as he said mar-teen. "He translates and astounds, and I get to play the befuddled American magician who can't keep up with his younger friend. It's a good gig, and the seafood—wow!" Gavriel rocked back in his seat, tossing his cuffed hands as best he could. Though when he leaned forward, his cracked eyes were heavy once again. "But I take it you didn't find me just to shoot the breeze."

Sophie nodded. "It's bad, Ron."

"I gathered," he muttered. "The beanstalk in the tan getup was talking a lot of hoopla about murders." Shaking his head, Gavriel rubbed his eyes. "Every time I come back to this river, it's trouble."

"Hey," Sophie softly jibed. "This is where you met me, remember?"

"Ha," Gavriel snorted. "Like I said, trouble."

He was just like she remembered. Older. Maybe a bit stiffer in the joints. But the kindness was there. The same grandfatherly affection. She'd once seen Gavriel chase a thief out of the backstage with a bit of loose pipe. When his hackles were up, he could swear a string that would make a soldier blush.

But he wasn't a murderer.

"Ron. I'm going to need some help." Sophie took the old man's hands, feeling the hard, strong callouses from a lifetime of work. The unique muscles that decades of sleight of hand had built. “This killer. He’s been taking women for the last year at least. I know you’ve passed through more than a couple times. We need to convince them you didn’t do this.”

Gavriel squeezed Sophie’s fingers, his jaw clicking as his mouth twisted in thought. “I don’t know what to tell you, Soph. I didn’t kill anyone. You know that. I can see it all over your face. When the riot broke out at the festival, I just clammed up. Locked my trailer, tried to make sure my props didn’t get stolen or burned.”

That wasn’t good. No alibi for Monroe’s disappearance meant Hitchens would be thrilled to keep Gavriel as a prime suspect.

“Tell me about this guy,” Ron said at last. “I can see it in your eyes. You’ve been at this a while now.”

“That obvious?” Sophie said.

“I remember a certain girl in overalls having the same expression on her face as she dropped the brass cups all over my rehearsal space. Probably the fiftieth time that day. She slammed the door on her way out and told me she was done with all of it.” Ron sat a bit straighter. “I also remember that she was back the next day. Picked them back up. And she tried again. You’re not a quitter, Sophie Quinn. You wouldn’t have survived my training if you were. Now, tell me about this killer.”

Sophie felt herself relax. Somehow, the old magician’s words made her feel at ease. Like she was being invited in from the cold and given a hot drink. Her breathing had slowed, like the gentle ticking of a clock or the tap of . . .

"Really, Ron?" Sophie gestured to where the man who taught her hypnotism was tapping a finger against the table, creating a steady, hypnotic rhythm.

"What?" Gavriel said innocently. "You were so tense. Now you feel better, right?"

Sophie put her head in her palm, laughing silently. It had been too long since she'd been in the presence of someone else with her skill set. She immediately replayed the old man's words and tone, hearing the hypnotic suggestion behind them.

"Fair play," Sophie said at last. Then taking the strength of that moment of calm, she told him what she knew.

"We think the killer is a man. The profile indicates someone who works with their hands. The coffins the victims were found in were handmade. Well-fitted with tight seals. He seems to weigh them down and let the women suffocate or, if the seals break, drown."

"The man who brought me in seemed to make a big deal out of some ether?" Gavriel supplied.

Sophie nodded. "We think the killer knocks the women out, probably with a rag soaked in ether. Then he can seal them in the coffins before they wake up and take them out into the Delta. He seems to pick drop-offs in locations where alligator fishing is common. We thought there might be a connection with a gang that has control over poaching in the area, but nothing concrete."

Ron was thoughtful for a moment, steepling his fingers as he looked off into the middle-distance.

"The coffins," he said at length. "Why make the coffins?"

Sophie leaned forward on her elbows. "What do you mean?"

"He's tossing them in the bayou. Why bother with coffins? Why not just wrap them in a rug. It'd be cheaper. Easier."

It was true. The coffins would not only take money but time. Effort. Why go to the trouble?

"Maybe it's symbolic for him?" Sophie said. "Serial killers often use their killings as a way to process psychological trauma. Reliving some moment that they can't get past or getting revenge in their own minds."

Gavriel gave a skeptical shrug. "I don't know anything about that. That's for you and your fancy degrees to figure out. But"—here Gavriel tapped a finger on the table, more direct and insistent this time—"if it means something to him, if it's an obsession . . . this won't be the only place it shows up."

Sophie raised an eyebrow. "What do you mean?" she asked.

"I mean . . ." Gavriel's hands folded the air in front of him, trying to get the shape of his words. "I mean, people's obsessions don't just stay in one compartment of their life. They spill over into everything. Work.

Leisure. They can't help it. Like you, Sophie." Here Sophie started, but Gavriel gave her sly wink. "You haven't been on the stage in a spell now. But you're a performer. It's in you. You love to sway the crowd. To give them the trick with your hypnotism and mentalism. To astound them with illusions. It comes out in everything you do. Even in all this." Gavriel waved vaguely around the room.

"And it will be the same with this killer," Gavriel continued. "The coffin is a part of who he is."

Chapter 17

The determined bleed for what they want. The obsessed make everyone else bleed.

Determination, Sophie thought. That's what she'd shown at the Barefoot offices . . . right?

Sophie hadn't realized she'd taken the lighter from Charles Randy's office until she found it in the pocket of her suit pants. As she turned it over in her palm, she idly examined the sleek gunmetal veneer. Flicking the flame up and dousing it in time with the ticking wall clock of the FBI field office.

Ticking away the seconds until the killer's next victim was pulled from the river.

Flick. She wouldn't have really set him on fire, Sophie told herself. She was sure she'd joked over the years about doing just that to the sleazy agent, but she wouldn't *really* have done it. Flexing the tendons in her hand, Sophie took a steadying breath. No time for what was done. They had to move forward.

Beside her, Greer rubbed his eyes. He had pulled a chair over for her, and they sat at his office desk, and as he set his cell phone down, he scratched at his salt-and-pepper hair. "They aren't willing to release him. Hitchens thinks Mr. Gavriel is a 'flight risk.'"

Sophie closed the lighter, resignation weighing down her voice. "So the only way forward is to find the real killer."

Greer let out a solemn sigh.

"Unless you think we already have him," Sophie said flatly.

She flicked the lighter. Once. Twice.

"No," Greer said at last. Sophie paused, turning to meet his gaze. "I'm sorry, Miss Quinn. I was sure. I was *so* sure." Pulling up his laptop, Greer shook his head and turned his attention to the screen. "I was listening in on your conversation in the interrogation room. And I agree with you. We have the wrong man."

Sophie followed the handsome agent's eyeline. He was pulling up the FBI databases, but there was something beyond that. His pupil flitted between the screen and the middle-distance. The faraway look of memory.

There was more to his decision. Given another moment, Sophie may have even asked him about it. Instead, he tapped the laptop screen and said, "I think Mr. Gavriel has a point. About the coffins not just being a part of the killings, but a part of our killer's life."

Sophie sidled up alongside Greer, looking over his shoulder as he furiously typed away at the database, flipping swiftly between a few separate databases.

"Carpenters?" Sophie asked, reading the occupation line in Greer's search.

"Exactly," he confirmed. "I remember what you said about ether being a solvent. Well, I saw it at Mr. Gavriel's trailer. He was using it to strip old pieces of wood. A woodworker or carpenter might have it on hand for the same reason."

Sophie snapped the lighter shut, stowing it in her pocket. "If someone out there is doing woodworking with recycled wood. They could have everything they need to build these coffins and no one would think it suspicious."

"Right."

Multiple databases lit up with dozens of results. Greer clicked between them, turning on result filters and refining the search. "We'll start local and go out from there," he added.

Sophie blinked at the rapid-fire computer work. She was certainly computer literate, but Greer was on a whole other level. He had mentioned his work as a Marine engineer extended into cyberspace, but it was another thing to *see* that same precise military efficiency on a keyboard.

With a final snap of a keystroke, Greer leaned back, making room for Sophie to see.

"Arturo Padilla," Sophie read out. As she did, she plugged the name into her phone, sifting the results until a woodworking shop appeared with a picture of a smiling, sun-toasted man holding a wood-planer. "He has a workshop in Duncan. 'Eco-friendly. Patio furniture. Cabi netry.'"

Greer stood, pulling his tan suit jacket off his chair as he did. "Duncan. That's not far from here. A small town. Out of the way."

"But it's on the bayou, I know it," Sophie confirmed. She was about to pocket her phone when she paused, stopping in her tracks.

"What is it?" Greer asked, his brow furrowing.

Sophie turned her phone around, one finger lingering on the phrase "biodegradable coffins and caskets."

* * *

Sophie let her car door hang open for a moment. She didn't want the noise to alert Arturo of their presence. The heat of the day had settled in, the afternoon sun was arcing to the west, and a shimmer of humid heat warped the air around the open workshop.

The building appeared to be half home and half workshop with a detached garage and old farm-style house separated by a gravel path that merged into the driveway. Neighbors were spread out, and Sophie could only barely see other similar houses through the summer oaks and thick brush.

"Do you think he's here?" Greer asked, stepping around the white sedan to join Sophie.

The garage door was open, but no cars were present. As they approached, Sophie saw a variety of saws, drills, hand tools, and electric power tools, all neatly organized on hooks and shelves around the space. Two sawhorses held pieces of some light wood in place with a set of clamps. The smells of gasoline and oil, varnish and paint stripper all mingled.

Greer let out a quick whistle, and Sophie followed his pointing finger to a can of paint thinner on a nearby shelf. That wasn't incriminating in itself, but the reminder brought the memory of ether vapor to her nostrils, and Sophie instinctively wrinkled her nose.

A loud wooden clatter snapped their attention to the back of the garage. A white door opened, revealing a clear shot of the tall grass and swamp trees behind the house. A scrawny man with black hair and a mousy mustache gave them a startled look.

"Hello," he said softly. "I'm sorry, but the workshop is appointment only. You have to call first." As he spoke, the man leaned forward and back, looking around Sophie and Greer to their car and over his shoulder to the tree-choked bayou at his back, revealing a tangerine-sized spot of balding hair at the back of his head.

"Arturo Padilla?" Sophie asked genially. It was only a courtesy; the man matched the picture she found online as well as the image in the FBI database. "We have a few questions for you. Is it alright if we sit down and talk for a bit?"

He shook his head in agitation, distracted and in a way that told her he wasn't answering her question. "You have to call ahead, you can't just show up."

Greer held up his ID, flipping it open as he stepped forward. "I'm afraid we're not really asking, sir. FBI. We're here regarding the fates of several young women in the area. Now we can talk here or we can talk down at the station."

Arturo suddenly reached for his back pocket. Greer and Sophie both tensed, reflexively pulling their guns.

"Hands where I can see them," Greer snapped.

Arturo's eyes bugged as both hands shot up. One was holding a sweat-stained handkerchief.

"You're not supposed to be here. You have to call ahead," he whined, angling his hand and head awkwardly to dab the sweat from his forehead while keeping his hand above his head.

Sophie lowered her gun, her heart racing. Something wasn't right here. Was he tweaking on something? The man's behavior was odd and his movements jerked with involuntary cringes.

At that moment, their conversation was interrupted by the sound of tires on the gravel driveway behind them, and Sophie turned as Greer stepped back to keep one eye on Arturo, casting a quick glance at the SUV that had pulled up.

Four men piled out of the car, and Sophie's blood ran cold.

"Well, well, well. Looks like y'all had the same idea."

The voice came from a pot-bellied man sporting a black eye. William Landry, the leader of the Dream. In one hand he held a broad-headed hatchet, idly gesturing to Arturo as he spoke.

Car doors slammed, and Sophie shot a look to Greer. He had turned to face the Dream as well, tensely assessing the group as the lanky younger brother, Miles Landry, stepped to William's side, spinning a heavy wooden baseball bat in slow loops. The man with the shaved head from the previous night was with them, a heavy chain wrapped around one wrist, the links clinking as they hung down to his knee. The last to join was a sallow-faced man who had sided with William in the standoff at the clubhouse. He didn't have an apparent weapon but ominously carried a coil of rope on one shoulder.

"I reckon that we have things from here," William called out. "Why don't y'all take an early supper. Come back in a couple hours. I'm sure you'll find everything you need for your case report. Wouldn't want to complicate the situation any more, would we?"

With a commanding shout that made Sophie jump, Greer barked back, "Get in your car and drive on, Landry."

The rotund fisherman gave a dark look to the tall FBI agent. "Can't do that. Not after what he done to those women."

"Innocent until proven guilty," Sophie shot back. "A jury will decide his fate. Not you."

At this the man with the shaved head snapped his chain. "Step aside. It's time for some *real* justice."

Sophie thumbed the safety on her pistol, looking to Greer for a cue. But before either could act, the door at the back of the garage slammed the wall.

Arturo was making a run for it.

Chapter 18

The Freedom's Dream Fishing Club wasted no time, darting forward the moment Sophie glanced at the door.

A gunshot tore the air, and the charging men stopped short, the sallow-faced man with the rope even diving to the ground for cover. Greer hadn't looked away, instead firing a warning shot into the gravel.

With the painful blast still ringing in her ears, he shoved Sophie's shoulder, shouting, "After him! I've got these clowns."

Sophie hesitated. Gun or no, Greer was still only one man against four. But as he gave another shout of "go!" Sophie dashed through the back door.

Tall grass and a few scattered holly bushes filled out the unkept land between the oaks and the bayou's slow-moving bed. Sophie ran ahead, following the swaying brush as Arturo's lanky body disappeared into the trees. It didn't take long for trunks and limbs to close off the space around her, leaving her feeling like a child in a department store, barely able to pick out the evasive figure running between the chaotic rows.

Another gunshot rang out behind her, and Sophie froze, chancing a glance back to the workshop as fear gripped her.

The skinhead with the chain came around the edge of the building, dragging Miles Landry by the elbow. There was no sign of Agent Greer or the other two members of the Dream, and a cold stone of dread fell in Sophie's stomach.

"There she is!" Miles pointed with his baseball bat just as Sophie turned, barreling through the underbrush, the desire to catch Arturo now second to the need to get away from her pursuers.

Sophie had been naturally athletic as a child. The need to keep up a youthful, vigorous appearance for the stage and the harsh training she put herself through preparing to take the FBI field exams meant that her body was ready when the adrenaline started pumping. She was used to fight or flight by now. But the swampy ground was unforgiving. Burrs and snaring thorns on branches were leaving tears in her suit pants, and her legs were already burning with the effort.

She had lost sight of Arturo Padilla, but by a broken branch here and a scuff of bark there, she knew she was still on his tail. Just as her pursuers were still on hers, though with considerably more cursing and shouting.

Arturo was leading them parallel to the river. A smaller tributary that Sophie recognized from the maps Deputy Director Hitchens had tacked up during their initial meeting. These were poaching grounds. Waters connected to where the victims had been left. Would he have a boat stashed nearby? If he took to the river, would they have any way to catch him?

Sophie stopped. Looking around the brush, she couldn't find a sign of which direction Arturo had gone. She scrutinized the bark and looked for trampled brush. He had been following the river; had he turned off? Headed inland? There were small creeks branching off the sluggish river. Maybe he'd turned off to follow one of them? But no, something about that didn't feel right.

Suddenly remembering she was also being chased, Sophie glanced over her shoulder. There was no sign of the pair. Their stumbling and cursing was markedly absent. Not even the usual bayou insects were humming around her. Only the soft crack and rustle of branches overhead. She felt that should be a relief. Instead, her pulse only quickened as chill unease made the hair on her neck stand on end.

Licking her lips, Sophie crept in a wide arc, bowing inland then back toward the river. Her brow furrowed in concentration, she looked for any sign of Arturo. He couldn't have escaped. She was right behind him . . . where did he—

Suddenly, a small white flutter caught Sophie's eye. She hustled toward it, bending low to see what it was.

Arturo's handkerchief. The one he'd used in the garage. If he cut this direction . . . Sophie closed her eyes, drawing a map in her mind.

Huckston Bridge.

The intersection of a main road with a small pier the locals used for fishing. If he reached it, Arturo would be able to escape up the river.

A branch cracked behind her.

Sophie dove to the side just as a chain viciously whipped through the space she'd been standing, tearing away leaves and branches. She drew her pistol, rolling on her hip to come up to her knees. Her eyes barely had time to register the outline of the brutishly muscled skinhead as he brought the chain down on her arm.

Sophie screamed in pain, her fingers going numb as the pistol fell from her grip.

"Never shoulda come here," the man growled. "Should of minded your own goddammed business."

The chain lashed out for her again, and Sophie threw herself backward, falling on her back in the muck and obscuring sedge near the river's edge. She grabbed her forearm where she'd been hit, unsure if the bone had chipped or broken, her heels slipping in the wet ground as she tried to regain her feet. Fallen logs and detritus scratched at her back.

"Now you're gonna get what's coming to you." The skinhead raised the chain, whirling it overhead so fast the links began to whistle.

Sophie's mind raced. Maybe she could dive to the side, take cover behind this log. This—log?

The man took a lunging step forward, bringing down the chain like an executioner's axe. Sophie tossed herself to the side, rolling bodily over the rough log just as the chain came down. The metal slapped the wood with a meaty *thunk*.

And then it moved.

Because it wasn't a log. It was the tail of an eleven-foot Louisiana alligator.

A long dinosaur face full of serrated teeth turned to face the attacker, the maw open and hissing a warning that needed no words to make itself clear.

The skinhead with the chain yelped, dancing back and holding up both palms as if the gesture might help the ancient reptile see reason.

Sophie pressed against a tree, forcing herself to her feet with the one arm still obeying her, the furious gator left between her and her attacker as she darted into the trees.

Huckston Bridge was only a quarter mile at the farthest, and as she ran, Sophie felt sensation returning to her chain-whipped arm. She could feel the bruise already setting in, but her fingers could move. It wasn't as bad as she initially thought, though she grimaced as she remembered dropping her pistol and hoped it had sunk into the muck and out of the skinhead's sight.

Concrete. Through the trees, Sophie caught the wet gleam of a concrete and steel bridge over the river. She put on a burst of speed, imagining their suspect pulling away on a river skiff. Gritting her teeth, Sophie burst into the open space beside the bridge.

She was level with the small pier reaching a long wooden finger into the river's current. A steep upward slope separated her from the road fifteen feet above, and while she could see the railing of the bridge overhead and the concrete supports reaching down to where she stood, if any cars passed, she would be unable to see them. She took all this in within a second before her focus returned to the more pressing issue.

Arturo was there.

He hadn't approached the dock; instead, he sat with his back against the concrete of the bridge supports. His arms were wrapped around his shins as he pressed his head to his knees. He was shaking and rocking.

Sophie approached carefully. She didn't have her gun, but from the look of him, she wouldn't need it.

Her shoe crunched on gravel, and Arturo started, looking up at her in wide-eyed panic. His eyes were wet. Had he been crying? Sophie showed her palms. She didn't smile. A smile in this circumstance would only come across as suspicious. Instead, she gave a sympathetic look, trying to reflect the sadness she could see in his expression.

With a crash, a figure came out of the tree line, covered in wet leaves and grass stains.

"Greer." Sophie waved to her partner, relief sending a ripple through her as some of the tension fled her body. "I was worried for a second there." Then the anxiety of the moment returning to her, Sophie added, "Two of them came after me. What happened to William and the other?"

The lean former Marine wiped a stray leaf from his pant leg as he came up alongside Sophie. "I'll tell you later. Right now, they're after us. We have to get out of here now, and—"

Arturo scrambled to his feet.

"Easy, friend," Greer said, raising his pistol.

The woodworker froze. He looked between the pair in frightened confusion. When he spoke, his voice was high and tinny. "I don't understand what's happening," he said.

"Allow me to explain," a voice called out, snatching back Sophie's attention. William Landry, his jowls vibrating with barely holstered rage, waved his gleaming hatchet as he emerged from the brush.

The tree line rustled once more. This time, the Dream did not stop as Greer's pistol swung around to them. The four men spread out, quietly encircling the trio cornered against the bridge's foundation as Landry called out, spittle flying from his lips, "You are a murderer! You come here, attack and kill our women! You are the lowest of the low. You're a cockroach! A rabid dog that needs to be put down!"

"Enough, Landry," Greer snapped back. But his focus was split. The skinhead with the chain and Landry's brother Miles moved to surround him, making it impossible to keep a line of fire on both. In a whisper, he added, "Anytime you want to draw that gun of yours, Miss Quinn."

"I lost it in the chase," she replied with her own nearly telepathic whisper. "Chain over there ambushed me."

Greer grimaced but gave a nod that he understood.

Arturo stayed where he was, looking for anywhere to escape, but as the four men closed the half-circle around them, even the carpenter seemed to realize the only way out would be through the Dream's gauntlet of menacing weapons.

Squinting in the afternoon sun, Landry spat at the ground. "And you. Traitor." He glared at Greer. "For what you done, you deserve to hang right with him. I've a mind to let you walk on account of the lady." He gave an offhanded wave toward Sophie. "But if you insist. We brought enough rope for three."

With a jab of his fingers, Landry motioned to the slope. The sallow-faced man with the rope hustled up and out of sight.

A long silence stretched between them. Greer gave an appraising look to the wincing suspect, then to Sophie. She could see the struggle behind his eyes.

"You'll let us both walk. Me and her?" the tall agent said at last.

Sophie shot Greer an anxious look. No, she mouthed. But he wasn't looking her way anymore.

"For today," Landry replied languorously. "You and I have an account to settle. But we'll resolve it like civilized men." Unconsciously, Landry's black eye twitched. Sophie could imagine what he was thinking at that moment, and it sent a shudder through her.

"But today we can leave? You swear it?" Greer said. As he spoke, Miles took another step forward, and Greer spun to face him, pulling up the gun and forcing him to freeze. As he did, the chain-wielding skinhead took another step from the other side, moving in turns like a game of "red light, green light."

"I swear," Landry said, rolling his eyes. "But we have business with that one. He stays." The hatchet leveled on Arturo, shaking where he stood, the afternoon sun reflecting off the oiled metal.

A pregnant pause in the conversation swelled, broken only by the shuffling of feet on gravel.

Landry cocked an eyebrow, looking up to the railing of the bridge overhead. "Leroy, what's taking you? Just tie the rope and throw it over."

Silence.

"Leroy? Answer me when I'm talking to you, goddammit!"

A silhouetted figure appeared on the bridge. Then another. And another.

Sophie looked up in disbelief as a half dozen police officers spread out to flank a thin, older woman in dark sunglasses on the bridge, guns pointing down on the group below as more officers made for the slope down to the pier. Shouts of "put down the weapons!" and "hands where I can see them!" rang out, and in dumbfounded shock, Sophie watched as the three members of the Dream still beside them were cuffed for the second time in twenty-four hours.

Lidya Merchant leaned forward on the railing and called out, "Agent Greer, if you and your partner can avoid calling in the cavalry *every* day of this case, I would appreciate it."

When Sophie faced her partner again, he was grinning ear to ear. Running a hand through his hair, he shrugged and explained. "I wasn't sure I'd catch up, so I told her to home in on your phone."

Sophie nodded, giving back a grin of her own. But it fell as her gaze returned to the man they had chased here. Curled in on himself, his face pressed to the concrete, Arturo Padilla. A man they'd risked their own lives to save from a gang of murderers. Not because he was a good man, but because he might be a murderer himself.

Chapter 19

The weakness was on him again.

Jack closed his eyes, feeling the setting sun toasting his face as he let the dull current guide his skiff along the sleepy bayou. Crawdad song and bullfrog belches gave weight to the mossy, wet earth smells around him. And he imagined what it might be like to sink into that cool dark under the boat.

His lungs filling. He might even open his veins first. Blood replaced with dark river water. Soon every bit of him would disappear, consumed by the swampland. He wouldn't feel the pain anymore.

He'd prove them all right. Once and for all, show what a weak man he was.

But it would be over.

His hand coiled on the pilot wheel, the harsh callouses scraping the worn leather. He was getting stronger. Wasn't he? He had said things he had only dreamed to say before. Taken control. All he had to do was stay the course. The pain would disappear if only he could grow stronger. Tough enough to endure it.

The natural noise around him lulled for a moment, and Jack heard a still, small sound behind him.

A whimper?

Looking to the black tarp, he imagined the coffin beneath it. His craft had gotten better over the last year. They were stronger, more solid than those earliest attempts. Corpses didn't fight to get out of their final beds, but inevitably the women did, so he ensured each coffin, despite their rough appearance, was sealed like a drum.

No air in or out. No moisture.

Sacred. Like a grave should be.

The whimpering continued, a little louder now. Now Jack could hear there was a rhythm to it. Pulling the anchor, he let the skiff guide itself to a halt. It was a pain finding a new place for her. Cops were crawling all over the river, even crannies he would have sworn only he and men like him knew about. But he was patient. There was the sound again . . . He stepped closer, leaning down to listen.

She was singing . . . praying?

Certainly not the singer-songwriter ballads or folk reels he had been bombarded with at the festival when he took her. But a song nonetheless.

His eye twitched, confusion and anger suddenly welling as his throat clinched. Faintly, Jack heard the young woman in the coffin as her tremulous voice carried on softly about "peace," "endurance," and a refrain weakly insisting all was well with her soul.

Fear. He could taste the fear rising on those words. But she didn't scream. Didn't fight. She was going to her death with whatever peace she could rally.

Even she's stronger than you are, you pathetic worm, the voice in him said. Jack felt a sensation like a fist crushing down on his heart, and a moan escaped his lips as he felt his eyes begin to water. Then he clenched his teeth, huffing as the veins on his arms and neck popped with the sudden tension. He brought a fist down on the tarp, slamming the coffin beneath.

The girl gave a muffled shriek, and Jack could almost see the way her hands had flown to her mouth to cover the escaping terror.

"You think that'll save you?" Jack shouted, both hands clenching the tarp as if he might rip it off at any moment. "You can't whitewash all the pain you've caused by playing the saint now. You're not a martyr, you whore. I didn't pull you from a soup kitchen or a church. You're vain! You shit all over the people under you, lying, manipulating to get what you want. You wanted their eyes on you, you skank! Just like that woman on your shows. And now that it's caught up with you, now that someone finally called you on your lies, you try to play the victim. Well, you're not the victim, Annie! I am!"

For a moment, the only sound was his heated breathing as Jack stretched himself over the coffin. Inside, a sobbing voice said meekly, "I'm so sorry for whoever hurt you. If I—"

"No!" Jack smashed a fist against the wood, cutting open his knuckles. "I don't have to listen to your lies anymore."

He ripped the tarp off the coffin, revealing the metal slats holding it above the skiff's sidewalls as he tore at the straps. "I'm stronger than I was back then," he growled, his hands shaking. "Stronger than you. Stronger than your lies."

The singer screamed, high and sharp, as the coffin splashed into the bayou. Propping his foot on the skiff's wall, Jack watched as the weights dragged the buoyant box beneath the murk.

But there was no peace.

Snarling, Jack was already wondering if Sophie Quinn would scream or sing.

Chapter 20

For the second time that day, Sophie sat in an interrogation room at the Louisiana field office. In her mind's eye, she saw the ghost of Ron Gavriel across the table, and her fingers idly plucked at a starfish charm from the eclectic bracelet on her wrist.

Greer had just finished setting the recording and was settling into a chair beside her. With a meaningful slap, he dropped the manila case file on the table beside him.

Across from them, Arturo Padilla gingerly tugged at the cuffs on his wrists, shifting and pulling to try and find a comfortable position. In the five minutes since he'd been brought in, he had yet to find one.

"Do you know why we've brought you in?" Greer started.

Arturo didn't reply, still fidgeting with his cuffs.

"I don't like these. Please take them off."

"I'm afraid the cuffs have to stay," Greer replied coolly. "You already tried to run from us once today."

"I promise, I won't run. Please take them off, please." As he spoke, Arturo continued to squirm. Now his feet were moving too, and the table leg let out a small squeak as he gave the cuffs a tug.

Sophie could see a small flush creeping into his cheeks. There was no sense of concern over Greer or her. In fact, everything in Arturo's posture indicated he thought very little of their presence beyond their ability to remove the offending metal from his wrists.

They had rushed to his workshop so quickly, Sophie and Greer had very little opportunity to dig beyond the bare essentials of his file. But serial killers often showed psychopathic tendencies. Psychopathy could manifest in antisocial behavior, thinking of others as objects and not whole people with lives of their own.

Pausing, Sophie forced herself to slow down. She was letting her thoughts carry her away. Right now, all she *really* knew was that something felt off. She needed to observe, to listen before casting any judgments.

"Please," Arturo repeated, enunciating the word as if it were a password, bringing his face down and rubbing his thin, mousy mustache across the back of his hand.

"Why don't we focus on why you're here," Greer said. "Your name came up when we—"

"I really need these cuffs off," Arturo Padilla repeated, speaking over Greer as if the noise coming from him was no more significant than the rattle of the air conditioner kicking on.

The tall agent frowned, his eyebrow beetling as he leaned forward. "Mr. Padilla, I don't think you realize . . ."

Sophie didn't hear the rest of Greer's warning. Arturo was flushed, going still everywhere except his chest as his breathing picked up. He was moments from hyperventilating. Which meant either he was a masterful manipulator or his distress was genuine. Guilt? *No*, Sophie thought. Something else. A panic attack?

Sophie stood. Holding out a hand to Greer, she smiled to Arturo and said, "I think we can go without the cuffs for now. You're not going to try anything foolish, right?"

Arturo nodded absently.

Out of the corner of her eye, Sophie saw Greer tense, his frown deepening as he raised an eyebrow. She didn't explain herself. By disagreeing with Greer she'd already surrendered some of their authority in the conversation. If he continued to argue with her, their credibility might completely collapse. It was a bit of a gamble, but if things continued as they were going, their suspect might be too distraught to be of *any* use.

Finally, Greer passed over the keys with a sigh.

"Thank you," Sophie said cheerfully, unlocking the cuffs and taking them back with her to the other side of the table.

Arturo let out a visible shudder of relief, slipping back in his chair and rubbing his wrists.

"Now, Mr. Padilla," Sophie prompted. "Do you know why we're here?"

"No," the suspect replied curtly.

"I think you do," Greer said insistently. "Several women have disappeared over the last year. They've been turning up in wooden coffins. You make wooden coffins, don't you, Mr. Padilla?"

"I didn't kill anybody."

He said the words flatly. Bluntly. Sophie tilted her head, her brow pinched in focus. There was nothing evasive or anxious about the words. It was as if the man was infinitely more concerned with his wrists being free than he was of being accused of serial murder.

"But you have hurt people," Greer continued. "You hurt a woman. Jessica Minor. You have an assault conviction on record. Got off kind of light, though, didn't you?"

"I hurt her," Arturo said, his voice again completely guileless. "I didn't mean to. She was too close. I couldn't breathe. It's over. Elena said it was over."

"Who's Elena?" Sophie prompted.

"My sister. She lives with me. At the house. I'm always busy. I have trouble focusing, on the right things, sometimes. She helps me." The wiry man rubbed his mustache again. "She can explain things better than me."

Sophie looked down to Arturo's file on her phone, confirming what he said. Elena Padilla. His sister, about ten years older than him. Same address. She tapped her phone thoughtfully. Most serial killers were loners, isolated. They needed space to do what they did. There were exceptions, of course, but if Arturo was alone, he would fit the profile perfectly. If he had a roommate, a family member in particular . . . It was another piece that didn't quite fit.

"Does the name Addison Garnier mean anything to you?" Sophie watched closely for any sign of reflexive recognition, but Arturo only shook his head. "What about Evie Shipman?" Again, no. Sophie went through the list of identified bodies. But the man across from her only stared at his hands, nonchalantly gesturing that he knew none of them.

"I think you're lying to us," Greer said, tapping the manila case file with an accusing finger.

There was no reply to that.

"So, you're going to tell me you don't know anything about *any* of the women?" Greer prodded, snatching up the file. "Addison Garnier." Greer slapped a crime scene photo, the pale body of the most recent victim photographed in her coffin. Arturo blinked, looking at the photo impassively. "Evie Shipman." Another photo, this body in much worse condition from weeks or months in the swamp. "Jewel French." Another picture.

Arturo looked away, his eyes blinking a bit too rapidly. "I didn't kill anyone," he said again, confusion in his voice, as if Greer simply hadn't heard him the first time.

But the tall agent leaned forward. "Listen, you can make this a lot easier for yourself if you tell us what you did. Because if you want to keep playing dumb, I can just test the wood in your workshop against the wood in these coffins. What do you think? Will we get a match?"

At that moment, the door behind them slammed open. Greer stood as Sophie whirled in her chair. A black-haired woman in a T-shirt and worn blue jeans gave them both a glare. "What the hell do you think you're doing?" she demanded.

Behind her an officer made a grab for her shoulder, protesting, "Ma'am, you can't go in there!"

She narrowly avoided his fingers, ducking forward and hurrying to Arturo's side.

Suddenly the room was crowded, as the officer came in after her. Arturo cringed, recoiling, panic flaring in his eyes. The table shifted as the officer chased the woman, his hip banging the edge and drawing a cry of fear from Arturo as he quickly grabbed the woman, pulling her close, like a child reaching for his mother in a thunderstorm.

Sophie held up a hand, stepping between the pursuing officer and the new arrival with a quick, commanding "It's alright."

With a reluctant scowl the officer stopped. He looked between Sophie and the dark-haired woman, but with a dismissive wave from Sophie and a nod from Greer, he retreated to the hall.

Sophie turned her attention to the woman. "You must be Elena."

"Yes. And what do you think you're doing, questioning my brother without me?" Her eyes bored into Sophie, venomous and dark.

"Ma'am," Greer said, holding up a placating palm. "Your brother is a suspect in a murder investigation."

Following Greer's glance, Elena noticed the assembled pictures on the table. Her furious expression was suddenly tempered with a queasy look, and she swallowed as she defiantly glared back to Greer, putting a reassuring arm around her brother as he shuddered in her grip. "All the more reason *I* should be here. My brother didn't do anything, and I'm not going to let you frame him. Anything you need to ask him, you can ask with me present. That's his legal *right*!"

His right?

All at once, the pieces came together. His odd behavior. His clinging to his sister.

"You're his caregiver," Sophie said suddenly.

Greer started, looking between the siblings then back to Arturo's file.

Elena nodded. "My brother is unwell. He had a head injury years ago. Since then, he doesn't understand how things work sometimes. He needs reminding. On his own, he forgets to eat. We were afraid he would starve himself without help." Shaking her head, Elena rubbed Arturo's shoulders reassuringly. "He can't go to the bank or to the grocery store. He panics in crowded places, so I handle most of his life except for his work. Sometimes woodworking is the only thing his mind can handle, but he's a genius at it."

Sophie and Greer looked to each other, and he nodded to the file on his phone, silently confirming what Arturo's older sister said.

"Be that as it may," Sophie said cautiously, "we have to follow the leads we have. These women have been found on the riverways connected to the bayou running behind your home. They've been found in coffins that look like Arturo's work. Carly Monroe was taken from the festival in New Orleans just last night, and testimony indicates it's the work of the same killer."

"Last night?" Elena asked, her mouth working in animated frustration, as she drew a deep, angry breath. "Well, then that guarantees Arturo is not your man."

"How's that?" Greer asked.

But Sophie's mind was already catching up with what Elena had told them. "He wouldn't go to the festival. It's a crowd, and even small groups make him panic." Sophie thought again about Arturo's initial insistence that they needed to make an appointment, and the Dream showing up in front of Arturo's garage. Four of them plus her and Greer. That was six strangers crowding around him unexpectantly. No wonder he ran.

Elena nodded. "I was with him. We watched a movie at home. I streamed it with some friends, they'll verify he was with us watching the movie . . . It's the only way Arturo can really 'hang out' with people."

And that was that.

Sophie and Greer took down the names; they asked a few more perfunctory questions to flesh out the siblings' story. They would have to follow up with the alibi. When it came to a string of murders spanning a year, having an alibi for one *potential* victim wouldn't be the end of the investigation . . . but it weakened the case against him.

Leaving the interrogation room, Greer grimly tapped the case file in his palm. "Well, what do you think, Miss Quinn?"

Sophie rubbed at her eyes. "Honestly?"

"Honestly."

"I don't think he's our killer." Sophie shrugged. Sweat was chilling on her skin as the heat of the day waned and was overtaken by the field office's chilly interior. "He could still be, but after what we've seen I don't think we'll find a solid connection."

Greer leaned against the hallway wall, considering her words. He seemed on the point of speaking when his phone began to ring.

"Deputy Director," Greer answered.

Hitchens. Sophie stiffened. No doubt calling to ask why they hadn't pinned this whole situation on Ron Gavriel yet . . . Sophie wondered if he had found out her connection to the old magician. He wouldn't need much of an excuse to kick her from the case after the way she'd embarrassed him in the briefing.

Greer looked to Sophie. "An update?"

Letting out a resigned sigh, Sophie braced herself.

Chapter 21

Would they find Carly Monroe's body tonight? Tomorrow morning? A year from now? Sophie turned the bracelet of beads sent to her by her mentor, thumbing it like a rosary as she stared into the setting sun.

She had claimed a bench in the small green space behind the field office where she could settle her thoughts. Hitchens had spent the better part of ten minutes on the phone tearing her and Greer to pieces. She'd only missed the first thirty seconds or so before the deputy director had demanded Greer put him on speakerphone.

"You have one job. Find the bastard responsible for these deaths and lock him up. I don't care if he's an arthritic magician, a carpenter, or your own meemaw in drag. Every second he goes uncaught makes us look like idiots. Every *day* of this investigation costs the bureau hundreds of thousands of dollars. And that's *before* we count your reckless waste of department resources arresting a goddamned *fishing club*!" Hitchens had latched onto the money argument for at least three of those ten minutes and in colorful language expressed how he might save some of that money by shipping Sophie back to Detroit and Greer with her for good measure.

Sophie took deep, meditative breaths, bringing her heart rate down. Hitchens was a distraction. They had a killer to catch and playing into his game would only waste mental energy she needed to find their next step. And there were no two ways about it, they were lost.

The crinkle of a paper bag and the hot grease smell of fast food brought Sophie out of her thoughts as Greer sat on the bench beside her. He silently passed her a chicken sandwich, pulling out a burger for himself.

He looked exhausted.

The setting sun exaggerated the dark lines under his eyes, and the tan-suited FBI agent's ordinarily orderly hair was falling around his eyes and ears in rowdy indifference. Nevertheless, he looked distinguished, and Sophie found herself staring at the strong cut of his jaw as he thoughtfully chewed.

"You didn't get much sleep last night, did you?" Sophie said sympathetically.

"I'll be alright," Greer murmured. "I'm sorry about Hitchens."

Sophie waved away the comment. "He's all bluster. I've worked with men like him before. We could have handed him the killer ten minutes after the briefing and he'd be demanding to know why it wasn't five minutes."

Greer gave a small chuckle at that, and the sound pulled a faint smile to Sophie's lips. Shaking his head, Greer turned and met her eyes, his pale blue irises a deeper hue in the twilight. "I'm sorry for how I behaved. Earlier." The day had been so long that for a moment Sophie only tilted her head in confusion until Greer elaborated. "I should have been with you to investigate the Barefoot offices. If I had, Mr. Gavriel wouldn't be in a cell."

Sophie nodded, pursing her lips. "That might be true," she said, keeping her voice steady and relaxed. "But the way it worked out means that we were there when Mr. Padilla was attacked by the Dream. I was rushing a bit because I thought you'd be there any minute. If we'd gone as carefully as I originally wanted to, we may not have reached Arturo in time."

"We did go there to accuse him of murder," Greer pointed out.

Sophie clicked her tongue, tilting her head in agreement. "Maybe. But it worked out in the end. He's alive, and the Dream are behind bars, to stay this time."

"Still." Greer leaned forward, his forearms on his knees. "It wasn't a tactical decision. I was . . . angry." Letting out an explosive sigh, Greer set aside what was left of his burger, giving a small, nearly silent clap of his hands. "As soon as we had that connection, between the missing singer and the riot and the killings . . . All I could think was how this murderer endangered my daughter. He murdered more than half a dozen other young women, but all I could think was how my little girl could have died that night. I wanted to get him, as soon as possible, and I wasn't going to let anyone slow me down or get in the way."

She could hear the feeling was still in his voice. Controlled but dangerous, like a penned bull.

"I've done it before," Greer said. "Rushed. Let my anger at the injustice in front of me drive me when I should have been thinking. I put an innocent man in prison." The confession came out in a pained staccato as Greer stared into the setting sun. "I couldn't live with myself. I put myself, my partners, the whole department on the chopping block with internal affairs just to get the evidence to exonerate him. Merchant is the only reason I still have a job, and for that, she's right along with me on Hitchens's shit list." Greer tilted his head toward Sophie. "Just like you."

The determined bleed for what they want, the obsessed make everyone else bleed. Sophie's mantra returned to mind, her own self-doubts now fully displayed in the man in front of her.

She reached out a hand, her palm hovering over her partner's shoulder, her fingers hesitantly curling as if she might pull back the gesture at any moment. Comfort. Why was something so basically human so difficult for her?

Because sometimes receiving comfort feels like giving up. It was a lie but one Sophie knew too well from experience.

Blinking away her uncertainty, Sophie laid her hand to Greer's shoulder. At first he stiffened, then she felt him release a long breath, and by degrees, the muscle under her hand unwound.

"Now we're even," Sophie said, a wry, knowing smirk spreading on her lips. "After the last case, I mean." It had only been a little more than a week since she'd driven off with Greer's car to follow a lead after a similar argument, that time with her on the shouting end.

Greer turned to Sophie, his face hopeful, though still heavy with regret. "All the same, I apologize. I know we have other things to worry about right now. I just want you to know, we're on the same team here. We're going to figure this thing out, and you don't have to worry about me going rogue on you again."

The earnestness in the older agent's expression was heartbreaking. Sophie clapped his shoulder, giving the big man a small shake and a smile before returning her hands to her sandwich.

"Alright, enough drama," Sophie said, the light enthusiasm in her voice driving away the heavier conversation. "Now that you've secured your Oscar nomination—"

"I meant what I said, Miss Quinn."

"I know you did," Sophie said, giving him a soft, knowing look. It vanished almost immediately behind a cheerful mask. "That's what made your performance so captivating. But we have a killer to catch. So, what do we have?"

Greer leaned back on the bench and rolled some of the stiffness from his neck. "Coffins. Handmade and pinewood."

Sophie bobbed her head. "Victims all young women. All dressed up as if going out. All found in waters known for alligator fishing and poaching."

"A smell of ether on the latest victim," Greer added. "And rumors that the killer was at the festival prior to the riot and is responsible for Carly Monroe's disappearance."

Sophie paused mid-bite. A rumor? No, a bit more than a rumor.

Jabbing forward with her sandwich, Sophie quickly added, "Camille." When Greer only looked at her quizzically, Sophie continued. "Kinsley's friend Camille was working the festival. She claims to have seen a similar coffin backstage right before Monroe's disappearance. That's a bit more than a rumor or speculation."

Shrugging, Greer replied, "But what do we do with that? Assuming it was the killer's and assuming he used it to kidnap Miss Monroe. Where do we go from here?" Scratching at the silver shadow beginning to show on his chin, he added, "I suppose we could start interviewing the festival crew. See who else might have seen something."

That would take forever. Sophie tapped the final bite of her sandwich against her palm. Who else would have seen something last night?

Suddenly she sat up. "Kinsley," she said.

"What? What about Kinsley?" Greer asked, his eyes intent on her.

Sophie was abruptly aware that she had come up against one of Greer's personal boundaries. She remembered how protective he'd been on that first car ride, in the kitchen of his home . . . She'd have to tread carefully.

"Look. I know you have your lines. You don't like bringing her into this part of what you do." Sophie couched her words carefully; she could already see Greer's broad shoulders tensing again. "But I think she saw something. When Camille was talking last night, Kinsley looked like she had something to add. She may not consciously remember it, she may have spent all day today subconsciously burying it."

"You want to interrogate my daughter," Greer said bluntly.

Sophie winced. "I'm not sure how much we'll get just asking her questions. It was a chaotic night." She let the words hang meaningfully, but Greer didn't respond immediately, instead waiting patiently, meeting her eyes. "I mean, I'll start by asking her, of course," Sophie continued. "But if she can't remember or her mind is too traumatized to recall . . . I want to put her under hypnosis."

Chapter 22

"What do you see?"

Sophie's brass pocket watch lay open on Greer's coffee table. The sun had set seconds before, and the dark living room was lit by the final dregs of daylight fleeing over the suburban neighborhood and a single tea candle between her and Greer's daughter. Kinsley lay back on the couch, her eyes closed, her hands at rest beside her. Sophie had done everything she could to make this a relaxing space.

Almost.

His leg jangling nervously, Greer sat to the side of the two younger women, looking anxiously between them. But when he had agreed to let Sophie hypnotize his daughter, with Kinsley's consent of course, he had specifically provisioned that he would get to watch. The big man's concern for his daughter was endearing, and the way his strong, determined personality melted away reminded Sophie of her own father. Always trying to help. Trying to be there for the people he loved.

Sophie felt a pang of distracting guilt. She still had not called her parents after their fight during her last case. They didn't even know she was still in Louisiana. The last they'd seen of her, they'd been hosting a church potluck in the trailer park's green. Music. Laughter. And their sole surviving daughter furiously arguing with them for daring to be happy while the man who had killed Molly walked free.

The look her father gave her wasn't too distant from the look Greer gave Kinsley now. Concern. Fear that the darkness she stared into would consume her.

It was sweet. Endearing.

And making Sophie's job more difficult by the second.

She shot Greer a pointed look, and he forced his leg to stillness. Breathing in, drawing on her own center of calm, Sophie repeated, "Kinsley, what do you see?"

A pause. The tick and pop of the candlewick. The dull scuff of Greer's shoe on the carpet as he adjusted his legs with the agonizing slowness of someone unwrapping candy at the theater.

"Is she asleep?" Greer asked in a whisper.

Sophie gritted her teeth, giving a wide-eyed glare to her partner, who held up a hand to withdraw the question. Feeling her frustration mounting, Sophie again closed her eyes, taking a steadying breath, forcing her heart and lungs to return to the calm baseline of the pocket watch.

"Kinsley," Sophie started. But there was no point. She could see the tension in the girl's shoulders, the way her eyebrow pinched slightly in thought.

The hypnotic trance was broken.

Sophie sighed. "Let's take a break."

"I'm sorry," Kinsley said softly, rotating back to a seated position. "I think I'm overthinking it."

"No," Sophie replied wearily. "It's not your fault. Take a moment to relax. Don't turn on the lights; we'll try again in a few minutes."

When they'd first arrived, Sophie and Greer had asked the usual sort of questions. Had she seen anything. Was there anything before the riot that was unusual or that drew her attention.

As Sophie expected, the chaos of the night was like silt thrown up in the clear water of the young woman's memory. She remembered fragments, small details. But nothing significant. There were gaps too. She knew she'd gone to find Camille but couldn't tell them anything about the moments leading up to finding her friend backstage. When Greer pressed her about how she got backstage, Kinsley landed on a story about pulling aside one of the fabric walls and sneaking through. But as she did, Sophie noted the tilt of her head, the slight dance of her pupils.

Perhaps as she slept or over the day while her father and Sophie were away, Kinsley's mind had concocted a cover story for this moment, plastering over something that she didn't remember or didn't *want* to remember. It might only be a small detail or it could be the whole situation, but there was more to this story. Something her brain was trying to protect her from or some shame her subconscious was trying to scrub away.

Was it what she had been on the verge of telling Sophie last night?

That was when Sophie had asked to hypnotize her.

Stepping into the adjoining kitchen, Sophie took the liberty of pouring herself a glass of water while Greer caught up.

"No luck, huh," he said, joining Sophie by the sink. "Do you think Kinsley might be . . . I don't know, immune to hypnosis?"

Sophie gave a wry laugh. "She might be. Some people are. But I wouldn't know."

"What do you mean?"

Sophie sighed. "I mean, I can't get anyone into deep trance while you're interrupting and . . ." Sophie gave a sweeping gesture up and down Greer's lanky body, "fidgeting."

"You want me to leave?" Greer said flatly, folding his arms. "While she goes back to the most terrifying night of her life, I'm supposed to—what? Go watch the game?"

Sophie pinched the bridge of her nose in frustration. "Right now, relaxation is key," she said at last. "You gave me permission to do this. So did she. You have to *let* me work."

Greer turned, pacing in a tight circle. "This was a bad idea."

"She knows something, Jonathan." Sophie put a hand on Greer's arm, stopping him where he stood. "I saw it when we were questioning her. But it's buried."

"Buried," he repeated. "What do you mean?"

"I mean"—Sophie took a moment, glancing at the doorway to the kitchen and lowering her voice—"I mean that it's likely something her mind doesn't *want* to remember. Something horrible or frightening or just . . . unsettling. Something that raises scary questions when her body is so flooded with stress that it's doing everything in its power to pretend it's all fine." Tightening her grip, Sophie stared hard into Greer's cold blue eyes. "But it's *not* fine. And pretending like it is . . . just moving on and trying to protect her is going to get other people killed."

Sophie's mind flashed with the memory of her mother, tears welling in her eyes, pleading with her to stop talking . . . to just enjoy the party. *It's so ugly. Please, mon petite, sit with me.*

Greer lowered his eyes, a heavy frown appearing on his face. "Kinsley is . . . spirited. She's sensitive and kind," he said softly. "You know she's a counselor at a summer camp? They love her there. The kids love her. She brings out the good in everyone. Brings out the good in me . . ." Taking a deep breath, Greer shook his head. "Maybe it's a good thing that her mind is burying this. I don't want her to lose her faith in people. I don't want her to end up like me, unable to sleep at night."

Sophie's mouth drew into a hard line. Sympathy and compassion grating against the urgent need to find the killer. But she didn't have a daughter. And the obvious decision suddenly wasn't so clear.

"Daddy."

Sophie and Greer both jumped, turning to the kitchen doorway. Kinsley was leaning against the frame, her cherry red curls half covering a downcast face.

"Sorry, baby-girl," Greer said, pulling his arm away from Sophie's grip. "You snuck up on us."

"I heard you talking," Kinsley said cautiously.

Greer stiffened, rubbing the back of his neck. He muttered an apology if they were too loud.

Kinsley waited for him to finish, absently tapping her knuckles on the doorframe. "Daddy, I know that you want to protect me. You're good at that. Since mom left . . . you've made me feel safe."

Here Kinsley met her father's eyes, standing up off the doorframe to face him. At her full height, Sophie realized for the first time that the girl was already an inch taller than her five-six frame. She was only seventeen too and might get taller yet, just like her dad.

"But I can take a few sleepless nights," she continued. "If it helps find him . . ."

"It's harder than you think," Greer said cautiously, his voice grave and full of warning.

"I believe you," she replied. "But how well do you think I'll sleep knowing there was something else I could have done? If another girl turns up dead and I didn't do everything I could?"

A silent, appraising moment passed between father and daughter, and Sophie again thought of her own parents. She'd never had a conversation like this with them after Molly was killed. Never talked through it. It had always been anger, resentment. There was pain here . . . but it was different. The kind pain of understanding.

Greer stepped to the doorway. He put a hand on Kinsley's shoulder and gave her a nod, a worried smile ticking at the corner of his mouth. Then, without a word, he pulled her against his chest, wrapping her in a warm hug. Sophie watched from a few paces back as Kinsley wrapped her arms around her father's back.

Then, between her father's shoulders, the teen girl gave her a secret thumbs-up.

Repressing a smile, Sophie nodded as Greer turned to face her again.

"I'm going to take a walk around the block. Call me when you're finished."

"Nothing can hurt you here," Sophie said quietly. "You are welcome to everything you see and everything you hear."

Ten minutes had passed since Greer left the house. In the darkness of the candlelit living room Sophie talked Kinsley down, her comforting voice controlled and gentle as she eased her mind into the cool dark of unbiased observation. The ticking pocket watch setting a rhythm and focus.

The gentle rise and fall of Kinsley's chest showed the hypnosis had worked. She was on the very edge of sleep, her unconscious and conscious mind allowing free passage between each other. She was ready.

"You want to find Camille. Why is that?"

Sophie tried to start as close to the false memory of sneaking backstage as she could without referring to it. She didn't know how much Kinsley's mind had made up, and in this susceptible state, false memories could easily become ingrained if she accidentally treated them as truth.

Kinsley's eyelids fluttered. Her voice was distant, almost as if she was talking to herself. "I'm frightened."

"What frightened you?" Sophie prompted.

"The people. The crowd. They're scared. Angry. I want to leave."

Sophie thought back to that night. Carly Monroe had gone missing, delaying the show. Then rumors had spread about the killer. This must have been before the rumors, just as the festival was getting rowdy.

"What will you do now?"

Kinsley flinched. "They won't let me backstage. I'm trying to tell them. I'm friends with Camille. I asked if someone can get her for me. No one is listening."

Sophie noted tics of tension, little bursts of remembered motion as Kinsley relived these moments. "What do you see?" she asked.

"Crew. Black T-shirts. There's a ladder. No one is watching it. It's short. I can pick it up on my own and put it against this trailer wall."

Sophie frowned. If she snuck over a trailer with a ladder, why would her mind replace that with a memory of sneaking through a fabric wall? She watched the semiconscious girl carefully.

There.

A twitch in her expression. A shudder.

"I'm inside now. I have to find Camille," Kinsley murmured.

"Hold on," Sophie said. The words weren't abrupt, a gentle prompt, like a parent holding a child's hand before crossing the street. "You're on the ladder. You're climbing the trailer."

Kinsley squirmed on the couch, an unconscious grimace spreading on her face as an unspoken sight passed behind her eyes. "I have to hurry," she said.

Her anxiety was spiking. Sophie knew that if it continued to climb, she might break out of the trance like someone startled out of a nightmare. But she needed answers.

"You're on the ladder. What do you see?"

She twitched, her nearly unconscious body involuntarily trying to turn away.

"You see something," Sophie prompted. "What is it?"

Kinsley winced, her voice trembling. "A ghost."

That wasn't it, but it would be close. Her mind was trying to protect her again, trying to plaster over the gaps.

"What do you see?" Sophie repeated.

"White as a ghost. A ghost car," Kinsley whimpered, her body curling in and turning to face the back of the couch. "Someone is dead. The ghost car takes them away to be buried."

Sophie bit her lip. She knew she shouldn't give details or prompts. Anything she added could be woven into the fantasy Kinsley's brain was trying to create. But she needed more. Something in this memory was painful for Kinsley, and at any moment the trance could be broken. She needed the right question. Ghost car? Taking someone to be buried?

Sophie closed her eyes and pictured the aerial view of the festival. The stages. The bayou on one side. Parking. Crew parking behind the backstage. If Kinsley was on a ladder by the backstage, then she could see the crew parking lot.

"Kinsley." Fighting to keep the urgency from her voice, Sophie asked, "Was anyone by the ghost car?"

With a shiver, Kinsley nodded, her voice fraying with new tears. "A man. He closed the funeral door. The coffin. He's turning this way. He-He's looking at me!"

The killer.

"Stay with me," Sophie urged. She felt her breath catch. They were close. Any detail could break the case wide open. She just needed a little more. "What does he look like?" Sophie asked.

Kinsley bolted upright. A sudden scream tearing from her throat.

Chapter 23

"It's a hearse?"

"A white hearse," Sophie said with a curt nod.

By the time Greer returned, Sophie had managed to talk Kinsley down from her panic. But the experience had clearly shaken her, and after a brief goodnight to her dad, she had gone upstairs to her room.

Sophie had been nervous at first that this would only cement her partner's fears. That he would resent her for the pain she'd put his daughter through. She'd even prepared herself for a cold shoulder by pulling out her laptop, diving straight into research.

The former Marine surprised her though. He wasn't exactly pleased about the tearstains on his daughter's cheeks when he returned, but Kinsley put on a brave face. When she had gone to bed, he asked Sophie one question: "Was it worth it?"

She told him yes.

And that was that.

Greer sat down, pulling out his own laptop as Sophie explained everything she learned.

"It's an odd sight," Sophie reaffirmed. "A white hearse. But when she started talking about the killer putting someone in the back, ready for burial, I assumed she must have seen the coffin, or at least the tail end of it. She probably didn't consciously register what she was seeing," Sophie said. "It would be strange, but not necessarily memorable when her mind was fixed on finding her friend. Then when the riot broke out, everything else was pushed aside."

The memory of Kinsley's scream returned to Sophie's mind, and she gave a forceful cough, sending away the small twinge of regret rising in her throat. She didn't think Kinsley had imagined the killer looking at her, though it may have only been a passing glance her general direction. The girl's subconscious may have been putting together a connection her conscious mind wasn't ready for. Sealing away the memory out of fear that he'd seen her and that he'd be coming for her next.

Greer murmured, "ghostly white," repeating what Sophie told him Kinsley had said. "You might be onto something."

Sophie shook her head. "I should have thought of it before. We knew from Camille's testimony that the coffin was on-site. If he's taking them out with him, he'd need a truck or a van or something capable of discretely hauling a box that big around. A hearse would be perfect for that."

"Don't beat yourself up," Greer said softly. "I didn't think of it either. Now all we can do is try to narrow our options."

And with that, the pair set to work.

A funeral parlor. That's what they were after.

Some place that offered custom coffins, maybe an unvarnished, biodegradable option like the sort Arturo Padilla built, but also offered hearse service as well as having white hearses available.

It was certainly possible that the killer personally owned a white hearse and built coffins out of his own basement, but Sophie pushed that possibility aside for now. For the time being, she let Ron Gavriel's advice guide her. This killer's obsession with the coffin wouldn't stay isolated into one aspect of his life, it would bleed into everything. He would seek work that let him be around them and build coffins.

No, a funeral parlor certainly fit the bill.

It was nearly nine o'clock when Sophie set down her laptop.

"I'm down to two options," she said briskly.

Greer stretched his arms overhead, his shoulders and back popping as he replied, "I've got one confirmed so far. What do you have?"

"I have Family Heart Funeral Services, south of the city," Sophie read back. "White hearse service and custom coffin arrangements."

"They're the one I have too," Greer agreed.

"The other," Sophie continued, "is Last Journey. They are based in New Orleans proper."

Closing their laptops, the pair stood, neither of them needing to confirm they would be looking into the parlors tonight. The image of the most recent victim was fresh in their minds, as was the fact that Carly Monroe had yet to be found.

Sophie pulled her phone, quickly scrolling to Lidya Merchant's number. The older woman answered on the first ring, her voice betraying no hint that she'd been in anything approaching sleep or rest.

"Quinn, what do you have for me?" Straight to the point as usual.

"Ma'am, we have a new lead. A white hearse. Agent Greer and I have two funeral parlors in mind. We're going to investigate one. Can you get a team out to the other?"

On the other end, Merchant shuffled some papers, and Sophie heard the click of a pen. "I can have a team out tomorrow. Hitchens is still bitter about your little displays, Quinn. He has a moratorium on backup calls for you two. Says 'no one gets sent unless someone's bleeding when they call.' Besides, we'll need time to get warrants together."

Sophie bristled, giving the sleeve of her suit coat a fierce tug. "The killer is accelerating. He could be taking another victim out to the river *tonight.* We don't have time for this!"

"I know," Merchant replied coolly. "Listen." Her voice lowered, and Sophie suddenly had the impression Merchant must still be at the field office and not wanting to be overheard. "Keep your phones on. I'll do what I can here. Find him and I'll get you backup. But for now, you're on your own."

Two locations. Two of them.

Merchant's warning echoed in Sophie's ears: no backup unless they're bleeding. Sophie looked to Greer, resolve hardening her expression.

They would go. After all, the determined bleed for what they want.

* * *

Sophie checked the new pistol in her underarm holster. She had replaced the one lost in her run-in with the Dream after the interrogation of Mr. Padilla, but she hoped she wouldn't need to use it. Then again, she was planning on finding the lair of a serial killer tonight. If she found him, alone, she would need all the advantage she could get.

In that moment, Sophie thought of her handsome partner. Right now, Greer would be arriving in the city, likely looking for a spot to park so he could get the lay of the Last Journey funeral parlor. There might be someone there to let him in or answer some questions if they kept "city hours." But as Sophie parked on the gravel lot outside Family Heart Funeral Services, the dark, single-story building lacking even an exterior streetlamp to illuminate the walk from the road to the door, she doubted she'd have the same opportunity.

A part of her wished they had stuck together, but as she had insisted to Greer back in the safety of his home, they might not have the time to search the parlors one after the other. If the killer was taking someone out of the river tonight, they might only have hours or minutes to catch him.

Squinting into the near pitch-darkness, the building held no signs of life and anyone driving by could be forgiven for thinking it abandoned if they spied the isolated building at all.

She didn't come here just to turn around, though.

Sophie clicked on her flashlight and stepped out of the car, walking lightly toward a set of cracked, white columns holding up a porch roof over a set of grim wrought-iron benches. She tested the door and found it locked.

"Perfect," she murmured, and for a moment she was tempted to pull out her lockpicks. Instead, she checked the adjacent windows. Better to see what she could before she risked making any noise.

The lobby and viewing area were visible. Clean and empty. Chairs neatly stacked against one wall, and doors into offices and other rooms visible but closed.

Sophie circled the building, checking each exterior window in turn. When she reached the rear, she found a wide metal canopy. Two hearses were parked beneath it. One the traditional black, the other a porcelain white.

She looked into the rear, pressing her face to the glass as she focused her flashlight within.

No coffins.

Forty yards or so behind the funeral parlor, Sophie caught the glimmering ribbon of bayou water. A wooden pier jutted out into the silent water, and she saw the flat-bottomed shape of a wide skiff.

A clang of metal striking stone sounded behind her, and Sophie spun around. Her heart drummed as she cast her flashlight around for what had made the sound. Finally, she landed on a narrow door angled between the wall of the funeral parlor and the gravel ground. It was a cellar door, but narrow in the way old stairways sometimes were, as if the steps behind it might only leave room for one foot on each. A padlock held the doorlatch in place. Free of rust and clinging dirt as if it saw regular use.

Whatever had made the sound was behind that threshold.

"Alright," Sophie said, tucking her flashlight in the crook of her neck. "Let's see what's behind door number one."

Despite missing a pick from her set, it only took a few minutes for Sophie to crack the padlock. The lock popped free and she pocketed it. There was no telling what had made that sound and the last thing Sophie wanted was to leave someone the chance to lock her underground.

After a few steps on the narrow stair, the descent began to curl, twisting like a spiral. The surfaces around the stairway were clotted with soundproof padding, already smelling of mildew. The sight sent a chill down her spine, and Sophie gripped her flashlight a bit tighter. Warm light began to shine up from the steps ahead, and Sophie killed her light. Stepping softly, not risking a single sound, she readied herself for whatever she would find.

"What in the world?" she whispered.

The concrete walls were bleeding with black mold. A flickering tanning lamp lay on one side of the room beside a pile of rumpled curtains and a small unmade bed. Bits of bone and gator skin were piled on a workbench beside a tv turned to static and bottles of clear liquid that filled the dank room with a biting, semisweet smell that irritated her nostrils.

Ether.

Stepping further in, Sophie found a heavy steel ring embedded in the concrete. A chain lay loosely coiled beside it, terminating in an open manacle.

A manacled chain through a steel ring.

Suddenly, Sophie was gasping.

No! she thought. *Not now!*

She couldn't breathe as the sensation of cold water filled her mouth and nostrils. She felt the manacle on her own ankle, though she could see it across the room. Then the walls around her fell away. The dark stains became heavy-limbed mangroves, and the hum of tv static turned to the swamp song of crawdads.

Find the key and one goes free. The callous, gravel-coarse voice of her sister's murderer filled her mind. *One key. One free. And one for me.*

It's not real, a small voice in her mind told her.

But Sophie began to hyperventilate. Her eyes fixated on the chain and steel ring. Now she was seeing the chain the River Killer had used, the way it fed through the anchor ring of his boat forcing Molly and Sophie into an underwater tug-of-war for who had enough slack to reach the surface and breathe.

The night her sister was murdered . . . the night Sophie survived.

She stumbled back, crouching low as her insides twisted in sudden agony. She could feel her vision blurring, her rapid, shallow breaths driving her toward the cliff of unconsciousness. She had to stop it. Had to get out. She could see him now, tall, strong. RK sauntering toward her, cigar in mouth, baseball cap pulled over his brow, obscuring his face.

It's not real.

Sophie's heart raced. She suddenly felt as if she were walking through winter snow, her skin clammy with icy sweat and her limbs violently shivering as she grabbed her phone. She squeezed her eyes closed, turning to face the wall, biting down on the inside of her cheek hard enough that she tasted blood.

The pain drove the flashback from her mind.

The swamp disappeared, seeping out through the walls like mist. She was back in the funeral parlor basement, one hand on her phone, the other flat against the wall as she took deep breaths.

She had to call Merchant. Greer. Someone.

Her thumb fumbled for the contacts list. Sophie closed her eyes again.

One breath. Another.

By degrees, her body ceased to shake. Her lungs calmed. She hadn't had a flashback that severe in a long time. So real it was as if RK were in the room beside her. She didn't have time to consider it, though.

She opened her contacts list.

And a strong, rough hand came down, smacking Sophie's phone to concrete.

Chapter 24

Calloused hands snatched Sophie by the shoulders. The attacker shoved her to the concrete wall, and Sophie barely got her palms up in time to brace for the impact.

"You shouldn't be here!" a strong, energetic voice shouted behind her. The voice of a man used to shouting.

Instinctively ducking, Sophie dodged a haymaker aimed to the back of her head and spun away from her assailant. One hand reached under her suit coat for her pistol as she shouted, "Freeze, FBI!"

The man bearing down on her gave no indication he'd even heard her, his furious eyes boring into hers.

He was only a few inches taller than she was. She might have even guessed him to be in his early thirties like herself. But that was where the similarities ended. His bare arms, emerging from a sweat-stained tank top, were thick with corded muscle. His hands were large, covered in veins like mountain ranges and callouses like Badlands earth.

As she drew her pistol, one of those coarse mitts snapped forward, clamping down on the slide of her pistol, jamming it open. Sophie's finger found the trigger, but the trapped weapon refused to fire as the man pressed Sophie back.

"It is you," the man growled, the smooth, pale skin of his face collapsing in furrows of confusion and rage. "How did you find me?" he demanded.

With a final shove, he slammed Sophie's back against the workbench. She let out a gasp of pain as gator bones rattled to the floor and the television rocked on its base.

"You can't wait for me to do things in my own time, can you?" the attacker raged, pressing his face inches from hers. "No! It's always about *you*. It's about when *you're* ready!" Spittle flew from frothing lips, speckling Sophie's face as he jerked the pistol from her hands.

Sophie paid the man's rambling no attention. Instead, she braced her hips on the table behind her and levered forward, driving her forehead into the man's nose.

He screamed in pain, one hand covering his now-bleeding nose, the other still holding Sophie's pistol by the barrel while he staggered back, into the center of the room. Sophie pressed the advantage. Closing with him, she braced one foot, the other shooting forward in a kick to the side of his knee. The man buckled and collapsed, the injured knee hitting the concrete just as Sophie twisted her hips, driving the heel of her hand forward in a shot across his jaw.

Dropping to all fours, the man spat a mouthful of blood onto the ground, and Sophie saw an older bloodstain on the concrete beside it, dyeing the floor between the killer and the chain embedded in the ground. But she couldn't let up.

He still had her gun.

Sophie sent a kick to his exposed ribs, but the man brought up a paw, catching her by the ankle. With an abrupt yank, Sophie fell to the concrete, her elbow jostling the coiled chain beside her as starbursts exploded behind her eyes.

She gasped as a heavy weight pressed against her stomach. The man was straddling her now, both of his hands scrabbling for her throat. He was going to strangle her.

Without stopping to think, Sophie's hand instinctively reached to the cold floor beside her. Her fingertips found cold metal, and with all her strength, Sophie threw a hook to her murderer's jaw. The chain in her hand followed, whipping against the man's temple, jaw, and throat with a wet snap.

He fell off her, howling an inhuman cry as his hands shook, hovering an inch from his freshly torn face.

Sophie dropped the chain, rolling and coming up to a crouch with her pistol in hand.

"Don't move!" she shouted.

One eye glared at her, the other covered by a palm that leaked red between the fingers. But the man did not rise.

Slowly. Careful not to take her eyes off of him, Sophie found her phone. It had slid to a pile of pinewood planks, nearly disappearing under them. She took it back, scrolling to Greer's number.

Then she stopped.

She looked at the planks. Two sawhorses were folded beside them. Carpentry tools hung on hooks behind a padlocked cage on the wall.

Tools.

Planks.

No coffins.

He's starting again, Sophie thought. *There's material but no coffins because he's just finished with one.*

"Where is she?" Sophie demanded.

"Whoring herself to some banker in Las Vegas," the man hissed.

Sophie frowned. A lie? Delusion? Sophie abandoned trying to understand the comment and shook the pistol insistently. "Carly Monroe. Where is she?"

"Oh," the man chuckled, gagging a bit before reaching into his mouth and removing a bloody tooth. He tossed it on the floor as he said, "The singer. Right. She's got what's coming. Taught her a lesson. Same lesson as all of you need to learn, it seems."

"I know what you do," Sophie said coolly. "There's no coffin here. But the last one you made was hers. I know she's in the bayou. Where?"

The man grimaced, shifting up onto one knee, wobbling a bit as he rose to a crouch.

"We took a sunset cruise," he said placidly. "To reflect on our lives. Our mistakes." He shook his head. "One last sunset cruise. I remember the spot."

Sunset.

Sophie checked the time on her phone. Nearly ten o'clock. She might have only been submerged two hours ago. Chewing her lip, Sophie recalled what they'd learned about the coffins. Well made. Tightly sealed. On their own they might last hours . . . before the air ran out.

That is, if the victim didn't fight. If they didn't panic and didn't try to break out. If they didn't make cracks that let the water in.

It was a lot of ifs.

Sophie looked to the bleeding man in front of her, her finger tensing on the pistol grip.

She waved the barrel to the stairs.

"Take me to her."

Chapter 25

"Bayou Aveu . . . Yes. Off the main river." At the back of the wide-bottomed skiff, Sophie had no choice but to shout as she coordinated with the FBI search and rescue teams over the old motor's angry muttering.

With one ear to the phone, the other tuned to the off-color ranting of her murderous pilot at the front of the skiff, she was growing wistful for the chill silence of her investigation at the funeral parlor.

The lazy river around her glittered with reflected moonlight, the trailing arms of willow trees and mangroves offering occasional shelter from the light drizzle that had started to fall. Rain was expected in Louisiana but seldom persisted more than half an hour, and when it did, you could expect days of it. Nevertheless, the light speckling had managed to sufficiently plaster her close-cut hair around her ears and forehead in auburn tiger stripes.

A dragonfly buzzed past her face, and Sophie couldn't help but feel a spike of frustration at being outpaced by the insect.

"Hey! Pick it up," she called to the front of the boat, pulling the phone away from her mouth for the moment.

The killer's shoulders pinched, and she saw his hands tighten on the wheel. But when he called back to her, his voice dripped with humor, like they were old friends sharing a joke. "I don't remember it by the place," he said. "I remember it by the *feel.* You know? If I speed up, it won't feel right. I might forget."

Sophie grimaced. The man was so erratic it was impossible for her to tell if he was toying with her or not. She knew it was a gamble to even trust him to be her guide, but . . . Sophie thought about Carly Monroe, the young woman trapped in a coffin somewhere on these rivers. Likely drowned or suffocated. Likely . . . not certainly. There were no guarantees here.

No guarantee that he was even taking her to the right spot. But Sophie found she believed him. When they found the first coffins, he hadn't stopped his killing, he'd started killing more and more aggressively. Like so many serial killers, there was a part of him that *wanted* to be caught. That wanted someone to see, to know. He'd take her to the right spot.

It was what he'd do when they got there that worried her.

Sophie stretched her stiff fingers, adjusting the grip on her pistol. She hadn't dared to lower it since leaving the funeral parlor basement. Not that she imagined it was the only thing keeping the killer from attacking her. No, that was also because he wanted her to see first. To know it was really him.

"Take it as fast as you can, anyway," Sophie called out.

The man bristled again, waving one meat-hook hand over his shoulder. "Yes, dear . . . whatever you say, dear."

Sophie raised an eyebrow, half listening as she continued to relay directions over the phone. There was that familiarity again. Why did he talk like that . . . like she was his wife? He'd done something similar in the basement, talked to her like they had met before.

Suddenly the killer stood, rising from the pilot chair and looking around the river like a prairie dog. "This is it," he said, the final word clearer than the rest as the motor behind Sophie died.

Risking a glance overboard, Sophie looked into the black water. Cricket song rose around her once more, the nighttime hiss of rushes on the river edge hinting at the passing of a snake or maybe a gator.

As her ears adjusted, Sophie heard the distant rumble of other motors. The FBI boats catching up with them.

"Yes, this is the place," the killer said again. And for a long moment, he simply stared into the murk under the boat. Perhaps into his own reflection on the water?

Sophie looked too, but only darkness and the lightning flashes of silver-bodied fish returned her gaze.

"Where is she?" Sophie asked. "I don't see her."

The man gave a derisive snort, glaring at Sophie like she was an unruly child who had just spoken out in church.

He leaned toward her, one foot stamping the floor of the skiff, setting it rocking as he snatched at a metal hook on the side wall.

"Hey!" Sophie said, pocketing her phone as she leveled the pistol at him. "No sudden moves."

The man brandished the hook, a heavy piece with a polearm grip used for hauling gators into the boat. "You want to see her or not? Make up your mind," he snapped.

Sophie nodded, gesturing to the waterline.

"Typical you," the man spat. "Never know what you want. Never say a word . . ."

Leaning low, the man sank the hook into the water, pulling and prodding. Sophie risked a glance around. The FBI boats had to be getting closer, but it was impossible to tell in the open air. She still couldn't see any other lights beyond the skiff's.

The boat tilted as the man groaned, yanking on the hook.

A brown streak appeared in the water. Wood.

The corner of a box spun, striking the bottom of the boat and setting them rocking. Now Sophie could see it standing upright, the top of the coffin just below the water's surface.

He must have removed one of the weights. There had to be some air still in the coffin, buoying it up. But hours had passed. How much fresh air had been in the coffin? Even if it didn't leak, how long could his victim have survived? Panicked. Hyperventilating in the cold dark.

Sophie's eyes narrowed as she looked closer, risking a step toward the man for a better angle. She looked for any bubbles, any sign of water leaking in. The water was too murky. The churn of the man's struggle with the hook was stirring up the water too much.

"Careful," Sophie said. "Don't break the seal."

"Telling me how to mind my own work now?" The man shot Sophie a glare. "That's just like you, Annie. Doesn't matter if it's gator fishing, my woodwork—even in bed, you always got *something* to say."

"Take it easy," Sophie said carefully. Making sure the gun was still in sight, she gave him a stone-faced stare in return. "The best way out for you is if she survives. Pull her up."

His hands white-knuckling the hook, the killer shook where he stood. "Don't patronize me, woman!" he shouted. "I've put up with your shit too long."

In the half second before it happened, Sophie saw the man's intent. The subtle shift of his footing, the adjustment to his grip. Throwing herself backward, Sophie heard the whistle of the gator hook as it sliced the air where her head had been.

The boat rocked from the sudden shift, and she slipped, falling on her back with a dull *thunk*. Then he was on her. Stalking forward, the man lifted the hook overhead for another swing. Sophie fired the pistol.

Muzzle flashes and the sharp bark of the gunshot lit up the darkness. Sophie faintly heard the man's cry of pain as she fired again. But as she did, the metal hook's rounded head struck her hand. A sudden sharp pain brought a scream to her lips as Sophie pulled her numb fingers to her chest. The pistol was gone. She thought she caught the dull glimmer of gunmetal spinning in the air before it disappeared over the skiff's sidewall.

Bracing her hands, Sophie shot a kick up and forward, aiming for the man's groin. But with a swing of the polearm grip, he knocked the attack aside, stepping over Sophie's legs and dropping hard on her stomach.

She threw a punch to his throat. Tossing the hook aside, the man's enormous hands came up, blocking her attack like baseball mitts as he pressed forward, using his heavier body to hold her to the floor of the boat.

His grip was like iron, years of work putting strength behind the fingers. First one, then the other, he pressed her wrists to the floor of the boat, using his knees to pin her as he reached for her throat.

She couldn't overpower him. She'd lost almost all mobility. If his hands reached her throat, he'd cut off her voice, her last tool. She needed to do something, anything to survive. These thoughts passed in a moment, wordless instinct driving her.

"You're right."

The man froze.

Sophie didn't struggle. Didn't move. Despite the part of her screaming to fight, she looked into his eyes. Into his broken, twisted mind as she looked for what he wanted to hear. The way he talked to her . . . the victims he chose . . . a picture formed in her mind. And she could only hope it was the right one.

"I never appreciated you," she said, softening her eyes, inflecting remorse in her expression. "Everything you do for us. You were never weak. I was."

Her bile turning as she tried to stomach the words, Sophie kept up her stare, looking into the man's furious eyes as she saw the effect of her words.

For several breaths, he only stared, his chest heaving. She could see blood seeping from his ribs, just below his left pectoral, another red spot blossoming on his shoulder. Her shots had hit.

"Annie . . ." he murmured weakly.

Sophie nodded. "You were doing your best. I couldn't see that. I do now."

Her heartbeat was hammering in her ears, the man's weight on her making it difficult to breath, difficult to keep the tranquil understanding on her face as she fought to keep his attention. Maybe if she could get a hand free . . .

"No," he said suddenly. His rough hands wrapped around her throat, fingers linking at the base of her skull as both thumbs pressed under her chin.

Sophie choked, all pretense gone as every muscle in her fought to break free.

"You're a liar," the man said grimly, his lips shaking, his eyes going unfocused. "You never say it when it matters. Never say anything until you're ready to scream."

Sophie's face flushed as the circulation in her neck was cut off by the man's grip. Her vision swam, the man fading and twisting above her like a malevolent specter.

"You're not Annie," he said, his voice ragged now. "But you may as well be."

A sudden crack.

Another. A blast and flare of flashing light.

Sophie could breathe again. Her eyes wide, she gasped, drinking in the heavy swamp air. The killer stood, his strong hands on his chest as his tank top turned bright red. His eyes rolled, and he wobbled on his feet.

"So weak . . ." he murmured. Then he fell back, water sprinkling over Sophie as he hit the river.

Sophie turned, weakly gripping the skiff's wall as a floodlight twisted to find her. Under the glare, with a pistol in hand and on a boat painted with the gold lettering FBI, stood Greer.

Chapter 26

"You're staring again."

"Sorry," Greer replied with an embarrassed cough. "It's barely noticeable, really."

Sophie stopped on the hospital steps and appraised the bruises on her neck in the sunlit glass of the lobby windows. She still had some of the makeup she'd bought for her infiltration of the Barefoot offices, but she didn't even want to look at it. All her throat needed was time. The dark marks would fade until they were only a memory.

Smirking up at her partner in the glass, Sophie shook her head. "You're a terrible liar, you know that?"

"So I've been told." Chuckling along with her, Greer took the lead down the steps toward the hospital parking lot, thumbing his key fob as he scanned the lanes for his car. "Still, better than the alternative."

Sophie made a small sound of agreement, wincing as the vibration hit her recently strangled throat. That would also take a while to heal. At least there would be no hospital stay for her this time.

Just painkillers and tea.

Casting a last look over her shoulder to the New Orleans hospital, Sophie asked her partner, "Think she'll be alright?"

Greer let out a long sigh, scratching at his neck uncomfortably. "She's young. Resilient. Besides, her monster is dead. I wouldn't be surprised if she's able to turn this whole ordeal into a record deal."

Sophie scoffed at that, then bobbed her head as she debated the idea. "I hope so," she said at last. "After what she's been through."

With Sophie treated, the pair had made a point to visit the final victim, Carly Monroe. The young singer-songwriter was surrounded by a sea of cards, flowers, and the odd stuffed animal. An officer had been posted outside her room, probably more for the girl's peace of mind than out of any actual fear of danger.

When they pulled her from the coffin, she'd looked half feral. Like an abused dog kept in too small a cage, shivering all over, her eyes wide in fear of everyone and everything around her.

But she was alive.

The air had lasted. Only an inch or so of water had seeped into the coffin over the three hours she was trapped.

A survivor.

But she hadn't seen the killer shot. Hadn't seen his dead body in the water. Would he haunt her dreams? Would the sight of river skiffs and pinewood coffins send Carly Monroe into flashbacks like her?

Lookin' for a friend, cher? mocked the demon in Sophie's memories. RK's voice, never far from her nightmares, taunted her. *Maybe you teach 'er your lil' trick with the cheek bite, eh? Saved you in that cellar? Non?*

"Miss Quinn?" Greer had stopped, a hand half extended as if he was afraid she might fall over.

Sophie felt her cheeks redden and gently waved the gesture away.

"I'm alright," she said. "Probably the painkillers."

Greer met her eyes, his sky blues gazing at her intently. Concern, compassion, and . . .

Sophie cleared her throat, wincing again at the pain in her trachea. "Sorry," she said in a voice barely a whisper.

Muttering a hurried apology, Greer fell in beside her as they walked a few more paces in silence. More comfortable than their first case together, Sophie found she'd developed an affection for the trimly muscled FBI agent. His polite way of speaking. His earnest compassion. The drive that she couldn't help but admire.

"Did you see Hitchens's press release?" Greer asked with a hint of cynical humor.

"No," Sophie admitted.

"There are clips online. But he's basically claiming everything short of shooting the killer himself."

"Huh," Sophie said with a shake of her head. "That sounds about right." Then pausing a moment, she asked, "Did you and Merchant find out who Annie was?"

Greer gave a clipped nod. "His wife. Ex-wife." Turning up a palm as he spoke, Greer said, "Once we knew he was at Family Heart Funeral, we looked into who owned it. Jackson Kearney. Wife, Anne Marie Kearney. No children. She filed for divorce about a year and a half ago. Claimed domestic abuse among other things." Greer let out a solemn sigh. "But she ended up dropping those charges. Likely a deal just to get the divorce done with and to move on. It happens sometimes. She's in Florida now. Alive and well, though we're having local PD confirm that."

"She is," Sophie said.

"How's that?"

Wrapping her arms around her elbows, Sophie glanced up at Greer. "The way he talked. What he said. I think he was getting revenge on her. She was gone. He couldn't reach her anymore, so he was indulging everything he wanted to say and do to her when they were together. He was paranoid. Thought *he* was the victim. I can't imagine what she went through if this was what he was repressing . . ."

"Christ . . ." Greer whispered, shaking his head.

Sophie's phone buzzed in her pocket.

"Ron," she said cheerfully, her injured throat croaking with the strain.

"Holy hell, Soph. You sound awful."

Sophie pulled her phone away from her ear, switching to speaker-phone as Ron Gavriel practically shouted into his own.

"Cut me some slack," she said. "I was just strangled."

"Strangled-shmangled. You need a lozenge and some licorice tea. What did I teach you, huh?"

In unison, the magicians said, "Excuses are for understudies."

"Damn right," the old magician chuckled. "But listen, now that I'm out of the clink, I've picked out a bar for that apology dinner you owe me."

"Oh?" Sophie raised an eyebrow, shaking her head to Greer who was stifling a smile of his own. "Let me guess, the—"

"The Merry-Go-Round, that's right," Gavriel finished. "And tell the beanstalk, we're talkin' three apology rounds, then we'll circle back to whether I forgive him or not."

Sophie turned her attention back to Greer. "What do you think? Sound fair?"

Epilogue

"You know you're asking me to risk my career, correct?" Lidya Merchant posed the question with the same cool indifference someone might use to enquire about the weather the previous week.

Sophie nodded. "I know. That should tell you how certain I am."

"It certainly tells me something," the older woman answered. Crossing her legs, she adjusted her grip on the umbrella she kept propped under one hand like a cane beside her chair and fell silent.

Their waitress passed the patio door, giving them a warm smile as she settled a cup of black coffee in front of Merchant and a steaming tea in front of Sophie, confirming their order before retreating once again.

"Licorice tea?" Merchant's eyebrow quirked over her sunglasses. Despite the shade of the café's patio umbrella, Lidya seemed to prefer the old school aesthetic and kept them on.

Sophie only shrugged in answer to the question, indicating her still-healing throat.

"Interesting," the silver-haired woman answered. "I'll have to keep that in mind." She lifted the still-steaming cup, inhaling the coffee's aroma and setting it back down. "If you get caught looking into this, it won't be like last time, Quinn." Merchant rolled her fingertips on the umbrella's decorative pommel. "It won't be a slap on the wrist and exile to Detroit. Hitchens is already looking for any excuse. Your association with myself and Agent Greer was enough, but now you've embarrassed him. He'll have you arrested."

"With respect, I'm onto something real here, ma'am," Sophie said quietly. "RK's killings go back further than any of us thought. There's a clear line to the Mississippi Line Minutemen—"

"A militia group," Merchant interrupted.

Sophie slowly nodded. "I've done my part, ma'am. You called and I answered. You said when this was over you'd help me."

"I did." For the first time since her return, something like real concern crossed Merchant's lips. She removed her sunglasses, her honey-brown eyes staring into Sophie as if reading the etchings of her retina. "If they know why you're there, they *will* kill you, Sophie. These off-grid groups do not play around. Even if the River Killer is no longer a part of their membership, you cannot expect easy answers or cooperation."

Sophie leaned forward, her jaw set with determination. "I only need a way in. I've been around dangerous men before."

"Not like this," Merchant answered bluntly.

Sophie hesitated, then nodded. "Not like this," she admitted. "But I can't let this go. Now that I know. I can't go back." She stood, pushing away from the table. "And if you can't help me, I'll have to find my own way."

Sophie started to walk to the patio gate when Merchant let out a sigh.

"You're like a spoiled child, you know that, Quinn?" Returning her sunglasses to her face, Merchant reached into the folded canopy of her umbrella and pulled out a USB drive. "I've scrubbed any trace of my involvement. Once you take this, I don't know anything about it. Do you understand?"

Sophie's heart was racing. Wiping her palm on her hip, she took the drive. In the same moment, Merchant stood. "Thank you for the coffee, Quinn. I'm sure I'll see you around the office."

For a moment, as the older agent's footsteps diminished and finally vanished, Sophie could only stare. This was it. Where RK was born. The place his evil took root.

Her next step.

Sophie slipped the drive into her pocket and was surprised to find her phone buzzing the moment she touched it. She lifted the screen, wondering who would be calling her this early. And her heart stopped.

Zadie Quinn.

"Mama?" Sophie murmured. Dread and guilt swelled in her healing throat. This was it. The call she hadn't had the courage to make. Her parents. But what would they want? To offer an olive branch? Or was this the moment they told her never to come back? A part of her felt she deserved it. Another part even craved it, that final end to a starved relationship.

Another buzz. Only one more and it would go to voicemail.

Sophie huffed a quick breath and answered.

"He-llo?"

Her voice stammered. The woman on the other end was sobbing from the moment the call opened. Her mother's voice, wracked with tears and choked crying. "Mama! It's Sophie, what-what is it? What's wrong?"

"Oh, Sophie!" Zadie wept across the phone line. "Sophie!"

"What is it?" Her own heart was pounding, she could feel her eyes wide with fear as she put a hand to her other ear, blocking out all other sound. "What happened?"

"Sophie, it's Noah! Your daddy, Sophie. They shot your daddy!"

What's Next for Sophie Quinn?

There is no greater pain than love and loss...

Three women are found dead in their homes, their nights of romance ended in violence. But as the men in their lives arrive one by one to confess the killings, FBI consultant Sophie Quinn suspects there's more to these eerily similar murders than meets the eye.

With her own father hospitalized and the final threads of family bonds stretched precariously thin, Sophie much choose between finding the man who took her sister from her and finding the truth behind this new violence.

And with no time to lose, Sophie faces a painful choice and the true extent of her obsessions.

Also by Georgia Wagner

Once a rising star in the FBI, with the best case closure rate of any investigator, Ella Porter is now exiled to a small gold mining town bordering the wilderness of Alaska. The reason for her new assignment? She allowed a prolific serial killer to escape custody.

But what no one knows is that she did it on purpose.

The day she shows up in Nome, bags still unpacked, the wife of the richest gold miner in town goes missing. This is the second woman to vanish in as many days. And it's up to Ella to find out what happened.

Assigning Ella to Nome is no accident, either. Though she swore she'd never return, Ella grew up in the small, gold mining town, treated like royalty as a child due to her own family's wealth. But like all gold tycoons, the Porter family secrets are as dark as Ella's own.

Also by Georgia Wagner

The skeletons in her closet are twitching...

Genius chessmaster and FBI consultant Artemis Blythe swore she'd never return to the misty Cascade Mountains.

Her father—a notorious serial killer, responsible for the deaths of seven women—is now imprisoned, in no small part due to a clue she provided nearly fifteen years ago.And now her father wants his vengeance.

A new serial killer is hunting the wealthy and the elite in the town of Pinelake. Artemis' father claims he knows the identity of the killer, but he'll only tell daughter dearest. Against her will, she finds herself forced back to her old stomping grounds.

Once known as a child chess prodigy, now the locals only think of her as 'The Ghostkiller's' daughter.In the face of a shamed family name and a brother involved with the Seattle mob, Artemis endeavours to use her tactical genius to solve the baffling case.

Hunting a murderer who strikes without a trace, if she fails, the next skeleton in her closet will be her own.

Want to know more?

Greenfield press is the brainchild of bestselling author Steve Higgs. He specializes in writing fast paced adventurous mystery and urban fantasy with a humorous lilt. Having made his money publishing his own work, Steve went looking for a few 'special' authors whose work he believed in.

Georgia Wagner was the first of those, but to find out more and to be the first to hear about new releases and what is coming next, you can join the Facebook group by copying the following link into your browser - www.facebook.com/GreenfieldPress.

GREENFIELD
PRESS

About the Author

Georgia Wagner worked as a ghost writer for many, many years before finally taking the plunge into self-publishing. Location and character are two big factors for Georgia, and getting those right allows the story to flow seamlessly onto the page. And flow it does, because Georgia is so prolific a new term is required to describe the rate at which nerve-tingling stories find their way into print.

When not found attached to a laptop, Georgia likes spending time in local arboretums, among the trees and ponds. An avid cultivator of orchids, begonias, and all things floral, Georgia also has a strong penchant for art, paintings, and sculptures. A many-decades long passion for mystery novels and years of chess tournament experience makes Georgia the perfect person to pen the Artemis Blythe series.

Printed in Great Britain
by Amazon

30775316R00145